# Progress in Molecular and Subcellular Biology

*Series Editors*: W.E.G. Müller (Managing Editor), Ph. Jeanteur, I. Kostovic, Y. Kuchino, A. Macieira-Coelho, R. E. Rhoads **33**



Springer

*Berlin*
*Heidelberg*
*New York*
*Hong Kong*
*London*
*Milan*
*Paris*
*Tokyo*

# Progress in Molecular and Subcellular Biology

*Volumes Published in the Series*

Werner E. G. Müller (Ed.)

# Silicon Biomineralization

## Biology – Biochemistry – Molecular Biology – Biotechnology

With 105 Figures, 17 in Color

Springer

Professor Dr. WERNER E.G. MÜLLER
Institut für Physiologische Chemie
Abt. Angewandte Molekularbiologie
Johannes Gutenberg-Universität
Duesbergweg 6
55099 Mainz
Germany

ISSN 0079-6484
ISBN 978-3-642-62451-3 ISBN 978-3-642-55486-5 (eBook)
DOI 10.1007/978-3-642-55486-5

Library of Congress Cataloging-in-Publication Data

Silicon biomineralization: biology, biochemistry, molecular biology, biotechnology/
Werner E.G. Müller (ed.).
p. cm. – (Progress in molecular and subcellular biology; 33)
Includes bibliographical references and index.
ISBN 3-540-00537-4 (alk. paper)
1. Biomineralization. 2. Silica. I. Müller, Werner E.G. II. Series.

QH506.P76no. 33



http://www.springer.de


Originally published by Springer-Verlag Berlin Heidelberg in 2003
Softcover reprint of the hardcover 1st edition 2003



Cover design: Meta Design, Berlin
Typesetting: SNP Best-set Typesetter Ltd., Hong Kong
39/3150WI-543210-Printed on acid-free paper

# Preface

In most multicellular taxa, but also in some restricted unicellular groups, either an organic or an inorganic skeleton is required for the stabilization of their body plan. Biomineralization is one method which has been successfully applied for the establishment of the form or the pattern of the organism. This process is dependent on the preceding formation of an organic matrix.

During evolution, silica deposition was used in diatoms and Radiolaria (Protozoa) as well as in sponges (Metazoa) as skeletal elements. It appears that the mechanisms for the formation of biogenic silica evolved independently in these three taxa. As in Protozoa and plants, biosilicification appears to be primarily driven by nonenzymatic processes and proceeds on organic matrices. In contrast, in sponges (phylum Porifera), this process is mediated by enzymes; the initiation of this process is likewise dependent on organic matrices.

In this volume, the role of biosilicas as stabilizing structures in different organisms is reviewed and their role for morphogenetic processes is outlined. The basic strategies of silica metabolism are summarized and also include novel and unexpected data. Finally, applied aspects of basic studies on biosilicification are pointed out – at present, it is impossible to estimate their value for future application in industrial processes.

It is hoped that this volume will attract new groups for studies on biosilica synthesis and metabolism and will help to exploit this treasure, which is provided by nature, in a sustainable manner.

W.E.G. Müller

# Contents

## Organisms: Diatoms

### Living Inside a Glass Box – Silica in Diatoms
F. Brümmer



### Components and Control of Silicification in Diatoms
Mark Hildebrand and Richard Wetherbee

**The Phylogeny of the Diatoms**
W.H.C.F. Kooistra, M. De Stefano, D.G. Mann, and L.K. Medlin



**Silicon – a Central Metabolite for Diatom Growth and Morphogenesis**
V. Martin-Jézéquel and P.J. Lopez

**Silicon in Plants**
D. Neumann



**Organisms: Sponges**

**Silica Deposition in Demosponges**
M.J. Uriz, X. Turon, and M.A. Becerro



**Molecular Mechanism of Spicule Formation in the Demosponge *Suberites domuncula*: Silicatein – Collagen – Myotrophin**
W.E.G. Müller, A. Krasko, G. Le Pennec, R. Steffen, M. Wiens, M.S.A. Ammar, I.M. Müller, and H.C. Schröder

## Biotechnology

**Biotechnological Advances in Biosilicification**
J.L. Sumerel and D.E. Morse



**Silicase, an Enzyme Which Degrades Biogenous Amorphous Silica: Contribution to the Metabolism of Silica Deposition in the Demosponge *Suberites domuncula***
H.C. Schröder, A. Krasko, G. Le Pennec, T. Adell, M. Wiens, H. Hassanein, I.M. Müller, and W.E.G. Müller

**Silicon Biomineralisation: Towards Mimicking Biogenic Silica Formation in Diatoms**
E.G. Vrieling, S. Hazelaar, W.W.C. Gieskes, Q. Sun, T.P.M. Beelen, and R.A. van Santen

# Organisms: Diatoms

# Living Inside a Glass Box – Silica in Diatoms

Franz Brümmer

## 1
## Introduction

Silicon (Si; name originates from the Latin, silicis = flint) is present in all living organisms and is required for the production of stable structural materials. This takes place not only in single-celled organisms, e.g. diatoms, but also in lower Metazoa like sponges as well as in higher plants. Silicon is one of the most abundant elements in the earth's crust, second only to oxygen. Some geologists estimate that nearly 90% of all known minerals are silicates in combination with other elements such as aluminium (andalusite, kyanite), calcium (wollastonite), iron (fayalite), magnesium (fosterite), zinc (willemite) or zirconium (zircon), just to mention a few.

Silica, or more precisely hydrated silica, often referred to as opal, is the second most abundant mineral type formed by organisms. Only carbonate minerals are distributed in a wider range and are more abundant.

The basic chemical unit of silicates is the ($SiO_4$) tetrahedron-shaped anionic group. The overall charge condition leaves the oxygens with the option of bonding to another silicon ion and therefore linking one ($SiO_4$) tetrahedron to another and another, and so on. Regrading the numbers of tetrahedrons and the different shapes results in the silicates being divided into subclasses, not by their chemistries, but by their structures, e.g. nesosilicates (single tetrahedrons), sorosilicates (double tetrahedrons), inosilicates (single and double chains) or cyclosilicates (rings).

The soluble form of silica is a monomer, orthosilicic acid, which is a silicon atom also tetrahedrally coordinated to four hydroxyl groups with the formula $Si(OH)_4$. Only this soluble form is biologically assimilable.

Several groups of marine organisms as well as their representatives in freshwater habitats such as diatoms, "radiolarians", choanoflagellates, sponges and higher plants take up $Si(OH)_4$ from water to build their opal – amorphous hydrated silica – skeletons.

Biologisches Institut, Abteilung Zoologie, Universität Stuttgart, Pfaffenwaldring 57, 70569 Stuttgart, Germany

Progress in Molecular and Subcellular Biology, Vol. 33
W. E. G. Müller (Ed.)

# 2 Silica in Protozoa, Sponges and Higher Plants

## 2.1 Phaeodaria

In the artificial grouping of radiolarians (Radiolaria), the Phaeodaria are oceanic protists encased with porous skeletons composed of biogenic opal with organic substances and traces of Mg, Ca and Cu (Lee et al. 2000). They have been observed near the water surface of the oceans down to greater depths of up to several thousands of meters. From a quantitative viewpoint, they are only important in the Pacific equatorial belt.

## 2.2 Choanoflagellates

The choanoflagellates are nowadays included within the protozoa as the order Choanoflagellida (Lee et al. 2000). They are a well-defined group of free-living, colourless monads. They are uninucleated and small, seldom more than 10 μm in size. Their most distinctive feature is the uniform appearance of the protoplast with a single anterior flagellum. Choanoflagellates are ubiquitously distributed in aquatic habitats. The 50 genera and more than 100 species are included in 3 families, one with basket-like siliceous loricae (Acanthoecidae) found only in marine or brackish water (for review see Leadbetter 1991).

## 2.3 Silicoflagellates

The silicoflagellates (Silicoflagellata) are known in botanical literature as the Dictyochophyceae inside the alga class of Heterokontophyta (van den Hoek et al. 1995). The known 30 extant species form a clade that is most closely related to diatoms. Silicoflagellates are small to medium-sized unicellular protists and are common in marine planktonic and benthic habitats. Silicoflagellates sensu stricto form complex siliceous external skeletons (Lee et al. 2000), for example, *Dictyocha speculum*. The skeleton takes the form of a flat basket composed of hollow, but robust tubes of silica.

Among the protists, there are several other groups including species forming tests of siliceous plates, e.g. testate amoebae, or showing siliceous endoskeletal elements, for example, the heterotrophic flagellate *Hermesium adriaticum* (Lee et al. 2000).

## 2.4 Sponges

Two, the Demospongiae and the Hexactinellida, out of the three classes of the phylum Porifera contain siliceous spicules. Demospongiae contains more than 85% of all living sponges with about 6000 valid species already described; Hexactinellida include about 500 described species, 7% of all Porifera (Hooper and van Soest 2002). The specific spicule characteristics are used for taxonomic purposes. From one to eight types of spicules may be present within one single sponge species. The spicules stand for two major functions: they provide mechanical support to the soft part of the sponges and second, the presence of hard spicules also provides some kind of defence mechanism discouraging predators from eating sponges. The spicules are built by specialised cell types, the sclerocytes (for review see Simpson 1984). Up to now, the how and where the secretion of spicules take place – intra- or extracellularly– have been intensively discussed (Garrone et al. 1981; Simpson 1984; Uriz et al. 2000; Schönberg 2001).

Siliceous sponges proliferated well on the shelves of the Jurassic Thethys Sea, building up huge sponge reefs (Leinfelder 1993; Leinfelder et al. 1994).

## 2.5 Plants

Within higher plants, the Cyperaceae (e.g. horsetails, Arthrophyta) and Gramineae (grasses) contain up to 10–15% silica, which can be found in the cell walls, inside the cell lumen as well as extracellularly on the outer cuticle. The opaline silica deposits are most commonly in the form of particles called phytoliths (Perry and Keeling-Tucker 2000).

In higher animals, the role of silica is not known; in human blood, however, 138 µmol/l silicate (as $SiO_2$) has been reported (Wissenschaftliche Tabellen Geigy 1979).

# 3 Living in a Glass Box – the Diatoms

The diatom lineage contains the most beautiful, delicate eukaryotic organisms that are usually classified as algae (for reviews, see Round et al. 1990; van den Hoek et al. 1995.

Diatoms are extremely abundant in both freshwater and marine ecosystems. It is estimated that 20–25% of all organic carbon fixation on our "Blue Planet" is carried out by diatoms made possible by the chlorophyll they contain. Diatoms are thus a major food resource for all imaginable micro-organisms and animal larvae and diatoms are a major source of atmospheric oxygen.

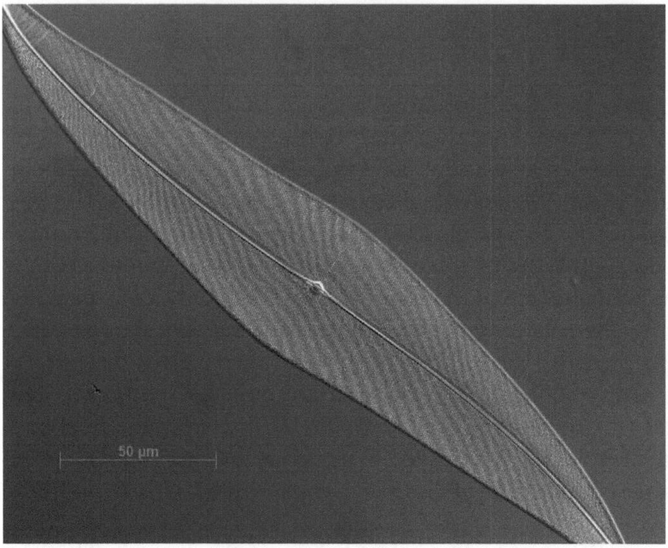


**Fig. 1.** *Pleursigma angulatum*, an example for a pennate diatom. In the middle of the cell, the raphe as well as the central nodule are visible

The oldest certain fossil diatoms are Lower Cretaceous (about 120 million years ago) in age. Most diatom fossils known are from Eocene and Miocene rocks, 35 and 5 million years old, respectively. The richest sources of diatom fossils are deposits of the skeleton known as diatomite or diatomaceous earth. This mineral was formed as ancient diatoms died and sank to the bottom of the oceans or lakes. Today, these large deposits of white chalky material are even mined and processed in cleansers, paints, filtering systems and abrasives. In addition, many toothpastes contain bits of fossil diatoms.

Within the Heterokontophyta, diatoms form the class Bacillariophyceae (Diatomophyceae = the diatoms). They are subdivided into two main groups (classes or orders) – the centrate diatoms (Centrobacillariophyceae; Centrales) and the pennate diatoms (Pennatibacillariophyceae; Pennales). The former are generally radially symmetrical, the latter show a typical bilateral symmetry. There are over 250 genera containing around 100,000 living diatom species.

All species of diatoms are unicellular or colonial coccoid algae. The cells secrete intricate skeletons of silica. The skeleton of a diatom, the frustule, is made of very pure silica coated with a thin layer of organic material. The presence of silica in the cell walls means that these tiny organisms live in a "glass house" or a "glass box". The skeleton is divided into two parts, one of which (the epitheca) is larger and older and overlaps the other (the hypotheca) like the lid of a box. Therefore, a more accurate description is that they live in glass Petri dishes (frustules). The top of the frustules, the epitheca, is perforated with many holes, arranged in a pattern characteristic of the species. The holes permit close contact with the environment and allow the diffusion of materials into and out of the cell.

Diatoms reproduce through cell division – one cell divides into two cells. First, the nucleus divides via mitosis and, second, two new valves are formed within the cell wall. The parental epitheca and hypotheca separate and new valves are laid down between them. The older valves fit over the newly formed hypotheca. Thus, each new cell contains an old epitheca and a new hypotheca resulting in a decrease in the mean size of a dividing population of diatoms. They become smaller and smaller with time. The decrease in size of progeny cells is called diminution and occurs until a certain size is reached. Fortunately, diatoms can also reproduce sexually in order to regain maximum size.

Diatoms are, as already mentioned, part of the drifting community, the plankton. With their rather heavy silica cell walls, planktonic diatoms are faced with the problem of how to remain in the uppermost layers ensuring enough light for photosynthesis. First, diatoms are well protected by their glass box and, therefore, able to stay intact in the turbulent mixing of the upper layers. Second, several diatoms can reduce their densities and become more buoyant by excluding heavy ions from their cell sap, thereby reducing their density to even less than the density of the surrounding sea water. Third, they can bear long spines or other protrusions which slow down their sinking rate ("living snowflakes").

On the other hand, for benthic forms the heavy frustule guarantees their remaining at the bottom, where nutrient concentrations are usually higher than in the water column. Also, in this case, the lucent frustule does not block photosynthesis and may serve for collecting and amplifying light.

Diatoms have also been recognised to produce toxins which infect shellfish and also humans along the food chain. The poisoning is called amnesic shellfish poisoning (ASP) and was first recognised in 1987 in Canada (Bates et al. 1989). To date, reports of ASP are mainly from North America and Canada, while only very low and insignificant concentrations have been detected in other parts of the world (Hallegraeff 1995; Bates et al. 1998). The symptoms include abdominal cramps, vomiting, disorientation and memory loss (amnesia). Most unexpectedly, the causative toxin is produced by a diatom (*Pseudo-nitzschia* spp.) and not by a dinoflagellate. The toxin responsible is called domoic acid which is a naturally occurring compound belonging to the kainoid class. For the mode of action, it is assumed that domoic acid is absorbed in very low rates through the gastrointestinal mucosa and transferred to the brain tissue, acting as a glutamate agonist in the brain and central nervous system, where it strongly binds to a special type of glutamate receptor (Wright and Quilliam 1995).

## 4
## Biosilicification in Diatoms

The cell wall of diatoms consists of polymerised silicic acid, an amorphous material without any crystalline structure. Each siliceous element, like the valves or girdles, is formed within its own flattened vesicle, the so-called silica deposition vesicle (SDV). The orthosilicic acid and other forms of silicon are transported to these SDVs and polymerisation occurs in order to create solid deposits of silica. The amorphous, non-crystalline silica of the diatom frustule, however, does not assemble itself to form the specific patterns that are so characteristic of this algae. There are also cytoplasmic structures and biochemical processes involved (Perry and Keeling-Tucker 2000). To make a long story short, Hecky and coworkers (1973) already proposed a hypothetical arrangement of organic layers in the diatom wall where a protein template (rich in serine, threonine and glycine) is proposed to interact with the mineral phase and an outer layer of polysaccharides acts as a buffer system. In the meantime, within the SDV, organic structures with the shape of individual components of the frustule were detected and interpreted as the matrix onto which silicic acid polymerisation can be conducted and other protein-containing materials like frustulins, HEP 200 proteins and silaffins were found and might also be candidates involved in the promotion of silicic acid polymerisation and precipitation of biogenic silica (reviewed by Perry and Keeling-Tucker 2000).

As silicon is the major limiting nutrient for diatom growth, the questions for the mechanisms of silicate uptake into the diatom cells must be addressed. Quite recently, not only silica acid transporter genes have been discovered (Hildebrand et al. 1997), but sizeable intracellular pools of soluble silicon have also been identified (Martin-Jezequel et al. 2000).

## 5 Conclusion

The investigations of diatoms provide some clear insights into the mechanism(s) of the nucleation and polymerisation of an amorphous material and also, what might be even more interesting, whose structures can function as templates and are, therefore, necessary for building up such beautiful and robust skeletons.

In sponges, the finding of Shimizu and coworkers (Shimizu et al. 1998; Cha et al. 1999) that the formerly termed "axial organic filament" is an enzyme, silicatein, which mediates the formation of the spicules, was a great step forward towards the molecular level. Quite recently, in the marine sponge *Suberites domuncula*, the gene expression of silicatein could be shown as well as the regulatory effect of silicate and myotrophin on this enzyme (Krasko et al. 2000).

Using molecular biology methods in combination with in vitro studies, it might be possible to further itemise the regulation system required for the formation of functional skeletons e.g. in diatoms and in sponges. And what about a silicon transporter in sponges?

## References


Bates SS, Bird CJ, de Freitas ASW, Foxall R, Gilgan M, Hanic LA, Johnson GR, McCulloch AW, Odense P, Pocklington R, Quilliam MA, Sim PG, Smith JC, Subba Rao DV, Todd ECD, Walter JA, Wright JLC (1989) Pennate diatom *Nitzschia pungens* as the primary source of domoic acid, a toxin in shellfish from eastern Prince Edward Island, Canada. Can J Fish Aquat Sci 46:1203–1215

Bates SS, Garrison DL, Hoerner RA (1998) Bloom dynamics and physiology of domoic-acid-producing *Pseudo-nitzschia* species. In: Anderson DM, Cembella AD, Hallegraeff GM (eds), Physiological ecology of harmful algal blooms. NATO ASI Series, vol G 41, Springer, Berlin Heidelberg New York, pp 267–292

Cha JN, Shimizu K, Zhou Y, Christianssen SC, Chmelka BF, Stucky GD, Morse DE (1999) Silicatein filaments and subunits from marine sponge direct the polymerization of silica and silicones in vitro. Proc Natl Acad Sci USA 96:361–365

Garrone R, Simpson TL, Pottu J (1981) Ultrastructure and deposition of silica in sponges. In: Simpson TL, Volcani BE (eds) Silicon and siliceous structures in biological systems. Springer, Berlin Heidelberg New York, pp 495–525

Hallegraeff GM (1995) Harmful algal blooms: a global overview. In: Hallegraef GM, Anderson DM, Cembella AD, Enevoldsen HO (eds), Manual on harmful marine microalgae. Intergovernmental Oceanographic Commission Manuals and Guides 33, UNESCO, Paris. pp 1–22

Hecky RE, Mopper K, Kilham P, Degens ET (1973) The amino acid and sugar composition of diatom cell walls. Mar Biol 19:323–331

Hildebrand M, Volcani BE, Gassmann W, Schroeder JI (1997) A gene family of silicon transporters. Nature 385:688–689

Hooper JNA, van Soest RWM (2002) Systema Porifera - a guide to the classification of sponges, vols I and II. Kluwer/Plenum Press, New York, pp 1–1754

Krasko A, Lorenz B, Batel R, Schröder HC, Müller IM, Müler WEG (2000) Expression of silicatein and collagen genes in the marine sponge *Suberites domuncula* is controlled by silicate and myotrophin. Eur J Biochem 267:4878–4887

Leadbetter BSC (1991) Choanoflagellate organization with special reference to loricate taxa. In: Patterson DJ, Larson J (eds) The biology of free-living heterotrophic flagellates. Clarendon Press, Oxford, pp 241–258

Lee JJ, Leedale GF, Bradbury P (2000) An illustrated guide to the protozoa, vols I and II, 2nd edn. Society of Protozoologists, Lawrence, pp 1–1432

Leinfelder RR (1993) Upper Jurassic reef types and controlling factors. Profil 5:1–45

Leinfelder RR, Krautter M, Laternser R, Nose M, Schmid DU, Schweigert G, Werner W, Keupp H, Brugger H, Herrmann R, Rehfeld-Kiefer U, Schroeder JH, Reinhold C, Koch R, Zeiss A, Schweizer V, Christmann H, Menges G, Luterbacher H (1994) The origin of Jurassic reefs: current research developments and results. FACIES 31:1–56

Martin-Jezequel V, Hildebrand M, Brzezinski MA (2000) Silicon metabolism in diatoms: implications for growth. J Phycol 36:821–840

Perry CC, Keeling-Tucker T (2000) Biosilicification: the role of the organic matrix in structure control. J Biol Inorg Chem 5:537–550

Schönberg CHL (2001) New mechanism in demosponge spicule formation. J Mar Biol Assoc UK 81:345–346

Shimizu K, Cha JN, Stucky GD, Morse DE (1998) Silicatein alpha: cathepsin L-like protein in sponge biosilica. Proc Natl Acad Sci USA 95:6234–6238

Round PH, Crawford MR, Man GD (1990) The diatoms biology and morphology of the genera. Cambridge University Press, New York, pp 1–774

Simpson TL (1984) The cell biology of sponges. Springer, Berlin Heidelberg New York, pp 1–662

Uriz MJ, Turon X, Becerro (2000) Silica deposition in Demospongiae: spiculogenesis in *Crambe crambe*. Cell Tissue Res 301:299–309

Van den Hoek C, Mann DG, Jahns HM (1995) Algae - an introduction to phycology. Cambridge University Press, Cambridge, pp 1–627

Wissenschaftliche Tabellen Geigy (1979) Volume II. CIBA-Geigy Ltd, Basle, pp 1–331
Wright JLC, Quilliam MA (1995) Methods for domoic acid, the amnesic shellfish poisons. In: Hallegraef GM, Anderson DM, Cembella AD, Enevoldsen HO (eds), Manual on harmful marine microalgae. Intergovernmental Oceanographic Commission Manuals and Guides 33, UNESCO, Paris. pp 113–133

# Components and Control of Silicification in Diatoms

Mark Hildebrand[1] and Richard Wetherbee[2]

## 1 Introduction

In nearly all publications relating to the synthesis of the silicified diatom cell wall, an inherent fascination with the physical beauty of the intricate and ornate structures that are formed and an intellectual curiosity about how these structures are made are expressed. Viewing electron micrographs of diatoms for the first time usually engenders an immediate interest.

Research in this field has advanced in conjunction with techniques available at a given time. First, and very shortly after the invention of the microscope (see Werner 1977 for a brief history of the early research on diatoms), visual observation and categorization began (Anonymous 1703; Baker 1753). Due to the large number of diatom species, estimated in the tens of thousands or greater (Werner 1977; Gordon and Drum 1994; Norton et al. 1996), this has continued into the present, although now with the use of the electron microscope (Pickett-Heaps et al. 1990; Round et al. 1990 for reviews). Diatoms were of great scientific interest in the nineteenth century; as mentioned by Werner (1977), by 1891 there were more than 1500 original research communications on diatoms. One pastime amongst microscopists during the Victorian era was to arrange individual diatom cells of different shapes under the microscope into pictures depicting larger objects such as flowers or landscapes – a skill that is still being performed (see www.diatoms.co.uk). With the advent of the electron microscope came investigations classifying detailed structures of the diatom cell wall, about which a comprehensive terminology has arisen (Ross et al. 1979). In addition to the categorization of completed structures, electron microscope images in time series, or at different stages of cell wall synthesis, have given insights into the process of formation of these structures. In the 1960s and 1970s, biochemical approaches were applied (mostly in the laboratory of Ben Volcani) to try to understand what metabolic changes occurred during cell wall synthesis, and to isolate and characterize the cellular components involved. In 1985, the first demonstration that diatom gene

[1] Marine Biology Research Division, Scripps Institution of Oceanography, University of California, San Diego, 9500 Gilman Dr., San Diego, California 92093-0202, USA
[2] School of Botany, The University of Melbourne, Parkville, Victoria 3010, Australia

Progress in Molecular and Subcellular Biology, Vol. 33
W. E. G. Müller (Ed.)

expression was regulated by silicon was presented (Reeves and Volcani 1985), which opened the door for the application of modern molecular techniques. Subsequent studies have utilized gene and protein isolation and characterization (Hildebrand et al. 1993, 1997, 1998; Kröger et al. 1997, 1999, 2000, 2001; Vrieling et al. 1999a; Kröger and Wetherbee 2000), immunolocalizations (Kröger et al. 1997; van de Poll et al. 1999; Kröger and Wetherbee 2000), identification and localization of cytoskeletal elements (Pickett-Heaps 1998b, van de Meene and Pickett-Heaps 2002), direct probing of the intracellular organelle in which silicification occurs (Li et al. 1989; Vrieling et al. 1999b) and examination of the form of silicon in the diatom cell wall by X-ray scattering (Vrieling et al. 2000) and atomic force microscopy (Crawford et al. 2001). These studies have identified and characterized some of the key molecular entities and microscale structures involved in silicification, leading to new insights into the process. This review is aimed at discussing our current knowledge and hypotheses regarding the function of molecular components involved in diatom cell wall silicification, highlighting what we do and do not know, and tying these insights into our understanding of macroscale cellular processes leading to the formation of the silicified diatom cell wall.

## 2 Features of Diatom Cell Walls and Terminology

The diatom cell wall, also called the frustule, is composed of two halves fitting together like a Petri dish (Fig. 1); the upper half is the epitheca and the lower the hypotheca. The upper and lower surfaces of the frustule are called valves, and the structures extending on the sides and overlapping the two halves of the cell are the girdle bands (Fig. 1). The valves can be perforated with pores called areolae (Fig. 1, center), and these can vary greatly in number, size, and structure. The girdle bands can also vary in structure, and their number is species-specific (Round et al. 1990). Diatoms fall into two general classes, based on the symmetry of their cell walls, the centrics and the pennates. Centric diatoms are radially symmetrical, although not all centric valves are circular as shown (Fig. 1, right), whereas pennate diatoms have bilateral symmetry (Fig. 1, center). There are diatom species with more complicated symmetries (Round et al. 1990). The valve surface in most pennate species is divided longitudinally by the raphe fissure (Fig. 1, center), a slit through which mucilage is secreted for cellular movement (Edgar 1983; Edgar and Zovortnik 1983; Edgar and Pickett-Heaps 1984a; Lind et al. 1997). The raphe fissure can run the length of the valve, or be interrupted by a solid structure called the central nodule (CN), and there are examples of pennate species with no raphe at all (Pickett-Heaps 1989; Round et al. 1990). When present, the raphe structure has a complex "tongue-in-groove" curvature (see Fig. 7B); it was proposed that this could prevent the valve from splitting under turgor or other stresses (Pickett-Heaps et al. 1979).

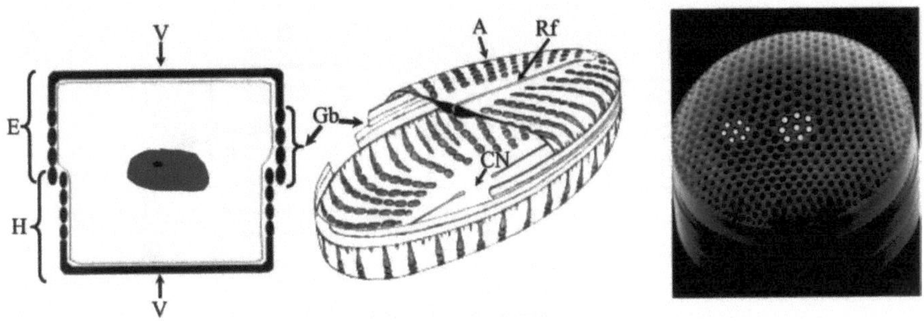


**Fig. 1.** Features of diatom cell walls. *Left* Diagram representing a lateral cross section through a diatom cell. The extreme upper and lower portions are the valves (*V*), and the upper and lower halves of the cell denoted by *brackets* are called the epitheca (*E*) and hypotheca (*H*), respectively. Structures on the sides and overlapping the two halves of the cell are the girdle bands (*Gb*). *Center* is a cutaway view of the frustule of the pennate diatom *Navicula pelliculosa*, from Reimann et al. (1966). The basic structure is as diagramed in the lateral cross section on left, but includes the areolae (*A*), raphe fissure (*Rf*), and central nodule (*CN*). *Right* Electron micrograph of a cleaned frustule of the centric diatom species *Thalassiosira oestrupii* from Fryxell and Hasle (1980). The *white dots* indicate the position at the center of the valve of an areola surrounded by seven others, and on the left an areola surrounded with the more typical six others. Areolae decrease slightly in size radially from the center

Vegetative cell division in diatoms occurs as shown in Fig. 2. Prior to division, the mother cell expands, but the two halves of the cell remain attached (Fig. 2A). After cytokinesis and cleavage, silicification of new valves begins on adjacent faces of what will become the daughter cells (Fig. 2B). The polymerization of silica occurs within the silica deposition vesicle (SDV), an intracellular organelle (Drum and Pankratz 1964; Reimann et al. 1966; Schmid et al. 1981) of unknown origin. This vesicle, bounded by a membrane called the silicalemma (Reimann et al. 1966; Schmid et al. 1981), always encases silica in forming valves. After valve formation, the SDV is exocytosed and becomes part of the daughter cell wall (Fig. 2C below, D above). Depending on the species, girdle bands form sequentially and then daughter cells separate (Fig. 2C, D below), or cells can separate prior to girdle band formation, resulting in daughter cells with girdle bands on the epitheca, but not on the hypotheca (Fig. 2D, above). In the latter case, girdle bands are added during subsequent growth and cell expansion (Fig. 2E).

# 3
# Transport of Silicic Acid into the Diatom Cell

The first demonstration that diatoms could take up silicon from the environment was by Lewin (1954, 1955). In numerous studies, silicon transport has been shown to be saturable (Paasche 1973; Azam et al. 1974; Martin-Jézéquel

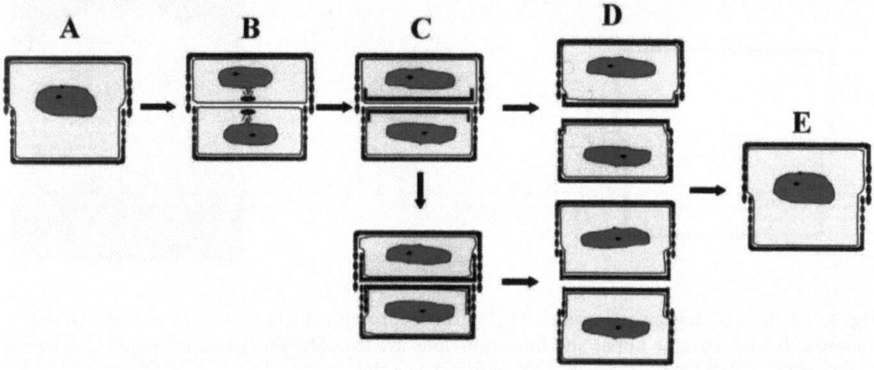


**Fig. 2A–E.** The process of cell wall synthesis and cell division in diatoms, shown in lateral view. **A** A mother cell that has expanded prior to cytokinesis, features are as shown in Fig. 1. **B** Cytokinesis and cleavage have occurred, and silicification is initiated at the center of the cleavage furrow in both daughter protoplasts. **C** *Above* The valve is completed, but has not yet been exocytosed, girdle bands have not formed. *Below* The valve is completed and has been exocytosed, and girdle bands have formed. **D** *Above* Valves in daughter cells from **C** (*above*) have been exocytosed and cells have separated. Girdle bands are still not formed. *Below* Daughter cells from **C** (*below*) have separated and are expanding. **E** Girdle band formation in cells from **D** (*above*), and continuing expansion of cells

et al. 2000), indicating a carrier-mediated process. Tracer techniques were developed to monitor transport using radiolabeled silicon and germanium (Azam 1974; Azam et al. 1974; Azam and Volcani 1974), of which the latter is a congener of silicon and has a more usable radioisotope. Competition experiments showed that the silicon transporting moiety was specific for silicon and germanium, with $K_s$ values of 4.5 and 6.8 µM for Si and Ge, respectively, and a $K_i$ value of 2.2 µM for Ge inhibition of Si uptake (Azam et al. 1974; Azam and Volcani 1974). In nature, germanium concentrations are so low as to not significantly affect silicon transport. It has only recently been demonstrated what chemical form of silicon is transported; Del Amo and Brzezinski (1999) showed that in most diatoms the unionized form of silicic acid [$Si(OH)_4$] was transported (which is the predominant form at pH 8.0 of seawater), although *Phaeodactylum tricornutum* apparently could transport the ionized form [$SiO(OH)_3^-$] as well (Reidel and Nelson 1985; Del Amo and Brzezinski 1999). Transport was sodium-coupled in marine diatoms (Bhattacharyya and Volcani 1980) and apparently sodium and perhaps potassium coupled in freshwater species (Sullivan 1976). Thus in marine species, the transporter acts as a sodium/silicic acid symporter, and indirect evidence suggested electrogenic transport with a $Si(OH)_4$:Na+ ratio of 1:1 (Bhattacharyya and Volcani 1980). Efflux of silicic acid from diatom cells has been measured (Azam et al. 1974; Sullivan 1976), which interestingly did not occur in the absence of extracellular silicon (Sullivan 1976). Since all types of transporters can function bidirectionally, it is likely that silicic acid uptake and efflux were controlled by the same moiety.

Sullivan (1977) showed that silicic acid transport was regulated during the course of the cell cycle in *Navicula pelliculosa*, being induced only just prior to cell wall silicification in G2/M phase, and not at other times. He also showed that transport activity had the characteristics of a protein that was induced and turned over rapidly. $K_s$ and $V_{max}$ values varied over the course of the cell cycle and during cell wall synthesis (Sullivan 1977), suggesting changes in the number and affinity of transporters, consistent with changes in protein abundance and composition.

The molecular characterization of the silicon transport system began with the cloning of the first silicic acid transporters from the diatom *Cylindrotheca fusiformis* (Hildebrand et al. 1997). The original cDNA clone, named *SIT1*, encoded a protein of 548 amino acids (Hildebrand et al. 1997). There were ten hydrophobic regions in the sequence predicted as membrane-spanning segments, and a long hydrophilic carboxy-terminus (Hildebrand et al. 1997; Hildebrand 2000). Microinjection experiments in *Xenopus laevis* oocytes demonstrated that the encoded protein was a silicic acid transporter, with characteristics similar to those identified in diatom whole cell uptake experiments (Hildebrand et al. 1997). *SIT1* was used as a probe to isolate other copies of *SIT* genes in *C. fusiformis*, and it was discovered that these constituted a gene family, with five members (Hildebrand et al. 1998). The amino acid sequences of the five types of SIT were highly conserved in the transmembrane domain, but less conserved in the carboxy-terminal region (Hildebrand et al. 1998; Hildebrand 2000). The similarity of the transmembrane domains suggested that the mechanism of silicic acid passage through the membrane was likely to be the same for each type of SIT. The carboxy-terminal segment in all five SITs had a very high probability to form a coiled-coil structure, suggesting that this portion interacted with other proteins. In other types of transporters, the carboxy-terminus can be involved in interaction with other proteins (Carafoli 1994), can control transport activity (Katagiri et al. 1992; Olivares et al. 1994; Due et al. 1995), sometimes by interaction with other proteins (Carafoli 1994), and can be responsible for targeting intracellular localization (Verhey et al. 1993, 1995; Olivares et al. 1994). By analogy, it was proposed that SIT activity or localization may be controlled by their carboxy-terminal domains (Hildebrand et al. 1998; Hildebrand 2000).

Levels of mRNA for the five *SIT*s varied coordinately, although total mRNA amounts for the different *SIT*s differed greatly (Hildebrand et al. 1998). This inferred that SIT proteins played specific roles in the transport process (Hildebrand et al. 1998). Overall *SIT* mRNA levels increased fourfold 20–40 min prior to a period shown in later experiments (Fig. 3) to correspond to an increase in uptake rate. These experiments were done using synchronized cultures of *C. fusiformis*, in which the majority of cells progress through the same stage of the cell cycle at the same time (Darley and Volcani 1969). Data in Fig. 3B indicated that incremental uptake changed from an average of 1.72 ±

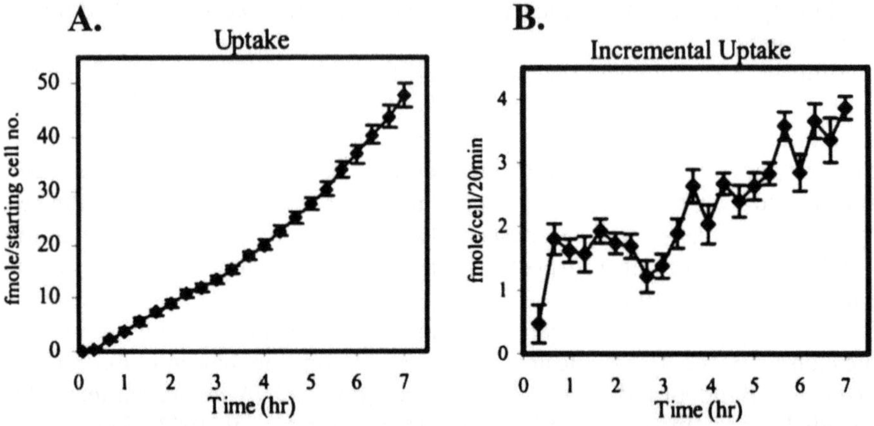


**Fig. 3A,B.** Silicic acid uptake in synchronized cultures of silicon-starved *Cylindrotheca fusiformis* after silicate replenishment. Sodium silicate was added to 100 µM at 0 h. Triplicate samples were measured, *error bars* denote the SEM. **A** Uptake, **B** incremental uptake over each 20-min period

0.048 fmol cell$^{-1}$ 20 min$^{-1}$ between 0.33 and 2.33 h (prior to *SIT* mRNA induction) to 2.86 ± 0.175 fmol cell$^{-1}$ 20 min$^{-1}$ between 3.33 and 7.0 h (after induction). When SIT mRNA levels were maximum at 2.67 and 3 h (Hildebrand et al. 1998), there was a distinct and reproducible decrease in incremental uptake (Fig. 3A, B). A reduction in uptake when transporter mRNA levels were maximum may seem odd, but a possible explanation will be presented after further discussion.

In most diatom species, silicic acid transport and cell wall silica deposition are temporally coupled (Chisholm et al. 1978), but there are exceptions (Chisholm et al. 1978; Binder and Chisholm 1980). In species where these processes are coupled, kinetic measurements have identified a mode known as internally controlled uptake (Conway et al. 1976; Conway and Harrison 1977) in which uptake rate is independent of external silicic acid concentration, but rather depends on the rate of intracellular utilization (i.e., cell wall deposition) of silicon. Results of these studies suggested that not only were transport and deposition temporally coupled, but that they were directly coupled through a feedback mechanism, proposed to work through the saturation of intracellular pools which then controlled uptake (Conway and Harrison 1977). The data in Fig. 4 support and extend this hypothesis. Silicon-starved synchronized cultures of *C. fusiformis* were replenished with 100 µM silicate in the presence and absence of 10 µM germanate (Ge). At this ratio of Ge : Si, germanate inhibits cell wall silica incorporation (Darley and Volcani 1969), but would not directly inhibit silicic acid uptake to any appreciable extent (Azam et al. 1974). Data in Fig. 4A show that Ge inhibited cell wall silica incorporation by 75% at the end of the experiment, and Fig. 4C shows an identical inhibition of silicic acid uptake. Soluble intracellular silicon pools increased more rapidly, and to greater levels, in the presence of Ge (Fig. 4B). An interpretation of these data

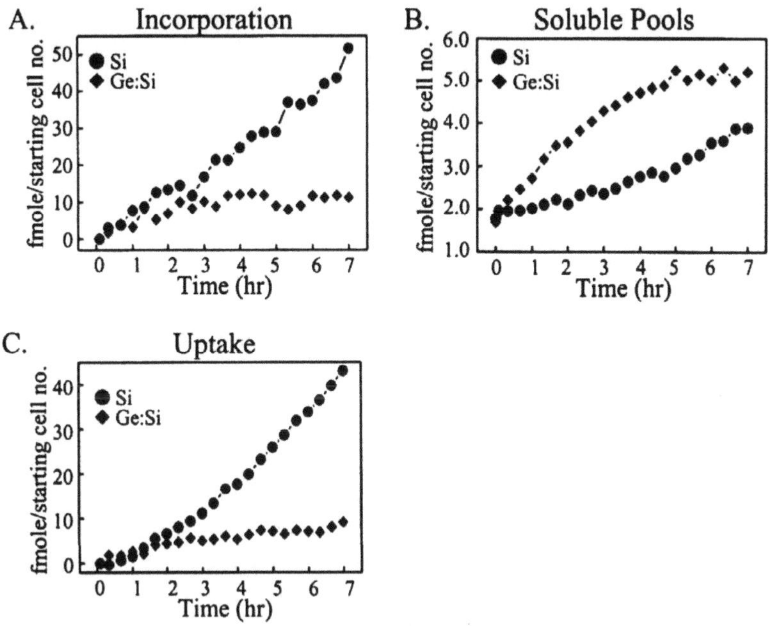


**Fig. 4A–C.** Response of synchronized cultures of silicon-starved *C. fusiformis* after replenishment with 100 μM silicate or 10 μM germanate:100 μM silicate. **A** Incorporation of silica into the cell wall, **B** levels of soluble intracellular pools of silicic acid, **C** uptake of silicic acid from the medium

(Hildebrand 2000) was that inhibition of cell wall silica incorporation increased intracellular pool levels, inducing a feedback mechanism that inhibited uptake, consistent with Conway and Harrison's hypothesis. However, since uptake was inhibited prior to intracellular Si levels reaching maximum (Fig. 4B, C), it was clear that pools did not have to be saturated for feedback to occur. This led to the proposal that regulation of uptake occurred not by the absolute level of silicic acid in the pools, but by the relative ratio of silicic acid to silicic acid binding components (Hildebrand 2000). There is consistent data in the literature suggesting that intracellular silicic acid is complexed with some cellular component (Azam et al. 1974; Sullivan 1986), perhaps as a means to maintain supersaturated levels of precursor material for the cell wall (Martin-Jézéquel et al. 2000; Hildebrand 2000). It was proposed that when excess intracellular silicic acid was present relative to its binding component, uptake was inhibited or efflux induced, and when excess binding component was present then uptake was induced (Hildebrand 2000). According to this model, silicic acid for the cell wall was drawn from internal pools as needed, and flux through the pools connected silicification with uptake.

The intracellular mechanism controlling transport (Conway and Harrison 1977; Hildebrand 2000) represents an important level of regulation over silicification. Rather than silicic acid being pumped into the cell and driving silicification, this mechanism coordinates uptake with requirements for deposition, allowing correct amounts of silicic acid to be taken up and preventing accumulation of excess intracellularly (Hildebrand 2000). With the exception of the SITs, it is unclear what the molecular components of this control mechanism are. We have only inferential evidence as to the nature of the intracellular silicon-binding components (Bhattacharyya and Volcani 1983; Hildebrand 2000), and no clear insights into how intracellular free silicic acid levels are actually sensed. However, the possibility that the SITs bind proteins at their carboxy-termini provides a reasonable explanation for how this mechanism may ultimately affect SIT transport activity, as shown in Fig. 5. In the example of the GLUT1 glucose transporter, deletion of the carboxy terminus locks the transporter into an inward facing orientation (Oka et al. 1990), suggesting that the C-terminus controls the direction of transport by inducing conformational changes in the protein. By analogy to this and other examples (Carafoli 1994; Oka et al. 1990), it was proposed (Hildebrand 2000) that SIT-interacting proteins (which could include the SITs through dimerization) regulated the direction of silicic acid passage through the membrane by inducing conformational changes in the SITs through their C-termini, favoring transport in one direction or the other (Fig. 5). It is possible that the SIT-interacting proteins are

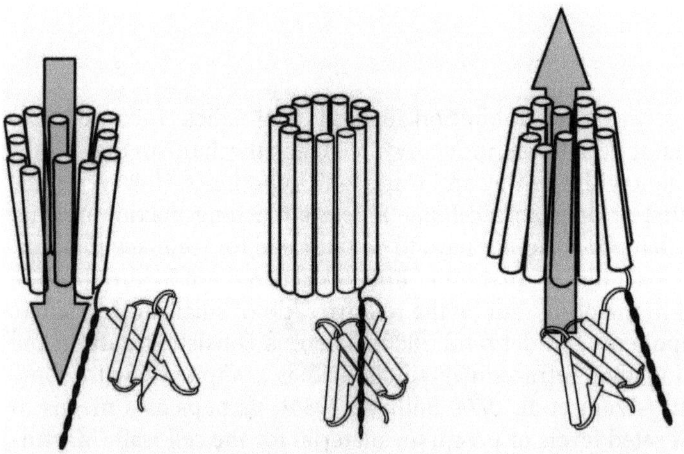

**Fig. 5.** Diagrammatic representation of a possible mechanism operating on the silicic acid transporters to control the direction of transport. The SIT is represented as a bundle of cylinders symbolizing possible helical transmembrane segments. The arrangement of cylinders is stylized and most certainly does not represent their actual arrangement. Extracellular would be at the *top*, intracellular at the *bottom*. At *bottom right* of each transporter is another protein interacting with the SIT carboxy-terminus through a coiled-coiled motif (see text). This protein regulates the direction of transport by altering the SIT's conformation. *Arrows* denote direction of transport. On the *left*, uptake is favored, in the *center* no transport occurs, and on the *right* efflux from the cell occurs

directly involved in sensing intracellular free silicic acid levels, or they may only be the penultimate step in a chain of events. Clearer understanding awaits their isolation and characterization. Returning to the question of why silicic acid uptake was reduced (Fig. 3) when *SIT* mRNA levels were maximum; we suggest that this was due to the removal of SIT-interacting proteins from plasma membrane (PM)-localized SITs by newly synthesized SITs that had not yet reached the PM. According to this hypothesis, removal of SIT-interacting proteins would diminish the uptake activity of some of the PM SITs, reducing overall uptake. The subsequent increase in uptake could result from synthesis of compensating amounts of SIT-interacting proteins.

The above data and hypotheses suggest that control of silicic acid transport occurs at many levels. Uptake, as well as deposition, is regulated during the cell cycle, and in most diatoms these processes are temporally coupled (Chisholm et al. 1978). Transport is also directly coupled to and controlled by the rate of silica deposition (Conway and Harrison 1977; Fig. 4) by a yet to be elucidated mechanism. Control at the transporter level is likely to occur by two mechanisms and perhaps three: (1) control of gene expression, measured thus far as mRNA accumulation (Hildebrand et al. 1998), but inferring changes in the amount and timing of appearance of SIT proteins (as in Sullivan 1977), (2) possible direct control at the protein level by SIT-interacting proteins, and (3) the as yet unsubstantiated possibility of control over intracellular targeting of the SITs. The multiple forms of SITs may have differing affinities or capacities for transport, and their regulated appearance may control uptake at specific times or under specific conditions (Hildebrand et al. 1998; Hildebrand 2000). Concentrations of external silicic acid are also likely to affect transport, since they do affect the overall extent of silicification (Martin-Jézéquel et al. 2000). The tight control over silicic acid transport at many levels may be due to the fact that excess free silicic acid in the cytoplasm could autopolymerize, which would be detrimental to the cell.

## 4 Intracellular Silicic Acid Transport

As recently reviewed (Hildebrand 2000), very little is known about the transport of silicic acid within the cell after uptake. Of course, silicic acid must make its way to and into the silica deposition vesicle (SDV), where polymerization occurs. Mehard et al. (1974) showed in *Nitzschia alba* and *C. fusiformis* that radiolabeled silicic acid copurified with a variety of organelles and subcellular fractions purified after treatment with the label. The lack of specific localization suggested that silicic acid could move about the cell freely. It had been proposed that transport through the cytoplasm and into the SDV occurred via "silicon transport vesicles", based on the identification of small vesicles fusing with the SDV and the presence of small spherical silica particles in newly polymerized silica, which were assumed to be derived from the vesicles (Schmid and Schulz 1979). However, there was no evidence that the vesicles contained silica or silicon (Schmid and Schulz 1979), and it was noted that such vesicles were not numerous enough nor could they carry the amount of silicic acid required for the cell wall (Li and Volcani 1985a). As we will discuss later, silica in solution naturally polymerizes in a spherical form, therefore the assumption that the shape of the diatom silica spheres was derived from spherical vesicles was not necessarily valid. In a little-cited paper (Rogerson et al. 1986), electron spectroscopic imaging was used to localize cytoplasmic silicon in *Thalassiosira pseudonana*. The element was found throughout the cytoplasm, and was not localized within vesicles. Although sometimes located on "lipid inclusions," silicon was present on the outside of these, and there also was an apparent association with ribosomes (Rogerson et al. 1986). The presence of silicic acid throughout the cytoplasm, and not sequestered in intracellular vesicles, was consistent with an unrestricted cytoplasmic transport mechanism. Perhaps the silicic acid binding components mentioned previously were involved in intracellular transport, but since these are not yet characterized this is unclear. Another consideration is the mechanism of transport of silicic acid into the SDV. Although it is possible that SDV transport is SIT-mediated, it also has been suggested that it could be electrophoretically driven (Hildebrand 2000), without the need for a specific transport protein.

# 5 Micromorphogenesis vs. Macromorphogenesis

The silicified wall components of diatoms are precisely sculpted in both time and space by two interactive mechanisms, micromorphogenesis and macromorphogenesis (Pickett-Heaps et al. 1990). Micromorphogenesis defines those processes, including silica polymerization that occur within the lumen of the SDV, on the inner surface of the silicalemma, or both. Diatom biosilica includes distinct nanostructures that develop in the lumen of the SDV, and the processes of nanofabrication cannot result simply from spontaneous autocondensation and accumulation of silica within the physicochemical environment and constraints provided by the SDV – hence the term micromorphogenesis (Wetherbee et al. 2000). Macromorphogenesis involves molding of the developing SDV by organelles and cytoskeletal components that interact with the cytoplasmic surface of the silicalemma to generate large-scale, complex shape.

## 5.1 Micromorphogenesis – the Nanostructure of Diatom Biosilica

There is an in-depth understanding of the chemistry of silica polymerization in aqueous solutions (Iler 1979). Silica is soluble to only about 2 mM at neutral pHs, and because polymerization is an energetically favorable process, a solution of silicic acid can be induced to polymerize either by increasing silicic acid concentration, or decreasing the pH of a supersaturated solution (Iler 1979). The process of polymerization is affected by pH and ionic conditions, and a defined series of intermediates have been identified leading to formation of either silica sols, or fused silica gel networks (Iler 1979). In either case, the form of individual units of polymerized silica is spherical. A net negative charge accumulates on the surface of polymerizing silica as a result of ionization of silanol groups (Iler 1979). Between pH 7–10 in the absence of salt, repulsion of negatively charged particles occurs, favoring formation of a sol with separated colloidal silica spheres of defined diameter (final product on the order of 100 nm). At pH <7, or pH 7–10 in the presence of salt, surface charge becomes neutralized and individual spheres can collide and fuse. Fused sol particles become arranged in chains, which can then form branches and networks, resulting in a three-dimensional gel network (Iler 1979).

The diatom silica deposition vesicle has long been thought to contain nucleating determinants in the silicalemma and/or its lumen, and enclose a specific chemical environment favoring polymerization (Sullivan 1986; Pickett-Heaps et al. 1990). The intralumenal pH of the SDV is apparently acidic (Vrieling et al. 1999b), which would favor three dimensional silica gel network formation. Other conditions in the SDV, such as ion concentrations, organic molecules, and the fact that it encloses a confined space, may also affect diatom silica polymerization.

In many studies, silica in diatom valves was observed to be initially deposited as fused globular particles or silica spheres (e.g., Chiappino and Volcani 1977; Schmid and Schulz 1979), much as seen in formation of silica gels in vitro. The "rough" newly deposited silica often appeared progressively smoother as silicification proceeded (Schmid and Schulz 1979; Schmid and Volcani 1983). It was thought that smoothing was due to the eventual filling-in of spaces between spheres with silica (Schmid and Schulz 1979). However, atomic force microscopic (AFM) examination of cross sections of the valve of *Pinularia viridis* revealed only spherical structures in the major portion of the valve (Crawford et al. 2001). The smooth surface visible by SEM only occurred as an outer layer at the interface between the silica and the silicalemma. This suggests that interaction with the lumenal surface of the silicalemma affected the silica micromorphology. Since spherical geometry is the state of lowest surface energy in aqueous solution, any aberration from this shape implies that some physical or chemical agent has affected the structure. A smooth surface layer forming at the silica/silicalemma interface is not universal however, since AFM analysis of the mature valve surface of a *Coscinodiscus* species showed the same rough structure as the underlying material (Wetherbee, unpubl.). Analyzing height data on spheres imaged by AFM suggested that diatom silica may be quite porous (Crawford et al. 2001), consistent with a small and large angle X-ray scattering study indicating pore sizes of 3–65 nm with homogeneous geometry (Vrieling et al. 2000).

SEM and AFM analyses have revealed different forms of silica in developing valves. Silica was found in three micromorphologies in *Ditylum brightwellii* (Li and Volcani 1985a), a thin, homogeneous base layer, siliceous microfibrils, and hexagonally packed columns. Silica microfibrils, identified in the initial examination of valve formation of *P. viridis* (Pickett-Heaps et al. 1979), are seen in valves partially dissolved by NaOH to consist of linked spherical silica particles (Fig. 6). Certain regions of valve silica appear compacted, corresponding to regions of surfaces of smooth silica more resistant to NaOH etching (Fig. 6). Lewin (1961) and Kamatani (1982) showed that spontaneous silica dissolution rates from acid-cleaned valves varied over time, consistent with the presence of different forms of silica. The different described forms of diatom silica may result from different micromorphogenetic processes.

A closer examination of silica structure in *P. viridis* reveals some fascinating details of silica deposition and micromorphology in this species. A thin cross section of a forming raphe (Fig. 7A) shows the previously identified (Pickett-Heaps et al. 1979) silica microfibrils, attached to the silicalemma with what appears to be fairly regular spacing. The extent of silicification is less near the valve-plane oriented silicalemmas and greater towards the raphe fissure and in the center of the SDV, although silicification is skewed towards the cytoplasmic side of the lumen (Fig. 7A). A similar arrangement was seen in the pennate *Navicula cuspidata* (Edgar and Pickett-Heaps 1984b). Examination of other cross sections suggested that silicification began where two silicalemma intralumenal surfaces were appressed, and as the SDV expanded and microfibrils lengthened, the central region became increasingly silicified (Pickett-Heaps et al. 1979). The microfibrillar structure is apparent in cross sections of

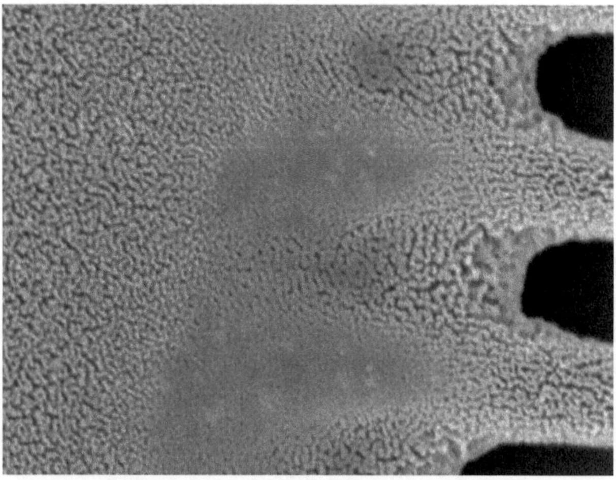

**Fig. 6.** Nanostructure of silica in the valve of *Pinnularia viridis* etched by NaOH treatment. Silica microfibrils can be seen especially at the *left* and *right*, and areas of smoother silica at *center right*. *Dark areas* at *far right* are areolar openings

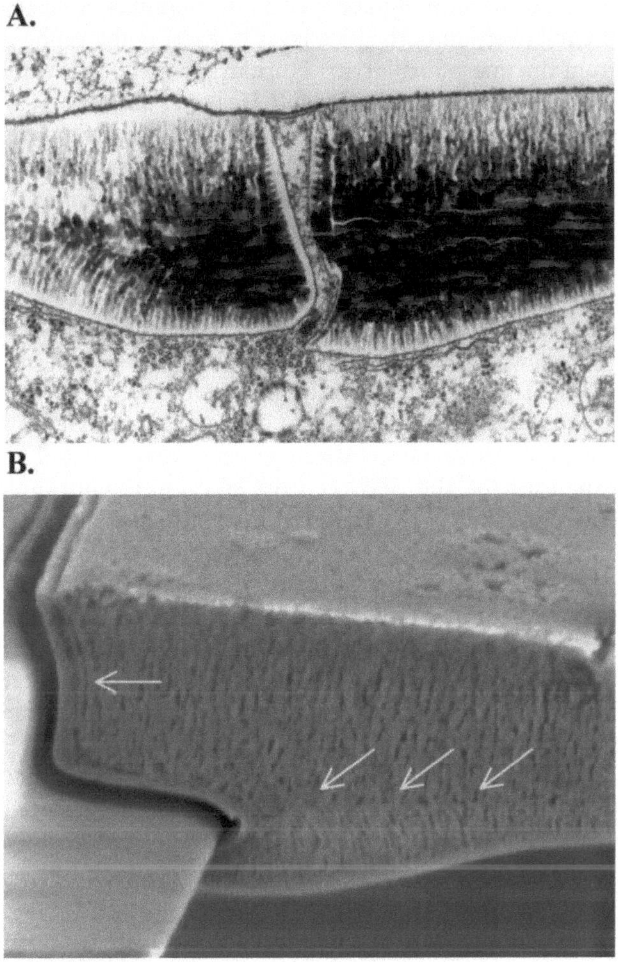


**Fig. 7A,B.** Cross sections of the raphe region of *P. viridis*. Cytoplasmic side is at the *bottom*. **A** TEM image of a silicifying raphe. Raphe fissure is at the *center*, the SDV surrounds the raphe. Dark material in the SDV lumen is silica, silica microfibrils are visible extending to the silicalemma. **B** SEM image of raphe region etched with NaOH. Raphe fissure is seen at *left*, forming a complex curve. Silica microfibrils are visible throughout most of the cross section. *Single arrow* at *left* denotes a region of compacted silica lining the raphe fissure. The three *angled arrows* denote a region where the silica microfibril structure is disturbed. This corresponds to a region of silica deposition where the two silicalemmas were initially appressed (see text)

fractured valves treated with NaOH (Fig. 7B); as in Fig. 7A, microfibrils are generally oriented in the direction of SDV expansion. Two other interesting features are visible in etched cross sections (Fig. 7B). The first is the apparent position of initial deposition where the two silicalemmas were closely appressed, visible as a "break" in the microfibrillar structure (diagonal arrows). The disruption in microfibrillar structure may result from the confined space between the initially appressed silicalemmas. A second interesting feature is the relatively smooth silica layer bordering the raphe fissure, which is distinct from the microfibrillar structure in the rest of the raphe (Fig. 7B). In the original study on *P. viridis* (Pickett-Heaps et al. 1979), it was suggested that silica deposition occurring initially along the raphe fissure prevented the fissure from filling in. In other words, the flexible structure formed by the SDV was rapidly converted to a "hard structure" by silicification. The compact silica micromorphology along the raphe, in contrast to that of most of the rest of the valve, may be indicative of the lack of expansion of the SDV in this region during polymerization. Based on this observation, the "smoothing" of silica observed in formation of other diatom valves (Schmid and Schulz 1979; Schmid and Volcani 1983) may also be a reflection of inhibition of expansion of the SDV. In Fig. 7A, relatively less silicification near the lumenal faces of the inner and outer valve surfaces of the expanded SDV may enable expansion of the SDV. These observations suggest that not only can the movements or expansion of the SDV control silica micromorphology, but also that the extent of silica polymerization may influence movements or expansion of the SDV.

Because the SDV is exocytosed after valve completion to form part of the mature diatom cell wall, investigators focused on isolating diatom cell wall proteins as a means of characterizing polymerization determinants within the SDV (Hecky et al. 1973). The assumption has been that key organic molecules involved in nanofabrication remain within mature walls. The amino acid compositions of diatom cell wall preparations were determined (Nakajima and Volcani 1969, 1970; Hecky et al. 1973) and an abundance of hydroxylated amino acids were identified. Based on these data, models of how hydroxylated amino acids could serve as nucleating centers to initiate polymerization were developed (Hecky et al. 1973; Lobel et al. 1996). However, since individual proteins were not isolated, the validity of these models was unclear. If other organic cell wall components were predominant, then the composition of cell wall preparations may not accurately reflect amino acids or other components involved in silicification.

A major breakthrough in our understanding of diatom silicification occurred when Kröger et al. (1999) isolated and characterized a set of polycationic polypeptides, called silaffins, tightly associated (released after hydrogen fluoride treatment) with cell wall silica in *C. fusiformis*. Tricine-SDS polyacrylamide gels revealed two bands with apparent molecular weights of 4 and 8 kDa (silaffins 1A and B, respectively), and one of 17 kDa (silaffin 2). The silaffin 1A fraction consisted of nearly identical peptides rich in regularly spaced lysine and arginine clusters, also enriched in serine. Silaffin 1B had a sequence similar to silaffin 1A, but which was repeated, hence its doubled molecular weight. The amino terminal sequence of silaffin 1B was determined and used to generate a probe, which was used to isolate a single open reading frame (ORF) encoding this sequence, the multiple repeats of the silaffin 1A sequences, and an acidic amino-terminal domain (Kröger et al. 1999). The ORF was called *sil 1* (Kröger et al. 1999). It was apparent that silaffins 1A and B were derived from proteolytic processing of the sil 1 protein, and a conserved tripeptide sequence at the end of the repeated domains was likely to be the cleavage signal. The amino acid sequence of silaffin 2 could not be determined, and so its relation to the *sil 1* ORF is unclear, however, if encoded by *sil 1*, its size suggests that it must either encompass the entire repeat region or the acidic domain and part of the repeats. The lysines in silaffin 1A were modified to ε-*N*, *N*-dimethyl-lysine, and lysine with 6–11 repeats of *N*-methyl propylamine attached to the ε-amino-group (Kröger et al. 1999). All three purified silaffins promoted the rapid polymerization of silica from a solution of silicic acid in vitro, and silaffin 1A did so optimally at pH 5–7, with some activity even at pH 4 (Kröger et al. 1999). Silaffin 1A formed 500–700 nm-diameter silica spheres that were connected or partly fused in random orientation, and an unfractionated mixture of all silaffins formed <50 nm-connected or partly fused spheres, also randomly oriented (Kröger et al. 1999). The latter were within the size range found in native diatom silica (Wetherbee et al. 2000). A synthetic peptide analog of one of the silaffin 1A repeated amino acid domains, lacking the modified lysines, did not promote polymerization at low pH, but could do so at pH 7.0. This result indicated that the post-translationally modified residues were essential for low pH activity, but also that some of the polymerization determinant was intrinsic to the polypeptide sequence. In vitro, poly-L-lysine and derivatives have been shown to precipitate silica (Cha et al. 2000; Patwardhan et al. 2001b), suggesting that similar amino acids in the synthetic silaffin peptide may have had the same ability.

Recently, Kröger and coworkers (2002) have extracted *C. fusiformis* silica with ammonium fluoride rather than hydrofluoric acid (HF) and showed that silaffin 1A was phosphorylated at all serine residues and at one trimethyl lysine group. The phosphates were essential for activity as it was shown that HF-extracted silaffins could not precipitate silica in the absence of phosphate buffer, whereas the phosphorylated ammonium fluoride-extracted silaffin 1A could (Kröger et al. 2002).

The silaffins coprecipitated with silica in vitro (Kröger et al. 1999; 2002). Thus, it may be likely that they were embedded in the silica spheres, although a direct measurement of this interaction has not been made. It is important to consider that "embedded" does not necessarily mean that organic molecules are completely enveloped by silica; partially embedded molecules may stick out from the surface, and consequently alter surface characteristics. The procedure required to isolate the silaffins did not allow determination of whether they were completely or partially embedded in vivo within diatom silica, or just tightly associated with the surface. In this regard, it should be noted that the HEP proteins (Kröger et al. 1997), now pleuralins (Kröger and Wetherbee 2000), appeared to be embedded within diatom silica and only released by desilicification in hydrofluoric acid (HF). Subsequently, the pleuralins were found to be surface adhesive-type molecules (Kröger and Wetherbee 2000), added externally to the silica or its organic casing. Swift and Wheeler (1992) identified proteins presumably embedded within silica of *Cyclotella meneghiniana*, by cleaning a cell wall preparation with NaOCl, then dissolving the silica with HF. The extent of cleaning was apparently carefully evaluated; the absence of residual organic material was checked by stains for glucuronic acids and primary amines, and also by SEM (Swift and Wheeler 1992). However, recent observations by AFM suggest that NaOCl treatment may not completely clean the surface of diatom silica (Wetherbee, unpubl.). It is, therefore, possible that the proteins isolated by Swift and Wheeler were not embedded in the silica, though one might expect that if the majority of the surface was clean (as suggested by their controls) that the predominant proteins isolated after HF treatment would be embedded. The coprecipitation of silaffins with silica (Kröger et al. 1999), and the formation of silica spheres in vivo and in vitro, are at least consistent with the possibility that silaffins are naturally partially or entirely embedded. None-the-less, to date the silaffins have not been localized within the SDV and mature valves. If they are entirely embedded and obscured from outside contact, they may prove difficult to localize using agents that rely on external binding such as antibodies.

Determination of the extent to which silaffins are embedded may be important in understanding the extent to which diatom silica is a composite organic/inorganic material. This has implications regarding its strength, stability, and resistance to dissolution. In the example of the calcium carbonate composite of the abalone shell, organic material comprises a few percent by weight, yet the material is 3000 times more fracture-resistant than the crystalline form alone (Smith et al. 1999). One consideration in the possible composite nature of diatom silica is whether the silica/organic interactions are due to covalent or hydrogen bonding. Covalently bonded silica/ABS copolymers had increased tensile strength relative to similar hydrogen-bonded material (Hsu and Lin 2000).

In addition to the silaffins, other proteins have been isolated from dissolved diatom silica. The pleuralins, isolated from *C. fusiformis* as previously mentioned, are one example (Kröger et al. 1997). Others were in the HF-released protein fraction characterized by Swift and Wheeler (1992), which was sized by HPLC to a range of 45–60 kDa. Vrieling et al. (1999a) observed two protein bands of 56 and 65 kDa by PAGE after silica dissolution by $Na_2CO_3$ treatment, which may correspond to Swift and Wheeler's HF fraction. Interestingly, bands of identical size were seen in the three diatom species examined (Vrieling et al. 1999a), which included two pennates and one centric, suggesting that these proteins were highly conserved. These proteins were apparently much less abundant in *C. fusiformis* than in the other species (Vrieling et al. 1999a). They were of much larger size (45–60 kDa) than the silaffins, and apparently did not have any silica precipitating activity (Swift and Wheeler 1992).

Recent results (Naik et al. 2002) suggest that amino acid types other than those already identified in diatom silica isolates could be involved in silica binding or precipitation. A combinatorial phage peptide display library was used to isolate peptides that specifically bound silica particles (Naik et al.

2002). In contrast to the silaffins, these peptides were generally greatly enriched in histidine residues, and depleted in lysines. Hydroxylated amino acids were present, but generally in lower amounts than in the silaffins. The authors concluded that the presence of hydroxy- and imidazole-containing amino acids, and a high cationic charge, was optimal for interaction with silica (Naik et al. 2002). Perhaps the amino acid biases in the phage display peptides and the silaffins were due to differences in pH conditions; silaffins operate under acidic pH (Vrieling et al. 1999b), whereas the phage display peptide experiments were done at pH 7.5 (Naik et al. 2002). The phage display peptides were also tested for their abilities to precipitate silica, and in some cases there was a good correlation between binding activity and precipitation, however, in others, the two activities did not correlate. This suggested that different amino acids or their arrangements might have been optimal for either silica binding or precipitation.

The results of Kröger et al. (2002) show that it is important to confirm that isolated silica-associated proteins have the same chemical features as in their native state, before drawing too many conclusions about possible functions and activities. Prior to this, it was questioned whether the hydroxylated amino acids in the silaffins are *O*-glycosylated, since HF treatment would cleave *O*-glycosidic bonds (Kröger et al. 2001), but ammonium fluoride extraction indicated that all serines were phosphorylated and that glycosylation did not occur in the native silaffins (Kröger et al. 2002). It has been shown in vitro that aliphatic polyols – sugar-like compounds, readily form stable five- and six-coordinated silicon complexes in aqueous solutions (Kinrade et al. 1999), suggesting that such interactions could occur in nature. Since sugar-containing glycoproteins are prevalent in other types of diatom cell wall preparations (Edgar and Pickett-Heaps 1982; Hoagland et al. 1993; McConville et al. 1999), the question of the role of glycosylation in silicification needs to be more carefully examined.

The initial characterization of the silaffins was followed up by surveying silica-associated material (also from HF-treated cell wall preparations, but not pre-cleaned) from a number of diatom species (Kröger et al. 2000). Polypeptides unique to each species, which were likely to be silaffins, were identified (Kröger et al. 2000). In addition, all species had a predominance of what were shown to be linear long-chain polyamines with different numbers (8–20) of repeated *N*-methyl-propylamine units, attached to methylation isoforms of putrescine. *C. fusiformis* was an exception, since free polyamines were much less prevalent, and as previously shown (Kröger et al. 1999), were mostly found attached to the modified lysines in the silaffins. Polyamines from *Nitzschia angularis* could precipitate silica from a silicate solution in vitro, polyamines of different size ranges formed different sized fused silica spheres, and the size and morphology of spheres varied greatly depending upon pH. The general trend was towards smaller spheres with increasing pH, consistent with the expected solution chemistry (Iler 1979), but the actual structures were far more complex, and often consisted of a mixture of fused larger and smaller spheres

(Kröger et al. 2000). Mixtures of silaffins and polyamines developed other distinct silica micromorphologies. A mixture of three silaffins from *N. angularis* formed large porous blocks made of dense aggregates of small silica particles quite distinct from what was formed by polyamines alone, but a mixture of these silaffins and polyamines resulted in a hybrid morphology (Kröger et al. 2000). Polyamines coprecipitated with silica, with a weight ratio of silica to polyamine of 1.25:1. Polyamines from a variety of diatom species ranged in size from 600 to 1500 Da (Kröger et al. 2000), which would correspond to molar ratios between 8–20:1 silica:polyamine. Since the polyamines contained 8–20 repeated *N*-methyl-propylamine units, the correspondence between this number and the ratio of silica:polyamine suggests a tight stoichiometry.

The mechanism of silaffin and polyamine-induced polymerization of silicic acid is not clear, but insight may be gained from studies on synthetic compounds and the more recent characterization (Kröger et al. 2002) of the diatom components. Mizutani et al. (1998) demonstrated that synthetic polyamines could precipitate silica. Interestingly, these experiments were done in the absence of phosphate, indicating that phosphates were not essential for silica precipitation in this system. Mizutani et al. (1998) showed that only polyamines (and not oligo or smaller amines) had silica precipitating activity, and that not only the number, but the relative arrangement of amino groups affected activity. They proposed that cooperative interaction of appropriately arranged amino groups could stabilize the transition state of a condensation reaction between two silicic acid monomers. Mizutani et al. (1998) also showed that synthetic polyamines were consumed in the reaction by coprecipitation with silica; thus they were not just catalysts, but were reactants as well. An apparent stoichiometric ratio of synthetic polyamine to silica was demonstrated, and IR spectra suggested hydrogen bonding between silanol and amine groups – indicating that the silica gel and amines were interacting at the molecular level (Mizutani et al. 1998). Along these lines, a recent NMR investigation provided evidence for the identification of a silicon/organic complex in the diatom *Navicula pelliculosa* (Kinrade et al. 2002), suggesting hexavalent silicon coordination to nitrogen. If polyamines act as nucleating centers for polymerization, then they may have a limited sphere of influence. As more silica was deposited, the activity of a given polyamine should become "snuffed out" due to the inability of substrate to reach nucleating centers. However, positively charged polyamines could be attracted to the newly formed negatively charged silica surface, thus providing new nucleation sites. In a study examining the influence of various parameters on polyallylamine-induced precipitation of silica using tetraethoxysilane (TEOS) as a precursor, as polymer concentration increased so did particle size (Patwardhan et al. 2002). Under lower concentrations of polymer the smaller particle surfaces were rougher and actually appeared striated (Patwardhan et al. 2002). Perhaps the striated appearance was due to polymerization on polyamines that were not obscured by further deposition, and the lack of further deposition was due to limitation of new nucleation sites. Smaller or larger polyamines with a constant density of polymerization determinants per unit length produced smaller or larger sized silica spheres, respectively (Patwardhan et al. 2002). This is consistent with diatom polyamines separated into lower and higher average molecular weights (Kröger et al. 2000), which because they consisted of repeated units, would have a relatively constant charge to length ratio. An apparent contradiction to these results occurred in the case of the silaffins from *C. fusiformis*. The smallest silaffin, 1A, produced much larger spheres than a mixture of all three classes of silaffins (Kröger et al. 1999). Perhaps this had to do with the effective concentration of components; a given weight of silaffin 1A would have had more molecules than the same weight of the silaffin mixture, and as shown (Patwardhan et al. 2002), a higher concentration of nucleating material formed larger spheres. Alternatively, self-aggregation of the silaffins (Kröger et al. 2002) could affect the size of the final product. The results of in vitro polyamine silica precipitation experiments suggest that the nature of the polyamine and relative concentration of polyamine and silica precursor must be important in the formation of well-defined and reproduced structures in the diatom, highlighting the need for coordinated and tight control over the amount of silicic acid and polymerization determinants in the SDV. An additional interesting result from the Mizutani et al. (1998) study was that the silica precipitate had a net positive charge (as judged by dye binding), proposed to be due to exposed amino groups at the surface. Silica in developing diatom cell walls apparently has a net negative charge since cationic rhodamine dyes are incorporated into it (Li et al. 1989), although it has not been shown what the net charge is at the silica surface. It would be interesting to determine the charge properties of silica precipitated in vitro by diatom polyamines and silaffins, as this may lend insight into whether polyamines or silaffins are completely embedded or whether portions of them stick out. Such information may relate to the mechanism of diatom silica polymerization, for example, this property could affect the fusing of silica spheres, or the interaction of the external silica surface with proteins.

Experiments using extracts from *N. angularis* indicated that the micromorphology of the final silica product could be dramatically altered by using different combinations of silaffins and polyamines, or by carrying out precipitations under different pH conditions (Kröger et al. 2000). The formation of different microstructures may relate to different microforms of silica identified in diatom valve cross sections (Li and Volcani 1985a), although the extent of SDV expansion is clearly involved (Fig. 7). In *P. viridis*, numerous tiny pores were found in a particular substructure of the valve, and these pores were not lined by the silicalemma during formation (Pickett-Heaps et al. 1979). Silica in sections of the valve of *Actinocyclus subtilis* had a "bubbly" structure, with numerous spaces of various sizes in random orientations, and these were also not lined with organic material (Andersen et al. 1986). Observations of such structures has led to the suggestion that the arrangement of intralumenal organic silica-precipitating components could be responsible for forming structures more intricate than those formed by molding of the SDV

(Pickett-Heaps et al. 1979, 1990). The native phosphorylated form of silaffin 1A self-assembles via electrostatic interactions (Kröger et al. 2002). Time-resolved analysis of silica morphogenesis identified an initial "plastic" structure of flat networks of a branched, irregularly shaped phase (Kröger et al. 2002). Later, spherical particles began to form, retaining their connections, similar to the connected silica spheres seen in the diatom valve. Such a mechanism could be involved in the formation of smaller silica structures inside the SDV.

It has not yet been demonstrated that silaffins and polyamines are directly responsible for formation of larger-scale structures in the cell wall, at least under conditions tested so far. Even the large irregular silica blocks produced by a mixture of silaffins and polyamines from *N. angularis* (Kröger et al. 2000) did not mimic the macromorphogenetic patterns found in the diatom cell wall, although their larger size suggests some relationship. This is in contrast to the example of silicatein proteins from silica of the sponge *Tethya aurantia*, which associate to form linear structures similar to the final silicified product (Shimizu et al. 1998; Cha et al. 1999). Of course, silaffins and polyamines may still influence larger-scale structure formation, since the chemical properties of these compounds may facilitate particular interactions or orientations during cell wall synthesis, and it seems reasonable that the nature of the building blocks will in some way affect the form of the final product. All silica in the wall of *C. fusiformis* (Reimann et al. 1965) is in the form of strands (even fibulae connecting the two halves of the raphe can be thought of as curved strands), which is not typical of most diatom species. In *C. fusiformis* there was also a dearth of free polyamines – these were mostly attached to the silaffins. Perhaps the peptide backbone or self-assembling ability (Kröger et al. 2002) of the silaffins can influence the spatial characteristics of polymerization, contributing to the macromorphology of the final silica product. Likewise, perhaps interactions between the peptide backbones in the mixture of silaffins from *N. angularis* were involved in the formation of the larger silica structures. It would be interesting to examine a diatom genus such as *Chaetoceros* (Pickett-Heaps 1998a,b), which has a more typical frustule, but also forms hollow long setae (spines), to see if there are differences in the relative timing of appearance of silaffins and polyamines as different structures are made.

Insights into possible macrostructure-directing activities of silica polymerization determinants in diatoms may be gained by examining in vitro silicification studies using synthetic compounds. The amphiphilic properties of micelle-forming surfactants have been widely exploited to generate mesoporous silica of different structure (Kresge et al. 1992; Monnier et al. 1993; Tanev and Pinnavaia 1995). In these examples, the arrangement, by a variety of means (e.g., Kresge et al. 1992; Trau et al. 1995; Lu et al. 1997; Tolbert et al. 1997), of polymerization determinants resulted in the formation of silica structures reflecting these arrangements or those derived from the polymerization process itself (Monnier et al. 1993). In an example not utilizing emulsions, Cha et al. (2000), synthesized several block copolypeptides with amphiphilic properties, containing a polycationic domain of polylysine, and a self-assembling domain of polycysteine, that precipitated silica from a solution of TEOS. One of these copolypeptides produced silica spheres when the reaction was carried out under nitrogen, but produced columns of silica when the cysteines were oxidized to promote polypeptide aggregation (Cha et al. 2000). What may be of particular interest to diatom silicification are examples where polymerization determinants are not statically prearranged, but are in motion. Subjecting a surfactant solution to flow has led to the formation of elongated or spirally twisted silica structures (Kim and Yang 2000), and polyallyamines assumed to be oriented by flow also formed elongated silica products (Patwardhan et al. 2001a). These are consistent with the dynamic movements and expansion of the SDV translocating polymerization determinants as polymerization occurs (Fig. 7). Although based on a limited number of unquantified observations, the microfibrils seen in diatom silica (Figs. 6 and 7) appear to be more elongated than silica precipitates formed in vitro by the silaffins and polyamines (Kröger et al. 1999, 2000), and their orientation in the direction of SDV expansion (Fig. 7), are consistent with this idea.

Although macrostructure-directing or -affecting properties of silaffins and polyamines (or other potential diatom silicification determinants) have not yet been directly demonstrated, a recently proposed model suggests a mechanism contributing to the formation of successively smaller hexagonally arranged pore structures in the valve of the genus *Coscinodiscus* (Sumper 2002), and also suggests possible mechanisms affecting macrostructure in other diatom species. Based on quantitative extraction from an aqueous solution by chloroform/methanol, polyamines isolated from *Coscinodiscus* silica appeared to have amphiphilic properties (Sumper 2002). In the initial stage of silica deposition in this species, the SDV was flat (Schmid and Volcani 1983), as occurs in *Pinnularia* (Fig. 7), and droplets of polyamines were proposed to self-arrange in an optimally packed hexagonal array (Sumper 2002). According to the model, silica polymerization was catalyzed at the outer hydrophilic surfaces of the droplets, leading to formation of the hexagonally arrayed silicified areolae of the base layer of the forming valve. After formation of the base layer, consumption of polyamines and creation of polymerized silica surfaces were proposed to be driving forces for a series of phase separation events in which larger polyamine droplets were broken successively into smaller ones, giving rise to the formation of smaller hexagonally arranged silica structures. This concept is complicated by the fact that the actual structures are separated in three dimensions (Schmid and Volcani 1983); after formation of the base layer, areolar walls must grow perpendicularly prior to resumption of the proposed phase separation events. Species-specific patterns within the genus *Coscinodiscus* could be explained by constraints on the wall-to-wall distances within the initial areolae (Sumper 2002). It is unclear whether these distances were determined by properties of the polyamines or by molding by the SDV; the relatively small variation in polypropylamine repeating unit size (16–19) in *Coscinodiscus* species (Sumper 2002) may suggest that the SDV is more responsible. It is acknowledged that the model cannot explain all of the features of deposition in *Coscinodiscus* (Sumper 2002), but the inconsistent features can be explained by movements of, and molding by, the SDV. It also may be likely that such proposed phase separation events do not occur in all diatom species (see discussion on *Pinnularia*, Fig. 7), however, the presented model is an elegant description of how the physico-chemical properties of polymerization determinants could contribute to macrostructure formation in some diatom species.

In light of the apparent regular spacing of the ends of silica microfibrils at the silicalemma seen in *P. viridis* (Fig. 7A), it is interesting to speculate that perhaps silaffins (or some other polymerization determinant) are attached with regular spacing to the silicalemma. According to this hypothesis, silicalemma-associated determinants could be responsible for initial deposition, and their regular spacing would ensure a coherent structure. The polyamines or small silaffins, being smaller and more diffusible, may catalyze polymerization between the initial deposition microfibrils. It is interesting to consider whether diffusibility is a factor in why the sil 1 protein is extensively proteolytically processed to form the smaller silaffins. A larger protein with the same density of nucleation sites should catalyze polymerization to the same extent, but constraints on the flexibility of the polypeptide backbone may lead to gaps in the final structure. Accumulating negative charge during silica polymerization would favor continuing transport of polyamines or small silaffins into voids in the polymerizing silica. The electrostatic properties of the silaffins and polyamines may differ, considering the high density of negatively charged phosphates on the silaffins at the low pH of the SDV. This may suggest that polyamines play more of a role in "filling in" than silaffins. Since out of six diatom species examined, only *C. fusiformis* had a predominance of silaffins over polyamines (Kröger et al. 2000), and since *C. fusiformis* only makes thin elongated silicified structures, perhaps polyamines play a lesser role in the "filling in" process in this species because the space to be filled in is correspondingly smaller. This discussion highlights the importance of further characterization of the largest silaffin in *C. fusiformis*, silaffin 2. If silaffin 2 contains the acidic domain identified in the *sil 1* ORF (Kröger et al. 1999), then perhaps this domain is involved in interaction with proteins of the silicalemma. It was proposed that the acidic pleuralins (Kröger and Wetherbee 2000) were bound to organic coat components of the terminal girdle bands (or to exposed silanol groups – but these are likely to also be negatively charged in silica); by analogy perhaps the acidic domain of the sil 1 protein has a similar function. A non-proteolytically cleaved form of sil 1 would be amphiphilic, with cationic features enabling its presence in polymerizing silica, and anionic features possibly enabling association with other proteins.

It is important to realize that our understanding of conditions inside the SDV is very rudimentary, and in vitro silica precipitation experiments done to date may not accurately mimic these conditions. In particular, the SDV represents a confined space (Wetherbee et al. 2000), is a dynamic structure, and the characteristics of the SDV lumen are generally unknown, all of which could influence the properties of silaffins and polyamines, and their roles in forming larger-scale structures. The possible involvement of silaffins or polyamines in macroscale structure formation may become more evident by, for example, carrying out precipitation reactions in emulsions (Kresge et al. 1992) that may mimic the confined conditions of the SDV, or under conditions of flow. Extensive modeling and observation has been done on reaction conditions and phase transitions that occur during silica polymerization in cationic emulsions (Monnier et al. 1993). Although modeled at high pH (Monnier et al. 1993), acidic pH conditions have been examined as well (Yang et al. 1998). In the interaction between silica and surfactant, as silica polyanions grow larger their diminished charge density drives a phase transformation from lamellar form to hexagonal, and reduces silica thickness as well (Monnier et al. 1993). Although there may be differences with regard to specifics, similar phase transitions as proposed by Sumper (2002) may be involved in diatom silicification. These transitions could either involve "free" polyamines or silaffins in the SDV lumen, or polymerization determinants attached to the silicalemma.Cytoskeletal movements of the SDV are clearly involved in molding the form of silicified structures, but in vitro silicification work (e.g., Kresge et al. 1992, Monnier et al. 1993) and Sumper's model (2002) suggest that intralumenal processes could contribute as well. Macrostructure formation could be driven partly by the physico-chemical properties of polymerization determinants, their local environment, and their interactions with the silicalemma, and partly by specific movements of the SDV to affect the shape or volume in which these reactions occur. Possible silicalemma-attached polymerization determinants could provide a direct link between intra- and extra-lumenal processes. Perhaps a dynamic tension exists between external cytoskeleton-directed SDV shape changes that alter microstructure, contribute to the directionality of polymerization, or induce possible phase transitions, and/or internal processes such as phase transitions or limitation of the extent of SDV expansion (Fig. 7) that may trigger changes in cytoskeletal movements. Much important work remains to be done to fully understand the properties of the silaffins, polyamines, and other possible polymerization determinants, and the contribution of these components to the in vivo silicification process on both a micro and macro scale.

Further characterization of the largest silaffin in *C. fusiformis*, silaffin 2, may be helpful in testing this hypothesis. If silaffin 2 contains the acidic domain identified in the *sil 1* ORF (Kröger et al. 1999), then perhaps this domain is involved in interaction with proteins of the silicalemma. It was proposed that the acidic pleuralins (Kröger and Wetherbee 2000) were bound to organic coat components of the terminal girdle bands (or to exposed silanol groups, but these are likely to also be negatively charged in silica); by analogy perhaps the acidic domain of the sil 1 protein has a similar function. A nonproteolytically cleaved form of sil 1 would be amphiphilic, with cationic features enabling its presence in polymerizing silica, and anionic features possibly enabling association with other proteins.

## 5.2 Control of Micromorphogenesis

Based on studies of silica precipitation by synthetic polyamines and other compounds, factors involved in the control of micromorphogenesis could include the relative concentrations of silica precursor and nucleating molecule, the size of the nucleating molecule, interactions between nucleating molecules, movements of nucleating centers, and possible phase transitions (Monnier et al. 1993; Mizutani et al. 1998; Kim and Yang 2000; Patwardhan et al. 2001a, 2002). Kröger and coworkers presented the hypothesis that combinatorial interaction between subsets of silaffins and polyamines expressed at different stages in valve morphogenesis could account for the formation of different silica morphologies (Kröger et al. 2000). One consequence of this hypothesis would be that changes in the relative amounts of silaffins and polyamines should be coordinated. Regulation of polyamine levels may involve control of key enzymes and genes in the multi-step pathway of their synthesis (Tabor and Tabor 1984). The silaffins could be regulated transcriptionally, at several expression steps post-transcription, or at the step of polyamine attachment to their lysines. The latter could provide a mechanism for polyamine regulation of silaffin activity, since without the modified lysines, low pH silica polymerization would not occur (Kröger et al. 1999). Since control of gene expression by polyamines is well documented in other organisms (Tabor and Tabor 1984), this represents another possible level of control by polyamines over the silaffins. Kröger and coworkers (2001) observed that the C-terminal tetrapeptide of silaffin-1A, which contains two arginines, was cleaved off during maturation of the protein. Arginine is the biosynthetic precursor of putrescine, and it was speculated that the cleaved tetrapeptide could be the precursor for eventual formation of polyamines that become attached to the silaffins (Kröger et al. 2001). If so, this suggests a possible control mechanism by polyamines over silaffins. However, one must consider the overall prevalence of polyamines in the cell (Tabor and Tabor 1984); therefore it may not be likely that their synthesis is limited by the silaffins, and also, this would not account for the predominance of silica-associated polyamines over silaffins in other diatoms. Comparison of silaffin sequences in species other than *C. fusiformis* would be helpful in validating this hypothesis.

In order to begin to elucidate in vivo mechanisms of regulation over silica polymerization, it is important to identify and characterize as many components in the process as possible. Other genes involved in silicification may be induced at the same time as the silaffin genes, and could be isolated using a subtraction-based approach (Hildebrand et al. 1993; Armbrust 1999). Proteins possibly involved in silicification induced at this time could also be examined by proteomics. A continuing investigation into the nanostructure of silica in diatom valves should be valuable in terms of correlating levels and timing of expression of polymerization components with the ability to form particular structures. This may also allow visualization of processes such as possible phase transitions. A major regulator in the synthesis of polyamines is ornithine decarboxylase (Tabor and Tabor 1984). This enzyme has fascinating levels of regulation, including a very rapid induction and turnover time in mammals, an antizyme protein that binds to and inactivates the enzyme in bacteria, an anti-antizyme that inactivates the antizyme, possible post-translational modifications that abolish activity, and cell cycle regulation (Tabor and Tabor 1984 and references therein). In general, polyamines are involved in many important cellular functions, and in diatoms, could provide interconnectivity between cell wall synthesis and other processes such as progression through the cell cycle. It would be of benefit to characterize ornithine decarboxylase and its gene in diatoms as a means of understanding the regulation of polyamine synthesis.

## 6
## Macromorphogenesis – the Formation of Large-Scale Silicified Structures in the Diatom Cell Wall

With the advent of microscopy came the dawn of fascination with the incredibly varied and detailed silica structures that diatoms make (Anonymous 1703; Baker 1753), a fascination that has deepened with electron microscopic investigations (Round et al. 1990). Examination of the stages in formation of the silicified wall have shown a meticulous control over the process, which stems from interactions at the molecular level, yet manifests in visible form. This is, as stated in a review, "... a profound and pervasive biological mystery – the creation of three dimensional structure from information contained in chemical form" (Pickett-Heaps et al. 1990). We will present several examples of the macro-scale development of the silicified cell wall in both pennate and centric diatom species to summarize our current understanding. It is important to emphasize that these are examples selected to illustrate general themes. In actuality, diatoms seem to be able to exploit every conceivable approach to form a silicified structure (e.g., Pickett-Heaps et al. 1990; Round et al. 1990). We will also examine synthesis of special silicified structures such as setae (spines) and other protuberances, because more recent studies are shedding light on the underlying cytoskeletal mechanisms involved in their formation. Much earlier work has been thoroughly reviewed (Pickett-Heaps et al. 1990; Round et al. 1990), but the reader is encouraged to read the original papers, because they offer detailed and fascinating explorations into the underlying cell biology.

Prior to the initiation of silicification, cytoskeletal movements occur as part of cell division, and it is apparent that the entire process of cell cleavage, mitosis, and cell wall synthesis is tightly coordinated (Pickett-Heaps et al. 1990).

One of the first electron microscopic examinations to follow the course of diatom silicification was done using synchronized cultures of the freshwater pennate species *Navicula pelliculosa* (Chiappino and Volcani 1977). When starved for silicon, *N. pelliculosa* arrests at the S/G2 phase boundary of the cell cycle (Coombs and Volcani 1968), and upon silicon replenishment, immediately proceeds with cell wall silicification. One characteristic of *N. pelliculosa* is that the epitheca has three girdle bands and the hypotheca only two (Chiappino and Volcani 1977). The initial event indicating the onset of cell wall synthesis, occurring prior to cytokinesis, is formation of a third girdle band on the hypotheca, which is made in a separate SDV, and targeted precisely to attach next to the second girdle band. The precise timing of synthesis and targeting of girdle bands is especially evident in *C. fusiformis*, which sequentially adds 20–30 girdle bands to the hypotheca in perfect concert with cell expansion after daughter cell separation (Kröger and Wetherbee 2000). In *N. pelliculosa*, after the correct number of girdle bands for the soon-to-be epitheca are present, cleavage occurs. A minute SDV, already containing polymerized silica, then becomes visible running the length of the cell, eventually forming one side of the raphe (Fig. 8A). Microtubules were present alongside the forming raphe strand (Chiappino and Volcani 1977). The enlarged silicified area in the center of the cell called the central nodule (CN), was shown in later

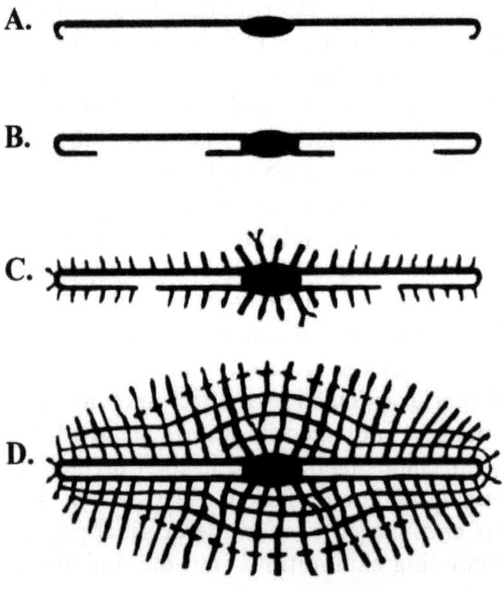

**Fig. 8A–D.** Process of valve silicification in *Navicula pelliculosa*, redrawn from Chiappino and Volcani (1977). **A** Initial deposition of one half of the raphe. Central nodule is at *center*, and the raphe strand extends the length of the cell and curves at the end. **B** Silica curves back and grows in towards the central nodule, and silicified strands progress outward from the central nodule. **C** The second side of the raphe is almost fused, and transapical ribs begin to form. **D** Raphe is complete, transapical ribs reach their longest extension, and cross extensions connecting the ribs are forming

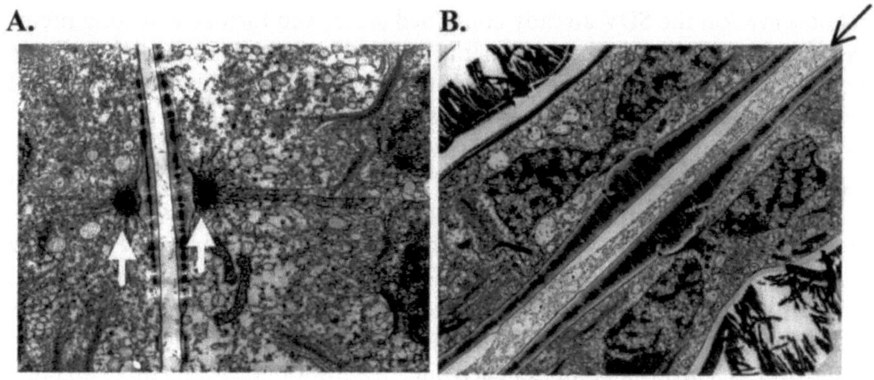


**Fig. 9A,B.** Cytoskeletal elements near the raphe fissure of *P. viridis*. **A** TEM of lateral cross section through the central nodule. *White arrows* locate the microtubule centers, which are located precisely on the central nodule. Microtubules can be seen radiating from the MC. **B** TEM of longitudinal cross section near central nodule. *Arrow* at *upper right* locates the raphe fiber

studies (Pickett-Heaps et al. 1979) to be the positioning site of the microtubule center (MC) during cell wall synthesis (Fig. 9A). Once the first half of the raphe reached the end of the cell, it curved, and then began to grow back in towards the CN (Fig. 8B). At the same time, a silicified strand grew outward from the CN towards the back-curving strand, and they eventually met and fused (Fig. 8B–D). The location of fusion is evident in mature walls (Chiappino and Volcani 1977), and is called the Voight discontinuity (Voight 1943). Transapical ribs began to form during raphe growth, but cross extensions connecting the ribs did not form until fusion of the two parts of the raphe. Cross extensions were formed sequentially, first near the raphe and then laterally from there, and during formation the two sides of the extensions grew in towards each other and met in the middle (Fig. 8D). After valve completion and exocytosis, the first girdle band was made, and only after it was released, did the second form and release, completing the hypotheca. Cells then separated.

The bidirectional growth of two silicified strands, propagating over a distance and then fusing precisely, is a feature seen in the formation of different structures in both pennate and centric species. This phenomenon is clear evidence of longer-range spatial control in macromorphogenesis, which in the case of raphe formation is likely directed by cytoskeletal elements.

An enlightening series of comparative examinations of silicification in pennates, which included monitoring movement of cytoskeletal centers, was carried out by Pickett-Heaps and colleagues (Pickett-Heaps et al. 1979; Pickett-Heaps and Kowalski 1981; Pickett-Heaps 1983). They compared *Pinnularia*, *Nitzschia*, and *Hantzschia*, which differed in the location of the raphe, and also *Navicula* (Edgar and Pickett-Heaps 1984b). After cytokinesis in *Pinnularia*, the microtubule center moved through a hole perforating the nucleus until it reached the cleavage furrow (Pickett-Heaps et al. 1979). As in *N. pelliculosa*, at first observation the SDV already contained silica, and formed as a long membranous sack sandwiched between the MC and plasmalemma. A bundle of microfilaments was visible first on one side and then the other of this initial rib; these eventually flanked and accompanied its extension and remained throughout morphogenesis. The MC remained precisely on the central nodule (Fig. 9A). The raphe rib became doubled and a "raphe fiber" appeared between the two ribs (Fig. 9B). The raphe fiber was a striated dense fiber found on the cytoplasmic side of the raphe fissure, running its length, and in *Pinnularia* included a complex curve at the end (Pickett-Heaps et al. 1979). Transapical ribs formed early and mitochondria became positioned between them. Later silicification created perforations in the valve called formena, and a mitochondrium was positioned under each one. In addition to ending up in an optimized location for respiration, the mitochondria apparently acted as barriers to the expanding SDV, which became molded around them (Pickett-Heaps et al. 1979).

This study showed that the SDV could act as a mold during silicification, by both expansion in conjunction with the cytoskeleton, and inhibition of expansion by cellular structures, in this case the mitochondria. Cytoskeletal elements such as the raphe fiber can also inhibit expansion. A study on *Navicula cuspidata* determined that the position of the raphe fiber coincided with either a compaction of the adjacent siliceous wall, or an indentation in the silicalemma, suggesting that the raphe fiber prevented silica deposition in its location when directly apposed to the SDV (Edgar and Pickett-Heaps 1984b). A hypothesis was presented that lateral movements of raphe fibers, coordinated by microtubules, controlled the formation of the raphe fissure by preventing the SDV from filling it in until the walls of the SDV were silicified (Pickett-Heaps et al. 1979; Edgar and Pickett-Heaps 1984b). Precise control was deemed likely in the formation of other silicified structures such as transapical ribs and their cross extensions (Edgar and Pickett-Heaps 1984b). The positioning of mitochondria between ribs in *Pinnularia* highlighted the coordination of other cytoskeletal-guided movements with the silicification process.

The pennate species *Hantzschia amphioxis* has a slight curvature to the cell, creating concave and convex sides (Fig. 10A). There is a distinctive "keel" where the raphe is located on the corner of the convex side, creating the "hantzschiod" morphology (Fig. 10A, arrows). After cleavage in this species, the MC moved to the center of the furrow (Fig. 10B), where the raphe was formed similarly as in *Pinnularia*, except that microtubules extended from the MC directly over the future raphe, upon which were organized rows of mitochondria (Pickett-Heaps and Kowalski 1981). After 1 h, the entire initiation complex, including the raphe, MC, microtubules, silicalemma, and mitochondria, moved to the convex side of the cell (Fig. 10C). The MC was then located in the keel, positioning the raphe there, and the remainder of the wall formed, resulting in hantzschiod morphology (Fig. 10C).

The movement of the initial silicification complex to one side of the cell was unexpected, but provided a possible explanation for cell wall symmetries

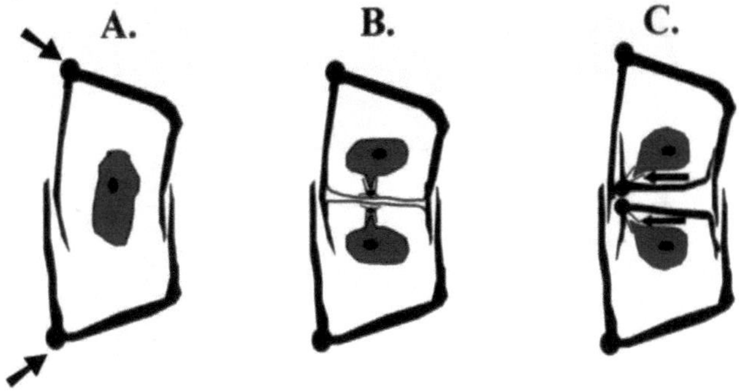


**Fig. 10A–C.** Cytoskeletal movements during cell wall formation of *Hantzschia amphioxis*, adapted from Pickett-Heaps and Kowalski (1981). **A** Vegetative cell. *Arrows* denote the keels where the raphes are located. **B** Cell after cleavage. Two daughter protoplasts are visible, and the microtubule center positions the start of silicification at the center of the cleavage furrow, and connects to the nucleus. **C** Movement of the initial silicification center to one side of the cell to position the raphe in the corner to maintain hantzschiod morphology. *Arrows* denote direction of movement

found in three *Nitzschia* species (Pickett-Heaps 1983). These species had two general morphologies, the nitzschiod, with keels at opposite corners of the cell, and hantzschiod, as described above (Fig. 11A, B). *Nitzschia sigmoidea* could assume both nitzschiod and hantzschiod morphologies (Fig. 11A and B), but *Nitzschia sigma* (Fig. 11C) and *Nitzschia tryblionella* (Fig. 11D) only nitzschiod. In *N. sigmoidea* and *N. sigma*, the silicalemma initially formed at the center of the cleavage furrow (Fig. 11A–C). In *N. sigmoidea*, after raphe formation, the initiation complex and associated organelles moved in either direction, but always in opposite directions, in the two daughter cells (Pickett-Heaps 1983). This explained how both nitzschiod and hantzschiod morphologies could form in daughter cells (Fig. 11A, B). In *N. sigma*, the initial complex also formed in the middle and moved in opposite directions in daughter cells, but in each cell always away from the location of the epithecal keel, thus always leading to nitzschiod morphology (Fig. 11C). In *N. tryblionella*, the initial complexes formed opposite each other on one side of the daughter cells (Fig. 11D). In the daughter cell where the complex was positioned appropriately for nitzschiod symmetry, the complex remained in that location. In the other daughter cell the complex moved, and did so to the opposite side to also maintain nitzschiod symmetry (Fig. 11D). Another fascinating observation from this study was that the fibulae (wall struts arching across the raphe canal) grew in specific directions. In *Hantzschia*, fibulae always grew in the direction from the valve to the girdle bands (Pickett-Heaps and Kowalski 1981). In contrast, in *Nitzschia*, valve to girdle directionality only occurred in one of the daughter cells, in the other growth occurred from the girdle to the valve (Pickett-Heaps 1983).

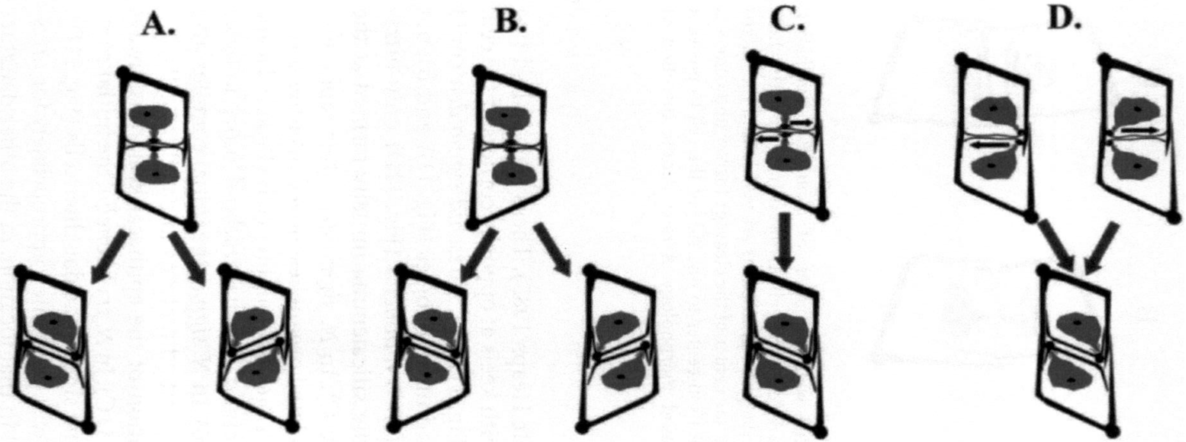


**Fig. 11A–D.** Cytoskeletal movements during silicification of three *Nitzschia* species, illustrating how particular cell morphologies are generated, adapted from Pickett-Heaps (1983). **A** and **B** *Nitzschia sigmoidea*, found in both nitzschiod (**A**) and hantzschiod (**B**) morphologies. Movement of initial silicification centers in either, but opposite directions in the daughter cells results in the generation of only nitzschiod or hantzschiod daughter cell morphologies from the nitzschiod mother cell (**A**, *below*), or a mixture of nitzschiod or hantzschiod from the nitzschiod mother cell (**B**, *below*). **C** *N. sigma*, found only in nitzschiod morphology. Initial silicification begins at the center of the cleavage furrow, and then always moves in opposite directions in daughter protoplasts to maintain nitzschiod morphology (**C**, *below*). **D** *N. tryblionella*, found only in nitzschiod morphology. The initial silicification center occurs at one corner of the daughter protoplasts, and in the cell where nitzschiod morphology can be maintained the center remains there. In the other cell, the center moves to the opposite side, maintaining nitzschiod morphology as well (**D**, *below*)

These comparative studies (Pickett-Heaps and Kowalski 1981; Pickett-Heaps 1983; Edgar and Pickett-Heaps 1984b) showed that specific movements of the initial complex controlled overall valve morphology in these pennates, and that the direction of growth of cell wall structures was also regulated. The cytoskeleton orchestrated a stepwise progression of events, which included positioning information, during silicification. Cytoskeletal elements not only controlled directionality within a single cell, but coordinated movements occurred in adjacent daughter cells.

In pennates, although the overall process of cell wall formation is similar, a more recent comparative study (Cox 1999) indicated that there were variations in different species in (1) the timing of areolae (pore) formation relative to other structures, (2) modifications of the internal central raphe fissure path, (3) modifications of the shape of the external raphe fissure, and (4) the predominant direction of silica deposition. Regarding the latter point, in most pennate species in this study (Cox 1999) initial deposition occurred predominantly at what would be the external surface of the valve (distal) and proceeded in towards the center of the cell (proximal), with the exception of one species where deposition was proximal to distal. In contrast, in centric species, proximal to distal deposition may be most common (Schmid and Schulz 1979; Schmid and Volcani 1983), but there are exceptions to this as well (Crawford and Schmid 1986).

We shall now describe examples of cell wall formation in two centric species, *Thalassiosira eccentrica* (Schmid and Schulz 1979), and *Coscinodiscus wailesii* (Schmid and Volcani 1983), highlighting the formation of truly three-dimensional valve structures (some pennates also make three-dimensional structures). The *T. eccentrica* valve consists of areolae in a hexagonal array occluded by what is called the cribrum sieve plate layer (which is more densely silicified) with foramen open to the exterior (Fig. 12d, e). *C. wailesii* has a generally similar arrangement, but with the cribrum on the outside, and the structure of the cribrum also differ (Fig. 13h). Just after cleavage in *T. eccentrica*, the initial silicification site is located at what will be the center of the hypovalve. The SDV and polymerizing silica extend radially outward from this area in the form of branching tubes, generating a two-dimensional system of a rough hexagonal "mesh" (Fig. 12a, b). Holes in the mesh eventually become filled in with silica (Fig. 12c). After this base layer forms, areolar walls grow perpendicular to it, and then become thicker (Fig. 12c, e). There is a radial gradient of maturation, with silicification more advanced near the center of the forming valve (Fig. 12c). After the areolar walls reach a particular height, growth then occurs perpendicular to the walls at the tops, in towards the center of the areolae (Fig. 12d, e). These foramen are not completely filled in – an open structure results, slightly smaller than the cribrum (Fig. 12d, e).

An interesting observation is that in several *Thalassiosira* species with a regular arrangement of areolae, the areolae are packed in an optimal hexagonal arrangement, but their diameters decrease with increasing distance from the center of the valve (Fig. 1). At the center, in the proximity of the initial

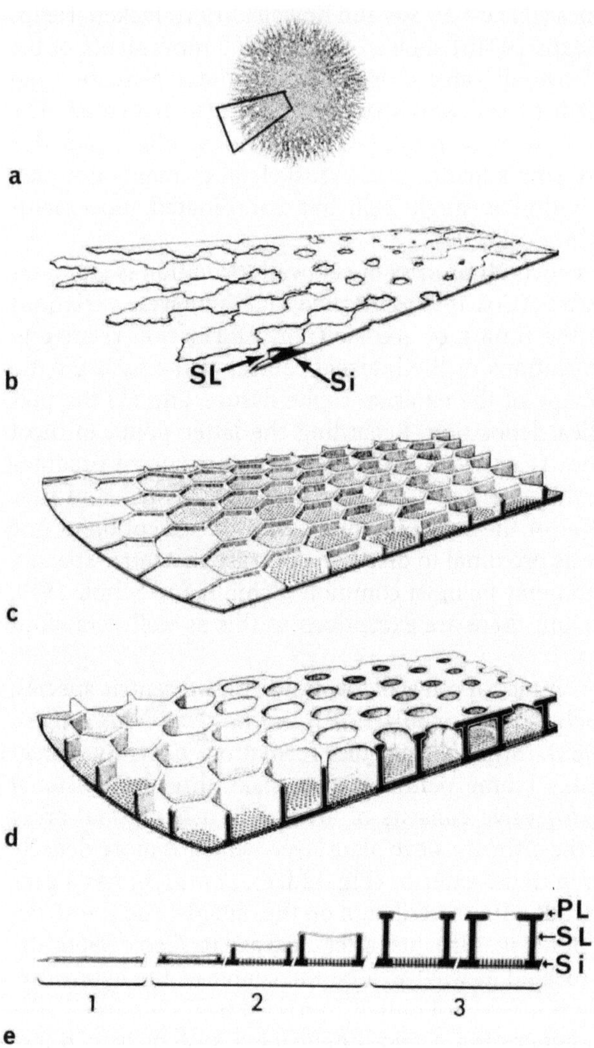


**Fig. 12a–e.** Valve formation in *Thalassiosira eccentrica*, from Schmid and Schulz (1979). **a** Stage 1: development of the base layer; silicification starts at the center of what will become the new valve, and silicified strands are deposited radially from the center. **b** Detail of **a**: radial structures become reticulate constituting the outline of the areolae. The finished base layer is a thin silicified plate, enveloped by the silicalemma. **c** Stage 2: development of the areolae walls and the "porous" plates. **d** Stage 3: development of the outer layer. **e** Cross section of the valve in developmental stages 1–3, and relative positions of silica (*Si*), silicalemma (*SL*), and plasmalemma (*PL*)

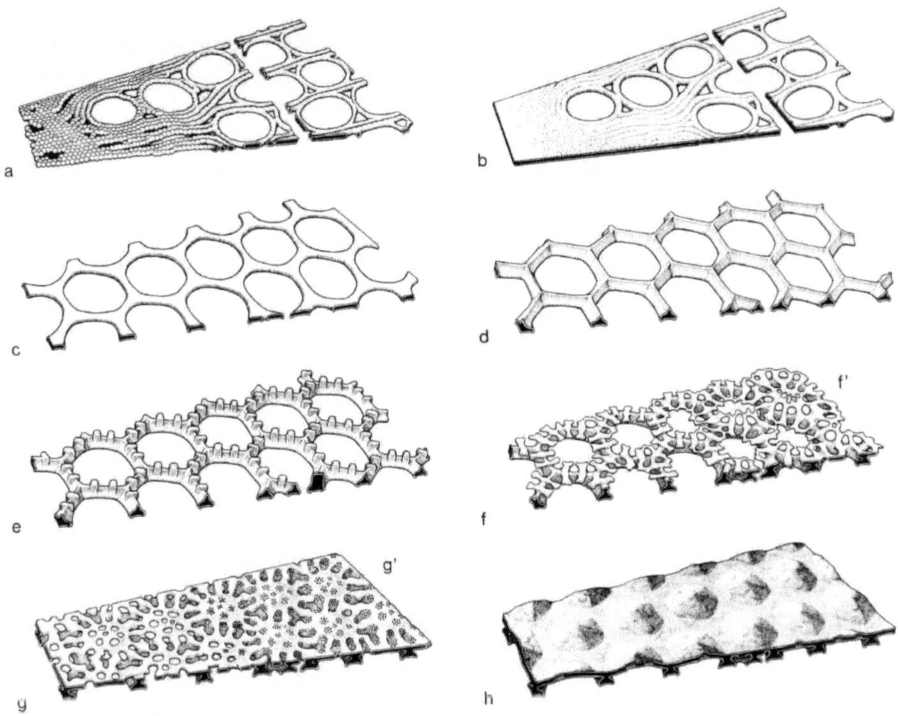


**Fig. 13a–h.** Schematic drawing of valve formation in *Coscindiscus wailesii*, from Schmid and Volcani (1983). **a** Initial deposition of silicified strands and formation of areolae in the base layer. Rough structure of silica is evident. **b** Initial smoothing of deposited silica. **c** Final smoothing of the base layer. Marginal curvatures surrounding each areola at the bottom are visible. **d** Perpendicular growth of areolar walls – note triangular cross section. **e** Growth of silicified "teeth" from the tops of the areolar walls. **f, f′, g, g′, h.** Growth of silica perpendicular to silicified teeth, forming the cribrum and cribellum of the outer valve surface

silicification site, is a single areola surrounded by seven others instead of six (Fig. 1). This arrangement constrains the rest of the areolae in the valve surface to reduce in diameter according to the −1/6 power of their distances from the center in order to maintain optimal packing (Longuet-Higgins 2001). Similar mathematical constraints may apply to other patterns in diatom valves (Longuet-Higgins 2001).

In *Coscinodiscus wailesii*, the initial silicification site is also at the center of the cleaved protoplasts (Schmid and Volcani 1983). A base layer is formed initially by deposition of tubular strands of connected "rough" (spherical) silica radiating outward from the center (Fig. 13a). Branching events occur, splitting one strand into two, and areolae are formed by the inward growth of adjacent strands, meeting in the middle (Schmid and Volcani 1983). The base layer becomes smooth (Fig. 13b, c), and marginal curvatures on the proximal side of each areola become visible (Fig. 13c). Growth then occurs perpendicular to the base layer, as in *T. eccentrica*, but the areolar walls have a distinctive triangular cross section (Fig. 13d). After the areolar walls are complete, small silicified "teeth" project upward from their tops (Fig. 13e). After reaching a specific height, growth then begins perpendicular to these teeth, filling in the areolae to eventually form the cribrum and even smaller cribellum (Fig. 13f–h). Formation of these structures also occurs by growth of silicified strands from each side meeting and fusing in the middle. As previously mentioned, the areolae, cribrum, and cribellum are hexagonally arranged (Sumper 2002). There is some question as to how these direct observations (Fig. 13; Schmid and Volcani 1983) correlate with Sumper's (2002) model, since the model seems to imply a pre-formation of hexagonally arranged polyamine droplets prior to silicification, which is in contrast to the "polymerize as you go" occurring in vivo.

These examples show the complete control that diatoms have over silicified form and direction of growth. The formation of specific structures is precisely coordinated and sequentially timed. Also apparent is a substantial amount of molding. As discussed previously, a smooth surface of silica, as occurs in the base layer of *C. wailesii* (Fig. 13c), indicates an area of appression between the silicalemma and depositing silica, which is an indication of molding. In a study on *D. brightwellii*, extending ribs in the base layer could either branch towards each other to form areolae, or could fuse together (Li and Volcani 1985a). Silica morphology differed in each case; when fusion was to occur, facing walls of adjacent ribs had a rough silica appearance, indicating the lack of intervening silicalemma, whereas in formation of areolae, facing rib walls were smooth, suggesting that the silicalemma was confining the ribs and only allowing branched growth at specific locations. If silicification were solely internal template-driven, it is not clear how such precise distinction could be made, nor how perpendicular growth from a completed base layer to form areolar walls could occur (Fig. 13). If polymerization determinants were in the SDV lumen or attached to the silicalemma, then expansion of the SDV could move these determinants in specific directions, allowing deposition on top of previously completed structures.

*T. eccentrica* and *C. wailesii* are large diatoms (as is *P. viridis*, Figs. 6 and 7), which makes observation of forming structures easier, but it is important to consider how these processes may relate to other, smaller, species. Many centric diatom species do not share the precise pattern of structures seen in the examples given here (Round et al. 1990).

The labiate process (LP), found predominantly in centric diatoms, is a tube that extends through the valve with a characteristic labiate inner surface. It is a channel through which mucilage can be secreted (Medlin et al. 1986; Pickett-Heaps et al. 1986). The LP is made in conjunction with a specialized organelle (Li and Volcani 1985a) called the labiate process apparatus (LPA). It is not clear whether specialized organelles or apparatii are involved in formation of other substructures in diatom valves. A fibrous plug in the SDV located immediately adjacent to the LPA is implicated in the formation of the labiate process, but the composition of the plug and details of its functioning are unknown

(Pickett-Heaps et al. 1988). Data support the hypothesis that the raphe in pennates was derived from the LP in centrics. Pennates arose later in evolution than centrics (Round et al. 1990). The pennate species *Diatoma vulgare* has no raphe, but has two small LPs located near opposite ends of the hypo- and epi-valve (Pickett-Heaps 1989). A homolog of the LPA, associated with formation of the LP, was clearly identified in this species, consistent with the derivation of the raphe from the LP (Pickett-Heaps 1989). The raphe also acts as an opening for the secretion of mucilage from the cell (Hoagland et al. 1993), involved in driving the characteristic gliding movement of pennates over surfaces (Lind et al. 1997).

The involvement of cytoskeletal elements in silicification has been demonstrated in numerous studies, which have included the use of cytoskeletal inhibitors (Schmid 1980; Blank and Sullivan 1983; Cohn et al. 1989). Recent work is beginning to clarify the molecular details of the activities and interactions of cytoskeletal elements. The genus *Chaetocerous* is characterized by formation of setae, which are long hollow external spines. In *C. rostratum*, two setae are formed extending from the valve (Li and Volcani 1985b). These are magnificent structures, especially when viewed from within at the base (Li and Volcani 1985b), made of longitudinal ridges of silica encircled by siliceous bands with small spines projecting out at regular distances, and tapering finely to the tip. Another species, *C. decipiens*, forms chains of cells attached to each other by fusion of setae at their bases. Setae in this case extend from between two girdle bands. The setae are quite long relative to the size of the cell, and it was apparent that the driving force for setae growth could not be turgor pressure (Pickett-Heaps 1998b). The extreme tips of growing setae were not silicified, and a transitory bulging of the cytoplasm was seen (Fig. 14). Just behind the tip was a fibrous band, apparently consisting of actin, that acted as a sleeve to define the diameter of the seta (Fig. 14). The fibrous band was appressed to the plasmalemma ahead of the silicification front, and overlapped into the most recently formed silica along the inner edge of the silicalemma (Fig. 14). It was proposed that tip extension was driven either by molecular ratcheting via a myosin-like motor mechanism (Fig. 14a), or by actin cycling in this sleeve (Fig. 14b), working as a molecular treadmill whose support base was the recently polymerized wall of the silicified tube (Pickett-Heaps 1998b).

An alternative mechanism of formation of protuberances was evident in the growth of structures called probosci in the centric *Proboscia alata* (van de Meene and Pickett-Heaps 2002). Surprisingly, instead of the structures growing outward from the cell, after formation of the tip, they actually resulted from valve growth back over the cell. Electron and fluorescence microscopy, coupled with drug treatments, revealed that microtubules moved away from the tip during valve formation, and that these were always coincident with the non-silicified portion of the valve and SDV (van de Meene and Pickett-Heaps 2002), suggesting that microtubules were involved in SDV expansion. A ring of actin was identified at the growing edge of the SDV, and the ring became larger as the valve grew. Thus, it appeared that the microtubules kept the probosci form

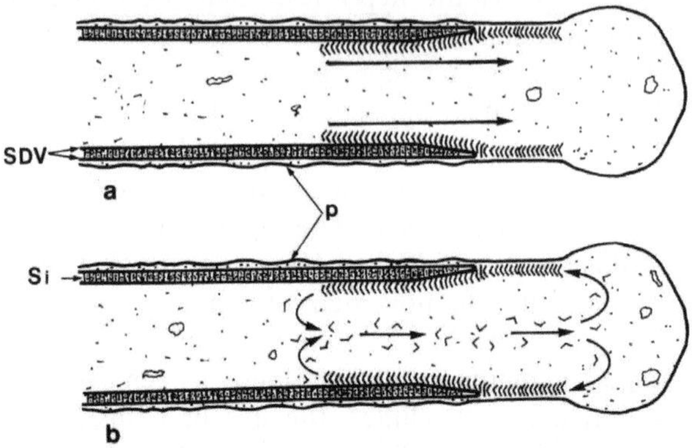


**Fig. 14a,b.** Two models for elongation of seta in *Chaetocerous decipiens*, from Pickett-Heaps (1998b). Silica (*Si*) within the SDV provides a rigid base and cylindrical template supporting a fibrous sleeve of actin. **a** The sleeve could move forward (*arrows*) as a single unit via a molecular motility system acting upon the just-formed wall; the sleeve molds the forming edge of the SDV just inside the plasmalemma (*p*). **b** A variation, with the components of the fibrous sleeve cycling like a molecular treadmill; subunits disassemble at the *left* and reassemble at the *right*

straight until they were silicified while the actin molded the conical outline of the valve by controlling the diameter of its base. The ability of the SDV to act as a mold was again apparent since probosci appressed against adjacent daughter cells were partly surrounded by the expanding SDV to form "claspers" – sites of attachment between cells. Molding across daughter cell protoplasts in general has been called "interactive" division, where if valves are nonplanar and tightly appressed, protrusions in one valve form depressions in the other (Mann 1984).

We have gained some insight into the cellular control of macromorphogenesis, but as fascinating as these few studies have been, it is important to realize that the species examined contain a minuscule fraction of the total number of existing silica morphologies (Round et al. 1990). The described mechanisms may be but a few options out of a wide variety that can be exploited by diatoms. Thus far, we have seen a direct involvement of the MC in positioning the silicification start site, and in some species movement of the initiation complex dramatically changes valve morphology. Microfilaments are found lining the edge of the expanding SDV in the raphe (Chiappino and Volcani 1977; Pickett-Heaps et al. 1979), and in the development of probosci in *P. alata* (van de Meene and Pickett-Heaps 2002), suggesting that these are pulling the SDV into a desired shape. It has been proposed that microtubule-driven lateral movements of raphe fibers control formation of the raphe fissure (Edgar and Pickett-Heaps 1984b). Actin plays a role as an apparent determinant of size; a ring of actin was proposed to define the diameter and provide a mechanism for extension of setae in *C. decipiens* (Pickett-Heaps 1998b), and an actin ring may determine the diameter of the forming valve in *P. alata* by gradually increasing its own diameter (van de Meene and Pickett-Heaps 2002). It would be valuable to continue electron and fluorescence microscopic studies on a variety of species, since it seems that with each new morphology examined comes insight into new mechanisms. One basic lesson learned from the published accounts is that the cytoskeleton is exerting specific control over the formation of silicified structures in the diatom cell; this is with the acknowledgement that as yet unidentified mechanisms of micromorphogenesis may contribute. A continuing investigation into these processes will not only shed light on silicification, but also be of general benefit to understand the cytoskeleton's role in cell biology.

In addition to cellular control of silicification, there are clearly environmental effects. The extent of silicification in a given diatom species can vary up to fourfold (Lewin 1957; Paasche 1980; Taylor 1985; Brzezinski et al. 1990). Growth rate has a significant effect; under environmental silicon levels not limiting for growth, silicification is reduced in populations of faster growing cells, which is likely due to less time being spent during the period of cell wall synthesis (Martin-Jézéquel et al. 2000). Under limiting silicon conditions, there is a direct correlation between growth rate and silicification; cells growing slower because of a lack of cell wall precursor material are less silicified (Martin-Jézéquel et al. 2000). Frustules that are less silicified may not only be thinner, but smaller as well, and siliceous spines can be diminished or absent (Paasche 1975; Davis 1976; Harrison et al. 1977; Brzezinski et al. 1990).

## 7 The Silica Deposition Vesicle – the "Black Box" in the Process of Silicification

The silica deposition vesicle contains determinants for silica polymerization, and likely provides a specialized microenvironment. Intralumenal conditions must be generated and controlled by transport activities in the silicalemma. The SDV also transmutes cytoskeletal movements into control over silicified form, and acts as a mold both actively and passively. The ability of the SDV to be stretched and molded must relate to the composition of the silicalemma, and a major question is how the integrity of the SDV is preserved while new membrane material is added in step with the rate of expansion.

Characterizing proteins of the SDV would clearly be of great benefit to understand the process of silicification. However, the SDV itself has not been isolated, making it difficult to even identify SDV proteins. After the completed valve is exocytosed, it is thought that the silicalemma becomes all or part of an organic casing surrounding the silicified components of the mature wall (Pickett-Heaps et al. 1990). In addition to this casing, however, there are other wall components. In some species an organic layer located under the silicified components called the diatopetum (von Stosch 1980) is present, and contains a complex mixture of sulfated polysaccharides, and a small amount of protein, suggesting the presence of proteoglycans (McConville et al. 1999). The organic casing of the silicified components of the wall and the diatopetum are clearly visually distinct (McConville et al. 1999; Kröger and Wetherbee 2000), and hence may be chemically so. There are also mucilage components of the wall, both coating the valve surface and secreted for cell movement or adhesion (Hoagland et al. 1993; Lind et al. 1997). Although the organic layer surrounding cell wall silica may contain some or all of the components present in the silicalemma, the presence of different cell wall components during extraction make it difficult to determine from which component extracted proteins have come. Also, AFM apparently did not identify a silica-coating organic layer in *P. viridis*, since uncoated silica was all that was left after scraping cell wall mucilage away with the tip of the AFM (Crawford et al. 2001). However, it is possible that a silica-specific organic layer underlying the mucilage was also scraped away by the AFM tip. It has been noted that the organic layer is so thin that it is not always visible in sections for TEM (Round et al. 1990). When visible, the documented examples of an organic silica casing have appeared convincing enough, but until this material is characterized, or specific probes are applied, there will still be some question as to whether it was derived from the silicalemma and/or is an entity separate from other cell wall material.

Data suggest that diatom cell wall material protects the silica from dissolving into the aqueous environment (Lewin 1961), since dissolution occurs at a minimal rate unless cell wall organic material is removed by chemical treatment (Lewin 1961; Kamatani 1982) or cells are treated with proteases (Bidle and Azam 1999). A strong correlation between ectohydrolytic protease activity and silica dissolution was shown, and no correlation with glucosidase, chitinase, and lipase activities (Bidle and Azam 2001), suggesting that the silica protective layer is largely proteinaceous. It is possible that cell wall material is highly cross-linked, since even after treatments that completely dissolve the silica, the casing has been observed intact (Reimann et al. 1966). Characteristics of the mucilage layer scraped away by AFM in *P. viridis* were consistent with this material being cross-linked (Crawford et al. 2001). If the proteins of the silicalemma are highly cross-linked, individual proteins may be difficult to isolate and characterize.

The above discussion highlights difficulties associated with isolating cell wall components as a means to characterize SDV proteins. There may be other options. One would be to isolate the proteins before they are incorporated into the SDV. A defined time period must elapse between synthesis of SDV proteins and their transport and incorporation into the SDV. The use of silicon-starved or synchronized cultures (Coombs and Volcani 1968; Darley and Volcani 1969) could provide a window of time in which to isolate these proteins. Inhibitors of Golgi vesicle movement may also be useful in this regard (Pelham 1991). SDV proteins could then be characterized, and their genes isolated. Gene isolation could more efficiently be done by using a subtractive cloning approach (Hildebrand et al. 1993; Armbrust 1999), again using silicon-starved or synchronized cultures to provide a defined time period when mRNA levels for these genes would be maximum. Cloning SDV protein genes would enable their expression in a heterologous host, providing sufficient amounts of material for characterization and antibody production. Antibodies could be used to confirm SDV localization and provide specific SDV probes. Fusion of possible SDV proteins with a marker such as GFP followed by diatom transformation (Zaslavskaia et al. 2000) may also facilitate localization.

The lack of understanding about components of the SDV allows us only to speculate about what reasonably may be there. This is one of the least known aspects of silicification, but the most important to be understood. Components of the SDV must confer the ability of silica polymerization, expansion, and molding. There must be proteins in the silicalemma that interact with the cytoskeleton. There must be transporters involved in maintaining the acidic condition of the SDV lumen, as well as other ionic conditions that are present. Water would be formed as an end product of silica polymerization (Iler 1979), therefore water channels may exist in the silicalemma. Silicic acid must be transported into the SDV, by an as yet unknown mechanism.

It has been suggested that the SDV may have been derived in an evolutionary sense from the lysosome (Zurzolo and Bowler 2001). Data consistent with this include the low pH of the lumen (Vrieling et al. 1999b), the accumulation of dyes such as rhodamine 123 (Li et al. 1989), implying an electrochemical gradient across the silicalemma, and the identification of silicatein, the silicifying protein in sponges, as a close homolog of cathepsin (Shimizu et al. 1998), a protease found in lysosomes. Extensive proteolytic processing of the silaffins also occurs presumably in the SDV or its precursor vesicles (Kröger et al. 1999). Confirmation of the possible derivation of the SDV from the lysozyme awaits characterization of SDV proteins, but perhaps application of lysozyme-specific probes could provide insight into the origin and workings of the SDV.

## 8 Conclusions and Future Prospects

Application of new research tools in the past decade have ushered in a new era of investigation and understanding of the process of silicified cell wall formation in diatoms. As described in this chapter, the isolation, identification, and characterization of the molecular entities involved are opening the door for mechanistic explanations of the in vivo process. We have isolated genes encoding the transporter responsible for silicic acid uptake into the cell (Hildebrand 2000; Hildebrand et al. 1997, 1998), and identified sequence features that may relate to mechanisms of transport control. We still know little about intracellular transport, but it is becoming clear that this occurs by a general transport mechanism without sequestration of silicic acid into vesicles (Mehard et al. 1974; Rogerson et al. 1986; Hildebrand 2000). Sequestration of another sort apparently does occur however, since several lines of evidence are consistent with intracellular silicic acid being bound by some compound or compounds (Bhattacharyya and Volcani 1983; Martin-Jézéquel et al. 2000; Hildebrand 2000). Insight into the molecular components involved in the maintenance and transport of intracellular silicic acid may clarify the mechanism of coupling between silica deposition and silicic acid uptake (Conway and Harrison 1977; Fig. 3). The first tentative direct identification of silicon bonded to a biological molecule in a diatom has recently been presented (Kinrade et al. 2002), and the demonstration that silicon readily forms five and six-coordinated structures in aqueous solutions with sugar-like polyols (Kinrade et al. 1999) suggests that common cellular compounds or their derivatives could be involved in diatom silicon processing. Considering this, it is a puzzle why identification of silicon/biological molecule interactions has been so difficult, since diatoms process so much silicon. One possibility is that most interactions are not covalent, but occur by hydrogen bonding. If so, detection of a silica/organic interaction could be accomplished by determining IR spectra (Mizutani et al. 1998), but measurements of a silicic acid/organic interaction, such as with the SITs or intracellular binding components, may be difficult due to the weak energetics and possible transitory nature of the interaction. In spite of the difficulties with direct detection, the cloning of a specific silicic acid transporter (Hildebrand et al. 1997) and identification of silica precipitating polypeptides and polyamines (Kröger et al. 1999, 2000) indicate that biological components do interact with, and catalyze reactions of silicon. We are gaining the first insights into the molecular determinants of silica polymerization (Kröger et al. 1999, 2000, 2001, 2002), but based on the literature of in vitro silica polymerization utilizing polyamines and surfactants, and observations of diatom silica microstructures, there is still much to learn about the diatom molecules. Regarding macroscale structure formation, it is apparent that cytoskeletal involvement is at least in the order of complexity as in mitosis. What is not yet clear is the extent to which intralumenal events contribute to macrostructure formation, and if they do, what the relative contributions of intra- and extralumenal events are. Electron microscopic examinations of the course of silicification and Sumper's model (2002) highlight the importance of understanding the relationship between micro and macro-scale processing. Application of imaging techniques using silica-incorporating dyes (Li et al. 1989; Shimizu et al. 2001) and electron and atomic force microscopy (Crawford et al. 2001), coupled with analysis of in vitro silica polymerization reactions (Mizutani et al. 1998; Patwardhan et al. 2001a, 2002), could provide insight into the spatial arrangements and movements of polymerization determinants during silicification, and glimpses into the mechanism of formation of various structures. Genetic approaches, as applied in elucidation of the eukaryotic cell cycle, would be powerful tools to apply to the investigation of diatom silicification. Mutants have been isolated in calcified scale formation in the coccolithophorids (Marsh 1999, 2000), and similar approaches may be fruitful in diatoms. The first steps to enable the manipulation of genes involved in diatom silicification has been achieved, namely the ability to introduce and express genes in diatoms by transformation (Dunahay et al. 1995; Apt et al. 1996; Falciatore et al. 1999; Fisher et al. 1999; Zaslavskaia et al. 2000), but much remains to be done.

The broad interest in diatom silicon research may be unique, since it spans such varied fields as cell biology, materials chemistry, global carbon and silicon cycling, nanotechnology, and it's own unique and emerging field of silicon biochemistry. We are now applying approaches that may allow us to finally understand the underlying molecular mechanisms of diatom cell wall silicification. Considering how intellectually intriguing and aesthetically pleasing the silicified cell wall is, one might expect that the molecular details of its formation will be just as fascinating. In one sense, this can be thought of as a culmination of over 300 years of investigation, but given the large number of diatom silica morphologies, and the already tangible hints at the multiple levels of cellular control over the process, we are likely to be intrigued for some time to come.

## References


Andersen RA, Medlin LK, Crawford RM (1986) An investigation of the cell wall components of *Actinocyclus subtilis*. J Phycol 22:466–479

Anonymous (1703) Two letters from a gentleman in the country relating to Mr. Leeuwenhoek's letter in *Transactions* no. 283. Communicated by Mr. C. Philos Trans R Soc Lond 23(288): 1494–1501

Apt KE, Kroth-Pancic PG, Grossmann AR (1996) Stable nuclear transformation of the diatom *Phaeodactylum tricornutum*. Mol Gen Genet 252:572–579

Armbrust EV (1999) Identification of a new gene family expressed during the onset of sexual reproduction in the centric diatom *Thalassiosira weissflogii*. Appl Environ Microbiol 65: 3121–3128

Azam F (1974) Silicic-acid uptake in diatoms studied with [$^{68}$Ge] germanic acid as a tracer. Planta 121:205–212

Azam F, Volcani BE (1974) Role of silicon in diatom metabolism. VI. Active transport of germanic acid in the heterotrophic diatom *Nitzschia alba*. Arch Microbiol 101:1–8

Azam F, Hemmingsen BB, Volcani BE (1974) Role of silicon in diatom metabolism V. Silicic acid transport and metabolism in the heterotrophic diatom *Nitzschia alba*. Arch Microbiol 97: 103–114

Baker H (1753) Employment for the microscope. Dodsley, London

Bhattacharyya P, Volcani BE (1980) Sodium-dependent silicate transport in the apochlorotic marine diatom *Nitzschia alba*. Proc Natl Acad Sci USA 77:6386–6390

Bhattacharyya P, Volcani BE (1983) Isolation of silicate ionophore(s) from the apochlorotic diatom *Nitzschia alba*. Biochem Biophys Res Commun 114:365–372

Bidle KD, Azam F (1999) Accelerated dissolution of diatom silica by marine bacterial assemblages. Nature 397:508–512

Bidle KD, Azam F (2001) Bacterial control of silicon regeneration from diatom detritus: significance of bacterial ectohydrolases and species identity. Limnol Oceanogr 46:1606–1623

Binder BJ, Chisholm SW (1980) Changes in the soluble silicon pool size in the marine diatom *Thalassiosira weissflogii*. Mar Biol Let 1:205–212

Blank GS, Sullivan CW (1983) Diatom mineralization of silicic acid. VI. The effects of microtubule inhibitors on silicic acid metabolism in *Navicula saprophila*. J Phycol 19:39–44

Brzezinski MA, Olson RJ, Chisholm SW (1990) Silicon availability and cell-cycle progression in marine diatoms. Mar Ecol Prog Ser 67:83–96

Carafoli E (1994) Biogenesis – plasma membrane calcium ATPase – 15 years of work on the purified enzyme. FASEB J 8:993–1002

Cha JN, Shimizu K, Zhou Y, Christiansen SC, Chmelka BF, Stucky GD, Morse DE (1999) Silicatein filaments and subunits from a marine sponge direct the polymerization of silica and silicones in vivo. Proc Natl Acad Sci USA 96:361–365

Cha JN, Stucky GD, Morse DE, Deming TJ (2000) Biomimetic synthesis of ordered silica structures by block copolypeptides. Nature 403:289–292

Chiappino ML, Volcani BE (1977) Studies on the biochemistry and fine structure of silica shell formation in diatoms. VIII. Sequential cell wall development in the pennate *Navicula pelliculosa*. Protoplasma 93:205–221

Chisholm SW, Azam F, Eppley RW (1978) Silicic acid incorporation in marine diatoms on light:dark cycles: use as an assay for phased cell division. Limnol Oceanogr 23:518–529

Cohn SA, Nash J, Pickett-Heaps JD (1989) The effect of drugs on diatom valve morphogenesis. Protoplasma 149:130–143

Conway HL, Harrison PJ (1977) Marine diatoms grown in chemostats under silicate or ammonium limitation. IV. Transient response of *Chaetoceros debilis*, *Skeletonema costatum*, and *Thalassiosira gravida* to a single addition of the limiting nutrient. Mar Biol 43:33–43

Conway HL, Harrison PJ, Davis CO (1976) Marine diatoms grown in chemostats under silicate or ammonium limitation. II. Transient response of *Skeletonema costatum* to a single addition of the limiting nutrient. Mar Biol 35:187–199

Coombs J, Volcani BE (1968) Studies on the biochemistry and fine structure of silica-shell formation in diatoms. Chemical changes in the wall of *Navicula pelliculosa* during its formation. Planta 82:280–292

Cox EJ (1999) Variation in patterns of valve morphogenesis between representatives of six biraphid diatom genera. J Phycol 35:1297–1312

Crawford RM, Schmid AM (1986) Ultrastructure of silica deposition in diatoms. In: Leadbeater BS, Riding R (eds) Biomineralization in lower plants and animals. System Soc 30:291–314

Crawford SA, Higgins MJ, Mulvaney P, Wetherbee R (2001) Nanostructure of the diatom frustule as revealed by atomic force and scanning electron microscopy. J Phycol 37:543–554

Darley WM, Volcani BE (1969) Role of silicon in diatom metabolism: a silicon requirement for deoxyribonucleic acid synthesis in the diatom *Cylindrotheca fusiformis* Reimann and Lewin. Exp Cell Res 58:334–342

Davis CO (1976) Continuous culture of marine diatoms under silicate limitation. II. Effect of light intensity on growth and nutrient uptake of *Skeletonema costatum*. J Phycol 12:291–300

Del Amo Y, Brzezinski MA (1999) The chemical form of dissolved Si taken up by marine diatoms. J Phycol 35:1162–1170

Drum RW, Pankratz HS (1964) Post mitotic fine structure of *Gomphonema parvulum*. J Ultrastruct Res 10:217–223

Dunahay TG, Jarvis EE, Roessler PG (1995) Genetic transformation of the diatoms *Cyclotella cryptica* and *Navivula saprophila*. J Phycol 31:1004–1012

Due AD, Zhinchao Q, Thomas JM, Buchs A, Powers AC, May JM (1995) Role of the C-terminal tail of the GLUT1 glucose transporter in its expression and function in *Xenopus laevis* oocytes. Biochemistry 34:5462–5471

Edgar LA (1983) Mucilage secretions of moving diatoms. Protoplasma 118:44–48

Edgar LA, Pickett-Heaps JD (1982) Ultrastuctural localization of polysaccharides in the motile diatom *Navicula cuspidata*. Protoplasma 113:10–22

Edgar LA, Pickett-Heaps JD (1984a) Diatom locomotion. Prog Phycol Res 3:49–88

Edgar LA, Pickett-Heaps JD (1984b) Valve morphogenesis in the pennate diatom *Navicula cuspidata*. J Phycol 20:47–61

Edgar LA, Zovortink M (1983) The mechanism of diatom locomotion. II. Identification of actin. Proc R Soc Lond B 218:345–348

Falciatore A, Cassoti R, Leblanc C, Abrescia C, Bowler C (1999) Transformation and nonselectable reporter genes in marine diatoms. Mar Biotech 1:239–251

Fisher H, Robl I, Sumper M, Kröger N (1999) Targeting and covalent modification of cell wall and membrane proteins heterogeneously expressed in the diatom *Cylindrotheca fusiformis*. J Phycol 35:113–120

Fryxell GA, Hasle GR (1980) The marine diatom *Thalassiosira oestrupii* structure, taxonomy and distribution. Am J Biol 67:804–814

Gordon R, Drum RW (1994) The chemical basis of diatom morphogenesis. Int Rev Cytol 150: 243–372

Harrison PJ, Conway HL, Holmes RW, Davis CO (1977) Marine diatoms grown in chemostats under silicate or ammonium limitation. III. Cellular composition and morphology of *Chaetoceros debilis, Skeletonema costatum*, and *Thalassiosira gravida*. Mar Biol 43:19–31

Hecky RE, Mopper K, Kilham P, Degens ET (1973) The amino acid and sugar composition of diatom cell-walls. Mar Biol 19:323–331

Hildebrand M (2000) Silicic acid transport and its control during cell wall silicification in diatoms. In: Baeuerlein E (ed) Biomineralization: from biology to biotechnology and medical application. Wiley-VCH, Weinheim, pp 171–188

Hildebrand M, Higgins DR, Busser K, Volcani BE (1993) Silicon-responsive cDNA clones isolated from the marine diatom *Cylindrotheca fusiformis*. Gene 132:213–218

Hildebrand M, Volcani BE, Gassmann W, Schroeder JI (1997) A gene family of silicon transporters. Nature 385:688–689

Hildebrand M, Dahlin K, Volcani BE (1998) Characterization of a silicon transporter gene family in *Cylindrotheca fusiformis*: sequences, expression analysis, and identification of homologs in other diatoms. Mol Gen Genet 260:480–486

Hoagland KD, Rosowski JR, Gretz MR, Roemer SC (1993) Diatom extracellular polymeric substances: function, fine structure, chemistry, and physiology. J Phycol 29:537–566

Hsu YG, Lin FJ (2000) Organic-inorganic composite materials from acrylonitrile-butadiene-styrene copolymers (ABS) and silica through and in situ sol-gel process. J Appl Polymer Sci 75:275–283

Iler RK (1979) The chemistry of silica: solubility, polymerization, colloid and surface properties, and biochemistry. Wiley, New York

Kamatani A (1982) Dissolution rates of silica from diatoms decomposing at different temperatures. Mar Biol 68:91–96

Katagiri H, Asano T, Ishihara H, Tsukuda K, Lin J-L, Inukai K, Kikuchi M, Yazaki Y, Oka Y (1992) Replacement of intracellular C-terminal domain of GLUT1 glucose transporter with that of GLUT2 increases Vmax and Km of transport activity. J Biol Chem 267:22550–22555

Kim W-J, Yang S-M (2000) Preparation of mesoporous materials from the flow-induced microstructure in aqueous surfactant solutions. Chem Mater 12:3227–3235

Kinrade SD, Del Nin JW, Schach AS, Sloan TA, Wilson KL, Knight CTG (1999) Stable five- and six-coordinated silicate anions in aqueous solution. Science 285:1542–1545

Kinrade SD, Gillson A-ME, Knight CTG (2002) Silicon-29 NMR evidence of a transient hexavalent silicon complex in the diatom *Navicula pelliculosa*. J Chem Soc Dalton Trans 307:1–3

Kresge CT, Leonowicz ME, Roth WJ, Vartuli JC, Beck JS (1992) Ordered mesoporous molecular sieves synthesized by a liquid-crystal template. Nature 359:710–712

Kröger N, Wetherbee R (2000) Pleuralins are involved in theca differentiation in the diatom *Cylindrotheca fusiformis*. Protist 151:263–273

Kröger N, Lehmann G, Rachel R, Sumper M (1997) Characterization of a 200-kDa diatom protein that is specifically associated with a silica-based substructure of the cell wall. Eur J Biochem 250:99–105

Kröger N, Deutzmann R, Sumper M (1999) Polycationic peptides from diatom biosilica that direct silica nanosphere formation. Science 286:1129–1132

Kröger N, Deutzmann R, Bergsdorf C, Sumper M (2000) Species-specific polyamines from diatoms control silica morphology. Proc Natl Acad Sci USA 97:14133–14138

Kröger N, Deutzmann R, Sumper M (2001) Silica-precipitating peptides from diatoms. J Biol Chem 276:26066–26070

Kröger N, Lorenz S, Brunner E, Sumper M (2002) Self assembly of highly phosphorylated silaffins and their function in biosilica morphogenesis. Science 298:584–586

Lewin JC (1954) Silicon metabolism in diatoms. I. Evidence for the role of reduced sulfur compounds in silicon utilization. J Gen Physiol 37:589–599

Lewin JC (1955) Silicon metabolism in diatoms. III. Respiration and silicon uptake in *Navicula pelliculosa*. J Gen Physiol 39:1–10

Lewin JC (1957) Silicon metabolism in diatoms. IV. Growth and frustule formation in *Navicula pelliculosa*. Can J Microbiol 3:427–433

Lewin JC (1961) The dissolution of silica from diatom walls. Geochim Cosmochim Acta 21: 182–198

Li C-W, Volcani BE (1985a) Studies on the biochemistry and fine structure of silica shell formation in diatoms. VIII. Morphogenesis of the cell wall in a centric diatom, *Ditylum brightwellii*. Protoplasma 124:10–29

Li C-W, Volcani BE (1985b) Studies on the biochemistry and fine structure of silica shell formation in diatoms. IX. Sequential valve formation in a centric diatom, *Chaetoceros rostratum*. Protoplasma 124:30–41

Li C-W, Chu S, Lee M (1989) Characterizing the silicon deposition vesicle of diatoms. Protoplasma 151:158–163

Lind JL, Heimann K, Miller EA, van Vliet C, Hoogenraad NJ, Wetherbee R (1997) Substratum adhesion and gliding in a diatom are mediated by extracellular proteoglycans. Planta 203:213–221

Lobel KD, West JK, Hench LL (1996) Computational model for protein-mediated biomineralization of the diatom frustule. Mar Biol 126:353–360

Longuet-Higgins MS (2001) Geometrical constraints on the development of a diatom. J Theor Biol 210:101–105

Lu YF, Ganguli R, Drewien CA, Anderson MT, Brinker CJ, Gong WL, Guo YX, Soyez H, Dunn B, Huang MH, Jink JI (1997) Continuous formation of supported cubic and hexagonal mesoporous films by sol gel dip coating. Nature 389:364–368

Mann DG (1984) An ontogenetic approach to diatom systematics. In: Mann DG (ed) Proceedings of the seventh international diatom symposium. Koeltz, Koenigstein, pp 113–141

Marsh ME (1999) Coccolith crystals of *Pleurochrysis carterae*: crystalographic faces, organization, and development. Protoplasma 207:54–66

Marsh ME (2000) Polyanions in the $CaCO_3$ mineralization of coccolithophores. In: Baeuerlein E (ed) Biomineralization: from biology to biotechnology and medical application. Wiley-VCH, Weinheim, pp 251–268

Martin-Jézéquel V, Hildebrand M, Brzezinski MA (2000) Silicon metabolism in diatoms: implications for growth. J Phycol 36:821–840

McConville MJ, Wetherbee R, Bacic A (1999) Subcellular localization and composition of the wall and secreted extracellular sulphated polysaccharides/proteoglycans of the diatom *Stauroneis amphioxys* Gregory. Protoplasma 206:188–200

Medlin LK, Crawford RM, Andersen RA (1986) Histochemical and ultrastructural evidence for the function of the labiate process in the movement of centric diatoms. Br Phycol J 21:297–301

Mehard CW, Sullivan CW, Azam F, Volcani BE (1974) Role of silicon in diatom metabolism. IV. Subcellular localization of silicon and germanium in *Nitzschia alba* and *Cylindrotheca fusiformis*. J Physiol 30:265–272

Mizutani T, Nagase H, Fujiwara N, Ogoshi H (1998) Silicic acid polymerization catalyzed by amines and polyamines. Bull Chem Soc Jpn 71:2017–2022

Monnier A, Schuth F, Huo D, Kumar D, Margolese D, Maxwell RS, Stucky GD, Krihsnamurty M, Petroff P, Firouzi A, Janicke M, Chmelka BF (1993) Cooperative formation of inorganic-organic interfaces in the synthesis of silicate mesostructures. Science 261:1299–1303

Naik RR, Brott LL, Clarson SJ, Stone MO (2002) Silica-precipitating peptides isolated from a combinatorial phage display peptide library. J Nanosci Nanotech 2:95–100

Nakajima T, Volcani BE (1969) 3,4-Dihydroxyproline: a new amino acid in diatom cell walls. Science 164:1400–1401

Nakajima T, Volcani BE (1970) Eta-N-trimethyl-L-gamma-hydroxylysine phosphate and its non-phosphorylated compound in diatom cell walls. Biochem Biophys Res Commun 39:28–33
Norten TA, Melkonian M, Andersen RA (1996) Algal biodiversity. Phycologia 35:353–365
Oka Y, Asano T, Shibasaki Y, Lin J-L, Tsukuda K, Katagiri H, Akanuma Y, Takaku F (1990) C-terminal truncated glucose transporter is locked into an inward-facing form without transport activity. Nature 345:550–553
Olivares L, Aragon C, Gimenez C, Zafra F (1994) Carboxyl terminus of the glycine transporter GLYT1 is necessary for correct processing of the protein. J Biol Chem 269:28400–28404
Paasche E (1973) Silicon and the ecology of marine plankton diatoms. II. Silicate-uptake kinetics in five diatom species. Mar Biol 19:262–269
Paasche E (1975) Growth of the plankton diatom *Thalassiosira nordenskioldii* Cleve at low silicate concentrations. J Exp Mar Biol Ecol 18:173–183
Paasche E (1980) Silicon. In: Morris I (ed) The physiological ecology of phytoplankton. Studies in ecology, vol 7. Univ California Press, Berkeley, pp 259–284
Patwardhan SV, Mukherjee N, Clarson SJ (2001a) Formation of fiber-like amorphous silica structures by externally applied shear. J Inorg Organometall Polym 11:117–121
Patwardhan SV, Mukherjee N, Clarson SJ (2001b) the use of poly-L-lysine to form novel silica morphologies and the role of polypeptides in biosilification. J Inorg Organometall Polym 11: 193–198
Patwardhan S, Mukherjee N, Clarson SJ (2002) Effect of process parameters on the polymer mediated synthesis of silica at neutral pH. J Inorg Organometall Polym (in press)
Pelham HRB (1991) Multiple targets for Brefeldin A. Cell 67:449–452
Pickett-Heaps JD (1983) Valve morphogenesis and the microtubule center in three species of the diatom *Nitzschia*. J Phycol 19:269–281
Pickett-Heaps JD (1989) Morphogenesis of the labiate process in the araphid pennate diatom *Diatom vulgare*. J Phycol 25:79–85
Pickett-Heaps JD (1998a) Cell division and morphogenesis of the centric diatom *Chaetoceros decipiens* I. Living cells. J Phycol 34:989–994
Pickett-Heaps JD (1998b) Cell division and morphogenesis of the centric diatom *Chaetoceros decipiens*. II. Electron microscopy and a new paradigm for tip growth. J Phycol 34:995–1004
Pickett-Heaps JD, Kowalski SE (1981) Valve morphogenesis and the microtubule center of the diatom *Hantzschia amphioxysis*. Eur J Cell Biol 25:150–170
Pickett-Heaps JD, Tippit DH, Andreozzi JA (1979) Cell division in the pennate diatom *Pinularia*. IV. Valve morphogenesis. Biol Cell 35:199–206
Pickett-Heaps JD, Hill D, Wetherbee R (1986) Cellular movement in the centric diatom *Odontella*. J Phycol 22:334–339
Pickett-Heaps JD, Wetherbee R, Hill D (1988) Cell division and morphogenesis of the labiate process in the centric diatom *Ditylum brightwellii*. Protoplasma 143:139–149
Pickett-Heaps J, Schmid A-MM, Edgar LA (1990) The cell biology of diatom valve formation. In: Round FE, Chapman DJ (eds) Progress in phycological research, vol 7. Biopress, Bristol
Reeves CD, Volcani BE (1985) Messenger RNA and polypeptide accumulation patterns in synchronized cultures of *Cylindrotheca fusiformis*. J Gen Microbiol 131:1735–1744
Reidel GF, Nelson DM (1985) Silicon uptake by algae with no known Si requirement. II. Strong pH dependence of uptake kinetic parameters in *Phaeodactylum tricornutum* (Bacillariophyceae). J Phycol 21:168–171
Reimann BEF, Lewin JC, Volcani BE (1965) Studies on the biochemistry and fine structure of silica shell formation in diatoms. I. The structure of the cell wall of *Cylindrotheca fusiformis*. J Cell Biol 24:39–55
Reimann BEF, Lewin JC, Volcani BE (1966) Studies on the biochemistry and fine structure of silica shell formation in diatoms. II. The structure of the cell wall of *Navicula pelliculosa* (Breb.) Hilse. J Phycol 2:74–84
Rogerson A, deFreitas ASW, McInnes AG (1986) Cytoplasmic silicon in the centric diatom *Thalassiosira pseudonana*. Can J Microbiol 33:128–131

Ross R, Cox EJ, Karayeva NI, Mann DG, Paddock TBB, Simonsen R, Sims PA (1979) An amended terminology for the siliceous components of the diatom cell. Nova Hedwiga 64:513–533
Round FE, Crawford RM, Mann DG (1990) The diatoms: biology and morphology of the genera. Cambridge Univ Press, Bath
Schmid A-MM (1980) Valve morphogenesis in diatoms: a pattern-related filamentous system in pennates and the effect of APM, colchicine, and osmotic pressure. Nova Hedwiga 33:811–847
Schmid A-MM, Schulz D (1979) Wall morphogenesis in diatoms: deposition of silica by cytoplasmic vesicles. Protoplasma 100:267–288
Schmid A-MM, Volcani BE (1983) Wall morphogenesis in *Coscinodiscus wailesii.* I. Valve morphology and development of its architecture. J Phycol 19:387–402
Schmid A-MM, Borowitzka MA, Volcani BE (1981) Morphogenesis and biochemistry of diatom cell walls. In: Kiermayer O (ed) Cytomorphogenesis in plants, vol 8, Springer, Berlin Heidelberg New York, pp 63–97
Shimizu K, Cha J, Stucky GD, Morse DE (1998) Silicatein $\alpha$: Cathepsin L-like protein in sponge biosilica. Proc Natl Acad Sci USA 95:6234–6238
Shimizu K, Del Amo Y, Brzezinski MA, Stucky GD, Morse DE (2001) A novel fluorescent silica tracer for biological silicification studies. Chem Biol 8:1051–1060
Smith BL, Schaffer TE, Viani M, Thompson JB, Frederick NA, Kindt J, Belcher A, Stucky GD, Morse DE, Hansma PK (1999) Molecular mechanistic origin of the toughness of natural adhesives. Nature 399:761–763
Sullivan CW (1976) Diatom mineralization of silicic acid. I. $Si(OH)_4$ transport characteristics of *Navicula pelliculosa.* J Phycol 12:390–396
Sullivan CW (1977) Diatom mineralization of silicic acid. II. Regulation of $Si(OH)_4$ transport rates during the cell cycle of *Navicula pelliculosa.* J Phycol 13:86–91
Sullivan CW (1986) Silicification by Diatoms. In: Evered D, O'Connor M (eds)_Silicon biochemistry. Wiley, Chichester, pp 59–89
Sumper M (2002) A phase separation model for the nanopatterning of diatom biosilica. Science 295:2430–2433
Swift DM, Wheeler AP (1992) Evidence of an organic matrix from diatom biosilica. J Phycol 28:202–209
Tabor CW, Tabor H (1984) Polyamines. Annu Rev Biochem 53:749–790
Tanev PT, Pinnavaia TJ (1995) A neutral templating route to mesoporous molecular sieves. Science 267:865–867
Taylor NJ (1985) Silica incorporation in the diatom *Coscinodiscus granii* as affected by light intensity. Br Phycol J 20:365–374
Tolbert SH, Firouzi A, Stucky GD, Chmelka BF (1997) Magnetic field alignment of ordered silicate-surfactant composites and mesoporous silica. Science 278:264–268
Trau M, Sankaran S, Saville DA, Aksay IA (1995) Electric-field-induced pattern formation in colloidal dispersions. Nature 374:437–439
van de Meene AML, Pickett-Heaps JD (2002) Valve morphogenesis in the centric diatom *Proboscia alalta* Sundstrom. J Phycol 38:351–363
van de Poll WH, Vrieling EG, Gieskes WW (1999) Localization and expression of frustulins in the pennate diatoms *Cylindrotheca fusiformis, Navicula pelliculosa,* and *Navicula salinarium.* J Phycol 35:1044–1053
Verhey KJ, Hausforff SF, Birnbaum MJ (1993) Identification of the carboxy terminus as important for the isoform-specific subcellular targeting of glucose transporter proteins. J Cell Biol 123:137–147
Verhey KJ, Yeh JI, Birnbaum MJ (1995) Distinct signals in the GLUT4 glucose transporter for internalization and for targeting to an insulin-responsive compartment. J Cell Biol 130:1071–1079
Voight M (1943) Sur certaines irrégularités dans la structure des diatomeés. Notes Bot Chim 4:1–5
von Stosch HA (1980) Structural and histochemical observations on the organic layers of the diatom cell wall. In: Ross R (ed) Proceedings of the sixth symposium on recent and fossil diatoms. Koeltz, Konigstein, pp 231–252

Vrieling EG, Beelen TPM, van Santen RA, Gieskes WWC (1999a) Diatom silica biomineralization as an inspirational source of new approaches to silica production. J Biotech 70:39–51

Vrieling EG, Gieskes WWC, Beelen TPM (1999b) Silicon deposition in diatoms: control by the pH inside the silicon deposition vesicle. J Phycol 35:548–559

Vrieling EG, Beelen TPM, van Santen RA, Gieskes WWC (2000) Nano-scale uniformity of pore architecture in diatomaceous silica: a combined small and wide angle X-ray scattering study. J Phycol 36:146–159

Werner D (ed) (1977) The biology of diatoms. Botanical monographs. Univ Calif Press, Berkeley

Wetherbee R, Crawford S, Mulvaney P (2000) The nanostructure and development of diatom biosilica. In: Baeuerlein E (ed) Biomineralization: from biology to biotechnology and medical application. Wiley-VCH, Weinheim, pp 189–206

Yang H, Vovk G, Coombs N, Sokolov I, Ozin GA (1998) Synthesis of mesoporous silica spheres under quiescent aqueous acidic conditions. J Mater Chem 8:743–750

Zaslavskaia LA, Lippmeier JC, Kroth PG, Grossman AR, Apt KE (2001) Transformation of the diatom *Phaeodactylum tricornutum* with a variety of selectable marker and reporter genes. J Phycol 36:379–386

Zurzolo C, Bowler C (2001) Exploring bioinorganic pattern formation in diatoms. A story of polarized trafficking. Plant Physiol 127:1339–1348

# The Phylogeny of the Diatoms

Wiebe H.C.F. Kooistra,[1] Mario De Stefano,[1] David G. Mann,[2] and Linda K. Medlin[3]

## 1 Introduction

Diatoms are among the most diverse groups of eukaryotic microorganisms and there are probably well in excess of 100,000 species (Mann and Droop 1996). The beauty and unbelievable variation of their intricately shaped and ornamented silica cell wall, called the frustule, have attracted taxonomists and naturalists alike (Round et al. 1990). More recently, ultrastructure, sexual reproduction, and the development of their unique type of zygote, called the auxospore, have also become the focus of research.

Even from the very beginning of diatom research, taxonomists have attempted to order the bewildering diversity of diatoms into discernible groupings and to infer some higher natural order among them, i.e. to unravel the evolutionary history of the whole class. Taxonomic approaches themselves have diversified as well over the last century (Schmidt et al. 1874–1959; Van Landingham 1967–1979; Round et al. 1990; Williams 1993; Hasle and Syvertsen 1996). Round et al. (1990) made ample use of electron microscopy and Medlin and her co-workers pioneered the use of molecular methods (e.g. Medlin et al. 1988, 1991, 2000). Yet, diatoms are certainly not the exclusive toys of taxonomists, as has been demonstrated by a stream of publications on an ever-expanding range of topics linked to these organisms.

Diatoms are of immense ecological significance and are present in almost every aquatic habitat imaginable. At present, they are ubiquitous in the plankton and benthos of marine and freshwater environments from the tropics to the polar regions as well as in temporarily humid surroundings, such as damp moss or rock (Round 1981; Mann 1999; Graham and Wilcox 2000). Several species have managed to carve out peculiar niches: e.g. brine channels of the sea ice, copepod talons, whale skins, hot springs or highly basic or acidic puddles (Round 1981; Round et al. 1990). A few species have even become endosymbionts of dinoflagellates (Chesnick et al. 1997) and foraminifera (Chai and Lee 1999, 2000). Insight into the evolutionary history of the group can unravel how the diatoms have invaded all these environments.

[1] Stazione Zoologica "Anton Dohrn", Villa Comunale, 80121 Naples, Italy
[2] Royal Botanic Garden, Edinburgh EH3 5LR, Scotland, UK
[3] Alfred Wegener Institute, Am Handelshafen 12, 27570 Bremerhaven, Germany

Progress in Molecular and Subcellular Biology, Vol. 33
W. E. G. Müller (Ed.)

Diatoms are virtually all autotrophic (for the few exceptions, see e.g. Li and Volcani 1987). As primary producers in the world's oceans, they form major players in the global cycling of nutrients. Just a few hundred species in the marine phytoplankton fix around a fifth of the total carbon and silica sequestered (Guillard and Kilham 1978; Goldman 1993; Hasle and Syvertsen 1996; Mann 1999). Because diatom primary production helps to fuel the marine food web and the biogeochemical cycles of the ocean, models that seek to predict the effects of increased atmospheric carbon dioxide levels (Falkowski et al. 1998; Smetacek 1998) will increasingly need to take the ecophysiology of these organisms into account. Insight into the geological age of the diatoms would thus give palaeoclimatologists a clue how long these organisms have already been modifying the biosphere.

Diatoms are used as indicator organisms, either for contemporary water quality, or as markers for historical or ancient changes in aquatic conditions (Kilham et al. 1996; Theriot et al. 1997; Mann 1999; Stoermer and Smol 1999). Diatom silica cell walls often preserve well, with their intricate microscopic details largely intact, and can be identifiable to species level after millions of years (e.g. Gersonde and Harwood 1990; Harwood and Gersonde 1990; Barron and Mahood 1993). Therefore, diatom fossils have been used for stratigraphic correlation (Baldauf 1992; Gersonde and Bárcena 1998; Stoermer and Smol 1999). Phylogenies permit critical assessment of the quality of morphological markers that are used to link fossil and extant species and to characterise palaeoconditions. In addition, phylogenies inferred from data obtained from the extant diversity can be compared with independently obtained phylogenies from fossil stratigraphy.

Last, but certainly not least, diatoms play an important role in the global economy. Fossil diatom frustules form an indispensable ingredient of products ranging from washing powder, car polish and toothpaste to dynamite, and they are widely used for filtration (e.g. of beer and wine, sugars, oils and water (Stoermer and Smol 1999). Extant diatoms are now being used for the production of food or valuable biochemical compounds (Falciatore and Bowler 2002). In addition, diatoms provide material scientists with models for the synthesis of silica-based materials at ambient temperatures and pressures (Kröger et al. 2000; Vrieling et al. 2000; Brott et al. 2001). Here too, phylogenetic knowledge will undoubtedly improve our understanding of how particular traits developed and evolved, leading to refined ontogenetic models, e.g. of silica sequestration in diatoms.

## 2 Approaches to Reconstruct Phylogenies

Before phylogenetic hypotheses can be developed to explain the evolution of diversity in diatoms, one needs detailed information about their diversity. Fortunately, the morphology and ontogeny of frustules, gametes and auxospores provide plenty of this and so does the ultrastructural organisation of the cells (Round et al. 1990; Droop et al. 1993; McLaughlin 1992; Pappas and Stoermer 1995). Phylogenetic hypotheses have been inferred from such sources (Simonsen 1979; Kociolek 1986; Kociolek and Stoermer 1986, 1988, 1989, 1993; Kociolek and Williams 1987; Kociolek et al. 1989; Theriot 1992). Wiley et al. (1991) give more general theoretical overviews of methodology.

Differences among homologous DNA sequences of organisms provide an independent data source for phylogenetic inferences. Various DNA markers change at different rates, so that markers can be chosen appropriate to the taxonomic level of comparison. For background information on phylogenetic inferences of sequence data, see Page and Holmes (1998) and contributions in Hillis et al. (1996). Molecular phylogenies of diatoms have recently been reconstructed from the nuclear small subunit (Medlin et al. 1991, 1993, 1996a,b, 1997a,b, 2000; Kooistra and Medlin 1996) and large subunit ribosomal DNA (Sörhannus et al. 1995; Lundholm et al. 2001); the internal transcribed spacer regions in the nuclear ribosomal DNA cistron (Zechman et al. 1994); the mitochondrion-encoded cytochrome c oxidase subunit I (*cox*A, Ehara et al. 2000); the plastid-encoded *tuf*A (Delwiche et al. 1995; Medlin et al. 1997a) and Rubisco large subunit (*rbc*L; Daugbjerg and Andersen 1997; Daugbjerg and Guillou 2001; Mann et al. 2001). Several computer programs have been developed to infer phylogenies from the data (e.g. PAUP*; Swofford 2000).

The general assumption behind phylogeny reconstruction is that distinguishable taxa come about through diversification, called cladogenesis, and that characters (traits, features) are acquired and change their state (in form, number, size) along the way. If so, one might expect that character states and taxa could always be ordered in nested sets. Yet, homoplasic "noise" of various kinds often obscures evolutionary signal and it is the way in which this noise – of conflicting data – is handled that determines the soundness of the method. For an explanation of general terminology used in phylogeny reconstruction (monophyletic, clade, lineage, paraphyletic, grade, polyphyletic, autapomorphy or unique character state, synapomorphy or derived character state, symplesiomorphy or ancestral character state, homoplasy or phylogenetically misinformative character state, convergence, parallelism) see various textbooks, e.g. Wiley et al. (1991).

Three types of approaches have been used to infer evolutionary relationships among taxa: the classical taxonomic approach, the cladistic phylogenetic approach, and the phenetic approach. A classical taxonomist defines a group based ideally on a set of character states that are unique to that group, regardless of whether those character states are in fact shared derived or shared ancestral. Then groups are gathered together into a hierarchical classification in which taxa at each level (species, genera, families, orders etc.) are based on shared traits.

The principal flaw in this approach is the way conflict among traits is handled. Generally, each specialist selects a list of "important" characters, that is, characters whose states discriminate among the taxa he or she believes in;

these are then used to justify the classification. There is nothing wrong with developing such classifications, as long as they are seen as working hypotheses (see discussion in Round et al. 1990 and Cox 1993). Specialists have often been able to develop classifications that are highly "natural" (phylogenetic). Frequently, however, the users of such systems, or even worse, the authors themselves, accept them without question as **The Truth.**

Another problem is that hierarchical classification needs an almost infinite and therefore, impractical number of levels (see Mann 1998, 1999). Furthermore, the taxonomic level assigned to a particular group, such as the diatoms as a whole, is largely arbitrary (Mann 1999). As an example, Round and Crawford (1990) and Round et al. (1990) treat the diatoms as a division (Bacillariophyta), whereas van den Hoek et al. (1995) treat them as a class (Bacillariophyceae); longer ago, they were considered a family ("natural order": Smith 1853). In their textbook, Graham and Wilcox (2000) avoid the problem by using informal names. We will use the classification in Round et al. (1990) as a convenient, but arbitrary reference.

Phylogenetics and phenetics handle conflicting information in a more formal way. The principal assumption of the phylogenetic approach is that the best topologies are the most parsimonious ones, i.e. those that need the fewest number of character state changes along their branches to explain the observed patterns in the available data. Maximum parsimony is, in fact, an application of Ockham's razor (Popper 1959; Sober 1990). Phylogenetics generally "works" because the various shared derived and shared ancestral character states that begot them generally reveal an ordered pattern of nested sets, whereas mis-informative ones are not expected to generate structure. A weakness of this approach is that it lacks an objective way of deciding which character state is shared derived and which one is shared ancestral. This problem is minimised by choosing a set of so-called outgroup taxa outside the ingroup taxa of interest to the researcher. All character states that are shared by some, but not all ingroup-taxa and one or more outgroup-taxa are assumed to be either ancestral or mis-informative. Those states that define single clades are automatically shared derived.

Phenetics is the study of relationships among a group of organisms on the basis of the degree of the overall similarity between them. No a priori and generally, no a posteriori value is given to particular character states: all information is incorporated to reconstruct relationships. A tree-like network expressing phenetic relationships is called a phenogram. Phenetics is particularly easily used with continuously varying, metric data (length, width, linear densities of ribs and pores, etc.), whereas phylogenetics does not handle such data very well. A problem with phenetics is, however, that similarity does not necessarily mean close relationship; some lineages and characters evolve slowly, others more rapidly. Similarity can arise also through convergence. Similar environmental pressures may force states of characters in unrelated species to become similar. A problem that lurks above any reconstruction of evolutionary relationships is that characters can affect one another's states

(Gould and Lewontin 1979), thus potentially giving rise to morphological highly similar, but evolutionary unrelated taxa. Fortunately, such effects abound less so in DNA sequences, but it does occur there too.

We will use a diatom phylogeny (see Fig. 3) generated from available nuclear SSU ribosomal RNA gene sequences to examine the natural status of taxonomic groups and the evolution of morphological features and aspects of life cycle, behaviour and ecology. We concentrate on relationships among taxa at various taxonomically perceived levels, on the origin of the group as a whole, and on how the modern diversity came about. Moreover, we evaluate the acquisition of various ultracellular and morphological structures and compare such insights with what can be gleaned from the fossil record.

# 3 The Diatom Silica Frustule

## 3.1 Morphology of the Silica Frustule

Diatoms are eukaryotic, unicellular or colonial microorganisms. Their hallmark is a unique type of silica cell wall, the frustule, which consists of two intricately shaped and ornamented compound structures called the epitheca and the hypotheca. Each theca is itself composite, comprising a large dish-like valve (epivalve or hypovalve; Figs 1H, I, 2C, I, L, M, N) and a series of narrow girdle bands (Fig. 2I; Round et al. 1990). Many valves show a characteristic division into a flattened upper surface, called the valve face, and its side, called the valve mantle). Although these elements are composed predominantly of silica, they do possess a matrix and an envelope of organic material. In addition to this organic cell wall material, most species also excrete organic material either as dissolved matter or in the form of gels, mucilage pads, stalks or chitin threads.

The valves and girdle bands consist of a system of silica ribs, which grow out from a circular (Fig. 1A), elongate or linear (Fig. 2I, P) pattern centre during formation (Round et al. 1990). Small lateral ribs or frets then grow out to link these primary ribs and the intervening spaces are filled up to form a primary sheet of perforated silica (Fig. 2I). In some unrelated genera, such as *Cylindrotheca* and *Papiliocellulus*, the valve has been reduced to a fragile network, whereas a few endosymbiontic species have dispensed with their cell walls altogether (Lee and Reimer 1984; Chesnick et al. 1997). In many species, a second and often very differently shaped layer of silica is laid down above, or occasionally below, the primary system of ribs and pores, complete with connecting walls, struts, flying buttresses and spandrels (Fig. 1K).

Generally, the frustule encases the protoplasm completely so that exchange with the surroundings must take place via a mesh of pores, usually distributed in a regular pattern all over the surface of the valves and girdle bands (Figs. 1 and 2). The pores may be covered by organic material only, or be occluded by

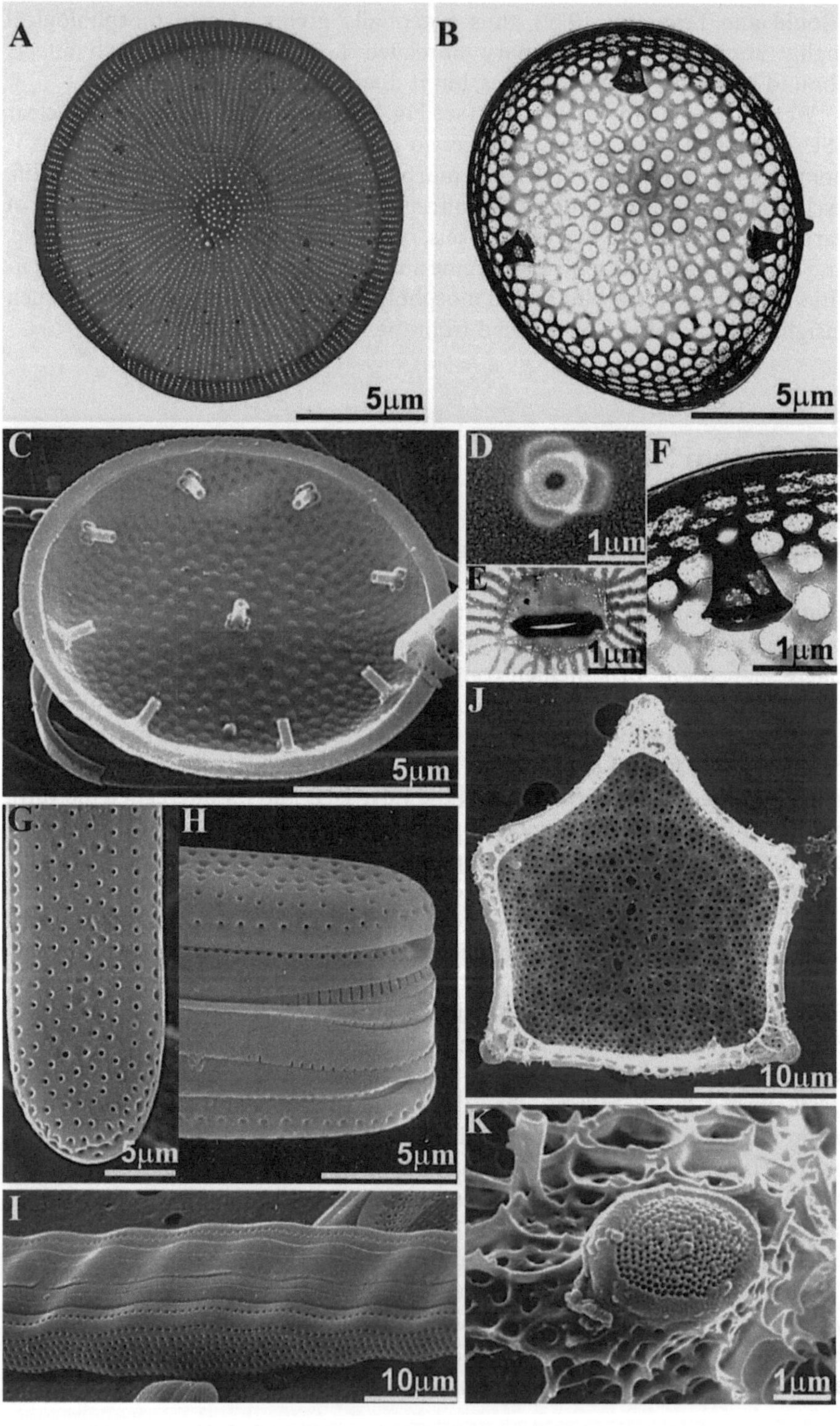
A
B
C
D
E
F
G
H
I
J
K
5µm
5µm
5µm
1µm
1µm
1µm
5µm
5µm
10µm
10µm
1µm

intricately shaped porous silica plates (Fig. 2O). For the detailed terminology used to describe such pores, see Round et al. (1990) or Hasle and Syvertsen (1996). In many species, unoccluded pores are present in clusters at the periphery of the valve face, forming specialised structures called ocelli (Fig. 1K), pseudocelli, ocellulimbi or apical pore fields (Figs. 2A–H), depending on their structure and the taxonomic group in which they occur. These structures are involved in the secretion of mucilage stalks or pads.

Many diatoms possess specialised tubular organelles (processes) that penetrate through the valve (Round et al. 1990; Hasle and Syvertsen 1996); they never occur on the girdle bands. The simplest one, the "tubular process", is nothing more than a simple tube. More common is a second type, the labiate process (rimoportula, Figs. 1B, E, F, 2B, D, F, G, H), which often reveals itself externally as a minute hole or tube (in some cases it is developed into a more elaborate structure, as for example in the long hollow spines of *Ditylum* and *Odontella*), but is flattened on the inside, resembling tightly pursed lips (Fig. 1F). Where present, labiate processes occur either in a circle along the rim of the valve face (Fig. 1B), scattered across the valve face, or clustered in the centre (Fig. 1E), or there may be only one or two, in particular characteristic places on the valve or mantle (Fig. 1C). In a few cases, there can be two different types of labiate process on the same valve. A third type of process, the strutted process or fultoportula (Fig. 1C, D), is involved in the secretion of chitin threads, which either link sister cells in chains or function as cellular "parachutes" for planktonic species, greatly increasing drag. A fourth one, the occluded process, consists of an external tube that has been blocked completely at its intersection with the valve face. The raphe (Fig. 2J, K) can perhaps be considered as a fifth kind of process, although it does not generally project from the wall as much, internally or externally, as the other types of processes. It generally consists of a pair of longitudinal slits through the valve, each beginning near the centre and extending towards the apex of the valve. The raphe shows a range of shapes and positions on the valve across species.

**Fig. 1A–K.** Examples of centric diatoms and their frustule features. **A** Radial centric diatom. TEM of *Leptocylindrus danicus*. Note circular pattern centre or annulus. **B** Typical radial centric diatom with large marginal labiate processes (detail of labiate process in **F**); TEM of *Roperia tessellata*. Note valve face with highly similar areolae in *Thalassiosira curviseriata* in Hasle and Syvertsen (1996, p. 64). **C** Secondary radial centric diatom from multipolar centric ancestry. Internal SEM view of *Thalassiosira œstrupii*. See large strutted processes (detail in **D**) and minute off-centre labiate process. **E** Central labiate process in multi- (bi-) polar centric *Chaetoceros* sp. **G–I** SEM images of *Toxarium undulatum*, a multi- (bi-) polar centric diatom originally believed to be a pennate diatom; **G** detail of apex in valvar view; **H** same detail in girdle view; **I** 75° tilted image of central region showing valve and girdle. Note the regular rows of pores in the perimeter of the valve face and mantle. **J** Internal view (SEM) of undescribed multi- (penta-) polar centric diatom belonging to Triceratiaceae. **K** Detail of large protruding ocellus of diatom in **J**

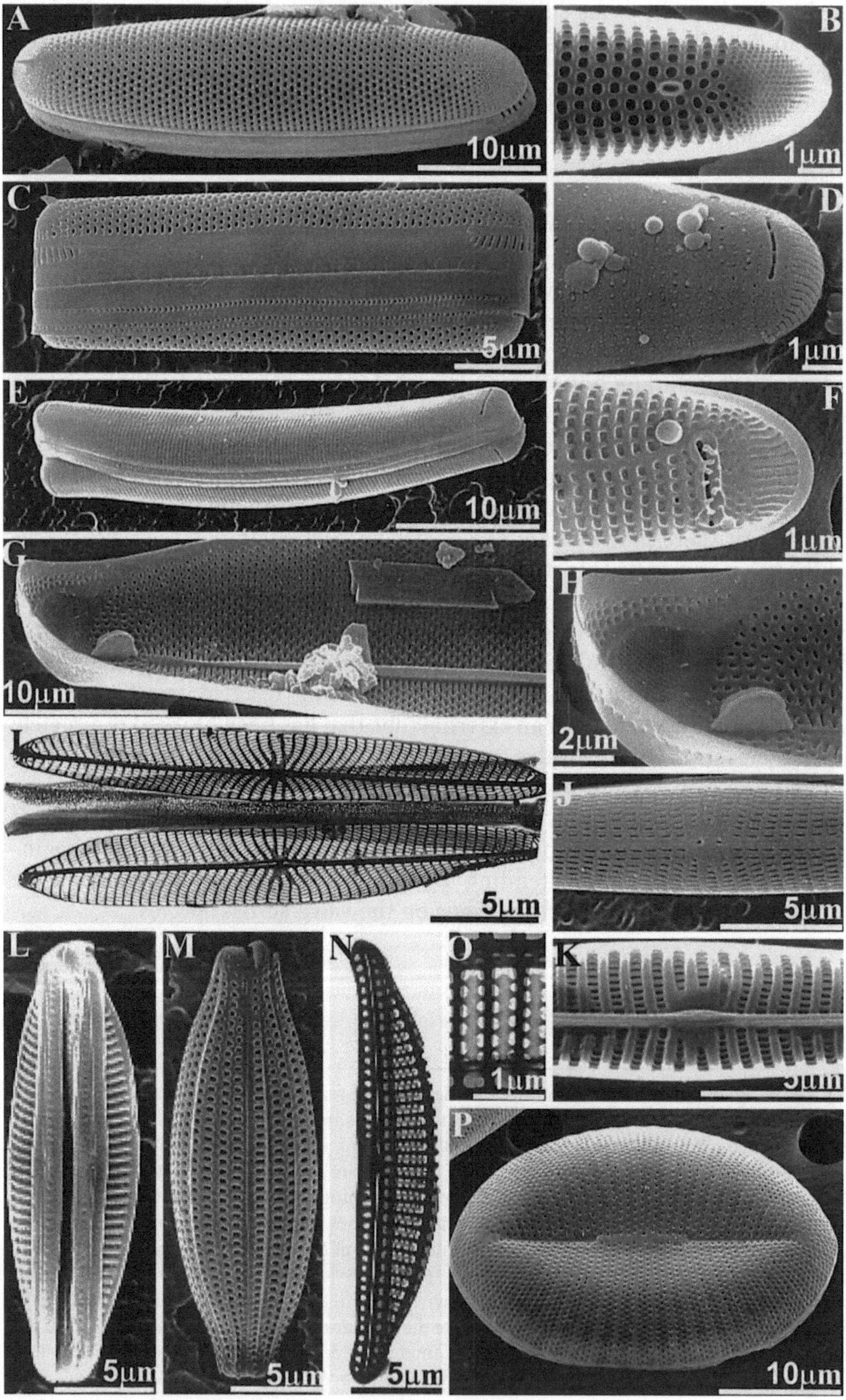
A
10µm
B
1µm
C
5µm
D
1µm
E
10µm
F
1µm
G
10µm
H
2µm
I
5µm
J
5µm
L
5µm
M
5µm
N
5µm
O
1µm
K
P
10µm

## 3.2 Taxonomy Based on Characteristics of the Silica Frustule

Diatom taxonomists are fortunate people, because the wealth of detail visible in the silica frustules provides a wealth of characters and states as a basis for delimiting taxa and inferring relationships between them. These features are always available for study (in contrast to characteristics of sexual reproduction) and they are relatively easily examined using electron microscopy (Round et al. 1990) or even only light microscopy (Hasle and Syvertsen 1996).

Taxonomists have separated diatoms into broad groups, based on the organisation of the pattern of ribs and pores on the valve face. In centric diatoms, the ribs can generally be seen to radiate from a ring (annulus). In the simplest case, in most of what might be called the "radial centrics", the annulus is more or less circular and central, and the valve outline is circular (e.g. Fig. 1A). In the "bipolar" or "multipolar" centrics, on the other hand, the annulus may or may not be central, it may not be circular (e.g. it is elongate or elliptical in *Attheya* and dumbbell-shaped in *Odontella*: Crawford et al. 1994; Pickett-Heaps et al. 1990). Radial centrics basically look like Petri dishes, whereas multipolar ones appear like soap bars or expensively designed perfume bottles. All of the different kinds of process, except the raphe, are found among the centric diatoms.

Another major group consists of the pennate diatoms, which have an elongate rib- or bar-like pattern centre, called the sternum (Fig. 2I, K), from which the ribs extend out more or less perpendicularly: they are organised like a feather (Latin "pinna" or "penna"). Pennate diatoms are usually elongate and their valves resemble ships. Pennates can be subdivided into two types: raphid pennates, possessing one or two raphe slits within or closely associated with

**Fig. 2A–P.** Examples of pennate diatoms and their frustule features. A–H show araphid pennates, I–P represent raphid diatoms. **A** SEM of *Grammatophora* sp. in valvar view. **B** Internal view (SEM) of *Grammatophora* sp. showing apical pore field and small labiate process. **C** SEM of *Grammatophora* sp. in girdle view. **D** SEM view of *Pseudohimantidium* sp. showing apical pore field and peculiar row of labiate processes. **E** SEM of *Pseudohimantidium* sp. in girdle view. **F** Internal view (SEM) of *Pseudohimantidium* sp. showing reduced lips of labiate processes. **G** Internal view (SEM) of *Striatella unipunctata* showing apical labiate process lined up with sternum. **H** Detail of labiate process and apical pore field of *Striatella unipunctata*. **I** TEM of *Navicula* sp. showing valve face with raphe-sternum, pattern of secondary ribs (interstriae), striae (rows of pores), and girdle bands. **J** External view (SEM) of same species showing striae and central raphe endings. **K** Internal view (SEM) of same species showing raphe-sternum, pattern of secondary ribs (interstriae) and striae. **L** SEM image of *Amphora* sp. in valvar view. The raphe sternum is not in the centre of the valve; one side of each valve is more developed than the other (dorsoventrality). **M** SEM image of same species in girdle view. Note large number of expanded girdle bands pushing the valves to one side of the cell. **N** TEM view of single valve of the same species. **O** TEM detail of the striae with occlusions. **P** External valve view of *Cocconeis britannica*, a monoraphid diatom. Note secondary filled-in raphe slit in the sternum valve their sternum; and araphid pennates, which lack such slits, but often instead have labiate processes at the apices or centre.

Some genera are difficult to fit into these broad taxonomic groupings. First, the border between the multipolar centrics and the araphid pennates is somewhat messy. Certain diatoms that are usually considered to be bipolar centrics, including *Dimeregramma* and *Subsilicea* nevertheless have parallel ribs on their valve faces and what appears to be a sternum, but they lack labiate processes at their apices, unlike the araphid pennate diatoms to which they are otherwise similar (Round et al. 1990). The bipolar centric *Helicotheca* lacks an annulus, although a similar diatom, *Neostreptotheca* has one. *Toxarium* (Figs. 1G–I) is considered a pennate diatom, but the best-known species *T. undulatum* lacks a midrib, parallel ribs, or any trace of labiate processes. However, it does appear to possess a pattern centre lying at or near the valve face–mantle junction. Essentially, the same type of pattern centre is visible in *Climacosphenia*, *Synedrosphenia* and *Ardissonea* except that here, the pores are organised into proper rows perpendicular to the apical plane. These species also possess two keel-like ribs slightly off-centre and curving around the apices. This structure is called a bifacial annulus (Mann 1984), because it subtends ribs on both sides, instead of only outwardly, as in the true annulus. Moreover, these species also lack labiate processes. Finally, *Psammodiscus* is circular and has radial symmetry, but appears to have neither an annulus nor a sternum. It has been suggested that it is related to one group of araphid pennate diatoms, but that the sternum has been reduced to a point (Mann 1984; Round et al. 1990)

Another inconvenient grey zone occurs also between raphids and araphids. *Pseudohimantidium* (Fig. 2D–F) possesses a short slit that resembles a raphe, but consists of a series of merged labiate processes (Simonsen 1970; Gibson 1979; Rivera et al. 1986; Round et al. 1990). Two short raphe slits are visible in the valve faces of *Eunotia*, *Actinella* and *Semiorbis* (Round et al. 1990) and *Eunophora* (Vyverman et al. 1998). The slits extend from the apex, but do not reach the central region of the valve face and they are not fully integrated into the sternum, as in other raphid diatoms. Labiate processes are present in some of these diatoms, whereas they are never found in raphid diatoms with a fully developed raphe system. Greater integration between pattern centre and raphe is present in *Peronia* and it has been suggested that the raphe originated from labiate processes (Hasle 1974) and only later became associated with and incorporated into the pattern centre (Mann 1984). The diatoms in the grey zones annoy classical taxonomists because they render their classifications very unbalanced. A taxonomic group with 10,000 species looks rather silly next to another one with only one species. Yet, these species are potentially gratifying to phylogeneticists and pheneticists because they may permit them to unravel evolutionary patterns.

## 3.3
## The Phylogeny Inferred from Nuclear SSU rDNA Sequences

A neighbour joining (NJ) phylogeny inferred from pair-wise maximum likelihood (ML) distances among nuclear small subunit (18S) nuclear ribosomal DNA sequence data is presented in Fig. 3. The tree has been cut in two to fit on two pages. Details of the analysis are given in the figure legend. The horizontal lines in this tree represent the internodes or branches, whereas the vertical lines stand for the nodes, stretched vertically to connect the internodes. Hence, the length of the vertical lines has no phylogenetic relevance, whereas the lengths of the horizontals are proportional to the amount of sequence divergence. Dichotomies obtaining a bootstrap statistical support less than 50% have been drawn in grey as observed in the NJ tree and have been collapsed into polytomies upon the first underlying clade that does get proper bootstrap support. Relationships among clades dropping onto such a polytomy remain unresolved; any combination of such clades can represent a monophyletic group. *Bolidomonas mediterranea* and *B. pacifica* were selected as outgroups because of their close relationships with diatoms (Guillou et al. 1999; Daugbjerg and Guillou 2001).

## 3.4
## Phylogenetic Relevance of Taxonomy and Frustule Characters

Figure 3 shows a basal polytomy composed of several clades with radial centrics and a moderately supported clade with the multipolar centrics and the pennates. The well-supported clade also consists of a polytomy containing several centric clades, primarily the multipolar centrics, a branch with *Toxarium* and a clade with all pennate diatoms. Within the basically multipolar centric group, there is a clade containing Thalassiosirophycidae. This is an especially interesting group because, like the basal, radial centric diatom clades, it consists of radially symmetrical species with a central round annulus; they have apparently retained the ancestral radial centric morphology only secondarily (convergence). Another important result is that the enigmatic pennate-like species *Toxarium* does not group with pennates. Within the pennate clade, the araphid pennates are paraphyletic, whereas the raphids are monophyletic.

Other gene regions have been used to address phylogenetic relationships above the generic level. In the *rbc*L tree, topology resolves not so well as with the nuclear SSU rDNA (Mann et al. 2001 and unpubl.). In cases where some resolution was achieved, the number of species and the taxonomic coverage are still rather low. Nevertheless, these preliminary results broadly support the phylogenetic superstructure inferred from SSU sequences (Kooistra and Medlin 1996; Medlin et al. 1996a,b, 1997a,b, 2000; this study).

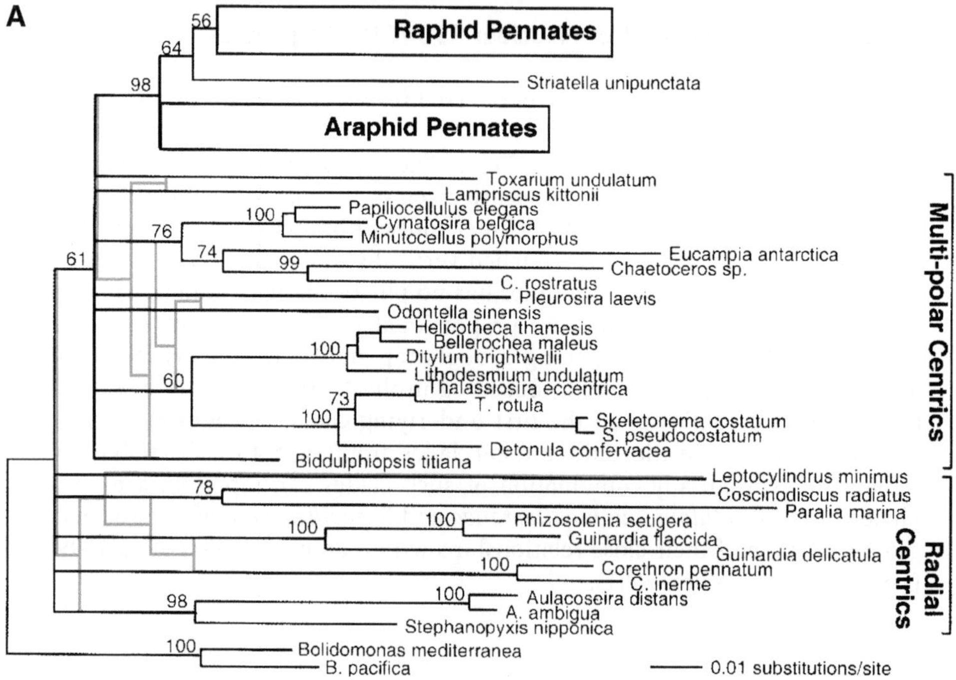


**Fig. 3A,B.** Phenogram inferred from an alignment of nuclear SSU rDNA sequences obtained from diatoms across the taxonomic diversity. The optimal base substitution model was estimated from the alignment using Hierarchical Likelihood Ratio Tests (hLRTs) in Modeltest Version 3.0 (Posada and Crandall 1998). Optimal parameters for base frequency: A = 0.2576, C = 0.1784, G = 0.2413, T = 0.3227; for base substitution frequency relative to G↔T = 1.000: A↔C = 0.9394, A↔G = 2.4904, A↔T = 1.1368, C↔G = 1.1279, C↔T = 3.7238; for proportion of invariable sites: I = 0.3360; for positional rate variation across alignment: $\gamma = 0.5716$. The neighbour-joining tree was inferred from pair-wise maximum likelihood distances constrained with the parameter settings as obtained with Modeltest in PAUP*. Nodal support was estimated using bootstrap analyses (1000 replicates). Bootstrap support values above 50% have been plotted above the clades. Clades obtaining less than 50% support (indicated in *grey*) have been collapsed into polytomies upon the first clade that did obtain proper bootstrap support. The branching order obtained from an NJ analysis of the sequences has been plotted in *grey lines* over the tree to show the resolution of the polytomies. *Bolidomonas mediterranea* and *B. pacifica* were chosen as outgroup taxa. The figure has been cut into two parts to accommodate presentation. **A** The pennate diatoms are collapsed into a box. **B** The radial centrics are collapsed into a box

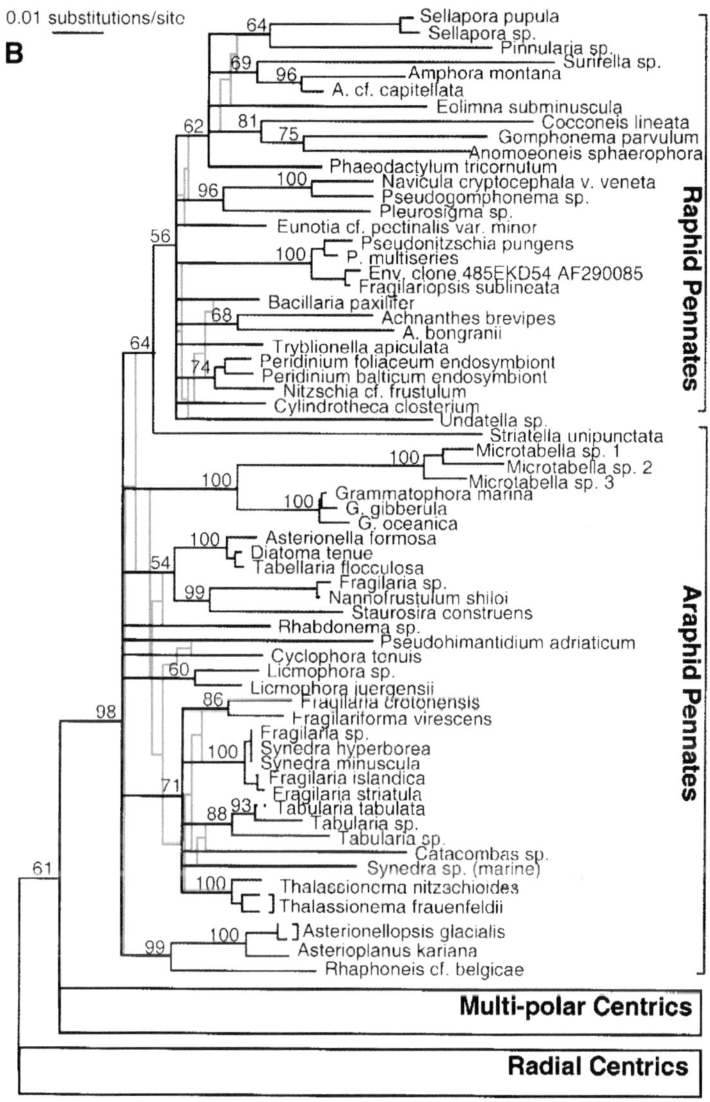


**Fig. 3A,B.** *Continued*

### 3.4.1
### *The Radial Centrics*

Radial centric diatoms (Figs. 1A–B) sprout as a series of clades in the basal polytomy of the tree in Fig. 3A. One clade contains *Coscinodiscus* and *Paralia*, both members of subclass Coscinodiscophycidae, yet they are classified in different orders. Another clade includes Rhizosoleniophycidae with their oddly twisted valves (*Rhizosolenia* and *Guinardia*). A third one contains *Aulacoseira* and *Stephanopyxis*, which belong to the Coscinodiscophycidae as well. The fourth clade comprises Corethrophycidae and the fifth one contains *Aulacoseira* and *Stephanopyxis*, both members of Coscinodiscophycidae. Although relationships among radial centrics remain unresolved, monophyly has not been falsified. The phylogram published by Medlin et al. (2000) suggests that the radial centrics are paraphyletic, *Paralia* forming a separate lineage basal to other genera. However, bootstrap support for this topology is not high. It should be noted that the set of radial centric taxa used in the analysis reported in Fig. 3A does not include *Melosira* or *Actinocyclus*. The SSU sequences of these taxa have highly elevated base substitution rates, in comparison to their relatives (Kooistra and Medlin 1996): such rates result in long branches, which can potentially distort the tree topology. However, even if these taxa are included in the present analyses, radial centrics still collapse into a polytomy.

Radial centric diatoms show a single pattern centre (Fig. 1A) from which the ribs radiate out to the valve perimeter. The annulus can be small and central, or it may lie at the junction of valve face and mantle, thus enclosing the whole of the valve face, as in some *Melosira* and *Aulacoseira* species. Labiate processes, where present, are located in a ring around the valve perimeter or scattered over the valve face. Several genera possess spines at the edge of the valve face, which interlock firmly with similar structures on the valve face of the adjacent sister cell.

### 3.4.2
### *The Bipolar Centrics (Fig. 3A)*

The sister clade to the radial centric diatoms clades consists of a polytomy with multi- (bi-, tri-, etc.) polar centrics (Fig. 1J), the other radial centric Thalassiosirophycidae (Fig. 1C), the supposedly pennate genus *Toxarium* and the true pennates. The fact that this polytomy also contains centrics shows that centrics are paraphyletic. Apparently, then, the centric type of organisation is ancestral and the pennates evolved from a particular group of centric diatoms, probably from those that have bi- or multipolar valves. Most members of this more advanced group of centrics and the pennates share some form of specialised mucilage secreting region in their cell walls. These regions take different forms in different groups, but in all cases they appear to be formed by modified valve pores in well-defined areas, rather than by modified labiate processes, strutted processes, etc. In the true pennates, these areas are generally positioned at the apices and are, therefore called apical pore fields (e.g. Fig. 2A), whereas in bipolar centrics they are referred to as ocelli (Fig. 1K) or pseudocelli. The Thalassiosirophycidae appear to have lost such structures and have retained the ancestral state for the valve shape.

Two distinct clades are recovered among the multipolar centrics. One of these (the lowermost in Fig. 3A) contains two well-supported clades. The first comprises the Thalassiosirophycidae (Fig. 1C), which are characterised by having strutted processes (Fig. 1C, D) a synapomorphy shared only among the species in this group); they also generally have a single large labiate process positioned on the valve mantle or near the centre of the valve. In some species (e.g. in the abundant marine planktonic diatom *Skeletonema*), the strutted processes are developed into long tubular extensions externally, which interdigitate and often fuse with similar tubes of the adjacent cell to link the cells into chains (see Round et al. 1990, pp. 140–141). These linking structures have evolved independently from those found in radial centrics. The equally well-supported sister clade to the Thalassiosirophycidae contains an apparently paraphyletic Lithodesmiophycidae, as well as *Bellerochea* and *Streptotheca* (Hemiaulales, Biddulphiophycidae). These diatoms are characterised by a centrally located labiate process, often adorned with an elongate exterior tube, and mostly also have a well-developed sheet-like ridge at the periphery of the valve face, called lamella. The latter structure generally connects with a similar structure protruding from the adjacent cells. Notably, such a central labiate or tubular process, located within the annulus, is also a feature common to the other clade in the multipolar centrics; the *Chaetoceros*–Cymatosiraceae group and might be a synapomorphy, relative to the radial centrics. Notably, *Biddulphiopsis*, *Lampriscus*, *Odontella* and *Pleurosira*, which collapse directly on the polytomy, do not have processes lying within the annulus.

The sister relation between the Thalassiosirophycidae and the *Lithodesmium*–*Bellerochea* cluster appears peculiar, given the radically different locations of structures in many Thalassiosirophycidae (especially the submarginal position of the labiate process in most species), the presence in the Thalassiosirophycidae of strutted processes, and the generally circular shape and radial symmetry of the valves. These features must be regarded as derived states for part or the entire group. It is particularly curious that the Thalassiosirophycidae exhibits more pronounced radial symmetry of valve outline and structure than any others of the "multipolar" group, and yet its usually single labiate process is found so oddly off-centre. There is no obvious explanation for this strange morphology. However, it should be noted that strutted processes (not labiate ones) do sometimes occur within the annulus in Thalassiosirophycidae (e.g. in *Thalassiosira rotula*, Syvertsen 1977), and that processes are missing from within the annulus in some valves of other multipolar centrics, including some valves of *Chaetoceros* and in *Biddulphiopsis*, *Odontella* and *Pleurosira*.

The other clade in the multipolar centrics consists of the orders Chaetocerotales (Chaetocerotophycidae), its sister *Eucampia* (Hemiaulales, Biddulphiophycidae) and their sister clade, the Cymatosirales (Cymatosirophycidae). Both generally have bipolar valves (the exception is *Bacteriastrum*, which is radially symmetrical – presumably a secondary evolution of radial symmetry or a retention of the ancestral state, as in the Thalassiosirophycidae). The Cymatosirales have a pair of simple pore fields, one at each apex of the elongate valves, which presumably were involved in mucilage secretion; similar areas of pores are present in some species of *Chaetoceros* at the bases of the long hair-like projections (setae) they bear at each apex.

The bipolar centrics *Odontella* and *Pleurosira* (Triceratiales, Biddulphiophycidae), which collapse directly onto the polytomy, possess well-developed ocelli surrounded by a distinct rim and have two (groups of) labiate processes in their valve face, one on either side of the centre. *Biddulphiopsis* (Biddulphiales, paraphyletic Biddulphiophycidae) and *Lampriscus* (Triceratiales, Biddulphiophycidae) do not group either though they share vaguely delimited apical pore fields (called pseudocelli). However, they show neither a trace of a central process nor a trace of subcentral groups of processes like those in *Pleurosira*. Instead, *Biddulphiopsis* possesses a row or ring of minute labiate processes near each pseudocellus, somewhat reminiscent of the arrangement in radial centrics.

Significantly, a strict phenogram representation from a NJ analysis (grey lines superimposed over the black lines in Fig. 3A,B) of the sequences to resolve the polytomies shows the *Biddulphiopsis* and *Lampriscus* branching off early in the multipolar centrics, suggesting that the ill-defined pore fields may have evolved earlier than the more differentiated pore fields (ocelli) of *Odontella*, *Pleurosira* (and probably of other multipolar genera like *Amphitetras*, *Cerataulus* and *Auliscus*), or the well-defined, but smaller apical pore fields of Cymatosiraceae and *Chaetoceros*. The setae of *Chaetoceros* may thus represent a further elaboration and elevation of what was originally an area of slightly modified pores. At some stage during the diversification of the multipolar centrics, the pattern of morphogenesis was altered in one or more lineages, such that one or more processes (labiate, tubular or strutted) came to be formed within the primary pattern centre, the annulus.

### 3.4.3
### *The Bipolar Centric* Toxarium *(Fig. 3A)*

Classically, *Toxarium* (Fig. 1G–I) has been considered to belong to the pennates because of its general shape and behaviour. However, its lack of labiate processes and a sternum rendered its position within pennates doubtful. Indeed, in the molecular phylogeny (Fig. 3A), this genus groups with multipolar centrics. In fact, the neighbour-joining tree (grey lines in Fig. 3) recovers the taxon as sister to the multipolar centric diatom *Lampriscus*. Results of an analysis with all multipolar centrics, the radial centrics *Stephanopyxis nipponica*, *Rhizosolenia setigera* and the pennates *Asterionellopsis* and *Asterionella*, revealed even proper bootstrap support for such a state of affairs (76%). In other words, *Toxarium* is not a pennate at all, but an extremely elongate bipolar centric related to *Lampriscus*.

Although the form of *Toxarium* could not be more different from that of *Lampriscus*, both genera lack processes of any kind. Because all their relatives do possess such organelles, the structures must have been lost secondarily in their common ancestry. Moreover, *Toxarium* does not have a sternum – the bar-like pattern centre characteristic of most pennates, which subtends two series of ribs, one on either side, as in a feather. Instead, the valve face bears a haphazard arrangement of pores (Fig. 1G). However, there are two longitudinal rows of evenly spaced, ordered pores at the junction of the valve face and mantle (Figs. 1I); at the apices these rows bend inwards towards each other and meet, apparently forming a complete ring (Fig. 1H). In fact, one could argue that the genus is really an extremely elongate bipolar centric diatom, with an extremely elongate annulus.

Although no molecular data are yet available to confirm the idea, it seems likely that *Toxarium* belongs to a small group of pennate-like genera with a similar construction. These may include *Climacosphenia, Synedrosphenia* and *Ardissonea*, all of which, like *Toxarium*, lack labiate processes and apical pore fields, despite the fact that all live attached to substrata by mucilage pads. In these three genera, it is even more evident that their pattern centre is an elongated annulus, because this is revealed by the arrangement of pores into four sets of orderly transverse rows, two associated with each straight longitudinal section of the ring. However, it is also important to stress that the annulus, at least in *Climacosphenia*, *Synedrosphenia* and *Ardissonea*, differs significantly from that of the centrics, because it subtends ribs (separating lines of pores) both outwardly and inwardly: the annulus is bifacial and, almost uniquely in diatoms, there is strong centripetal development of the valves (the other principal group in which this occurs is in the Surirellales, a very late-evolving lineage of raphid diatoms).

Whether *Toxarium* is to be considered a pennate or not is merely a question of definitions. Nonetheless, the extremely elongated form evolved at least twice through elongation of a bipolar centric. One pennate-like lineage developed a bifacial annulus, visible today in *Toxarium*, *Climacosphenia* and their allies, whereas the other, which we will conveniently call the true pennates, developed a sternum, possibly by extreme elongation of the unifacial annulus present in multipolar centrics. The multipolar centric *Odontella* exhibits an elongate dumbbell-like annulus (Fig. 40e in Pickett-Heaps et al. 1990), which is not far from a sternum in its form. In this scenario, the true pennates inherited their apical labiate processes from an elongated bipolar centric ancestor with apical pore fields and pericentral or apical processes (as in many bipolar centrics like e.g. *Odontella*).

### 3.4.4
### *The Araphid Pennates (Fig. 3B)*

The "true" pennates constitute a polytomy with several clades containing araphid pennates and a single clade containing all of the raphid pennates. In this topology pennates are thus monophyletic. The araphids appear to be paraphyletic because the araphid pennate *Striatella* appears as the sister group of the raphid diatoms.

Within the araphid pennates proper, a series of clades can be discerned arising from its basal polytomy. In the lowermost one, *Asterionellopsis* (family Fragilariaceae *sensu*; Round et al. 1990) and *Rhaphoneis* (family Rhaphoneidaceae) are found together with high bootstrap support. The next lineage includes a clade with *Fragilaria, Fragilariforma* and *Synedra* with a nearest sister clade holding *Tabularia* and *Catacombas* (all Fragilariaceae) and a next nearest sister clade with *Thalassionema* (Thalassionemataceae). The latter genus possesses what appear to be linking structures, like those in *Fragilaria*, but, whereas *Fragilaria* cells are linked into ribbon chains, *Thalassionema* forms zigzag and star-like colonies in which sister cells connect through mucilage pads; the linking structures have apparently lost their primary function. The close relationship and uncertain status of paraphyletic *Fragilaria* – which occurs in chains – and *Synedra* – which does not – was already noted by Lange-Bertalot (1980) and points to the need for further taxonomic changes beyond those advocated by Williams and Round (summarised in Round et al. 1990). Chain-forming linkage structures in these genera do not signify relatedness.

Next is a clade with two species of *Licmophora* (Licmophoraceae). The clade above this one contains *Asterionella, Diatoma, Tabellaria, Nanofrustulum, Staurosira* (all members of Fragilariaceae). The monophyletic genera *Microtabella* and *Grammatophora* also group together in a clade. This result is not surprising given their similar frustule, characterised by the presence of internal partitions within the cell, formed by inward extensions of the girdle bands, and their mode of colony formation. These two genera have sometimes been classified together with *Striatella* in Striatellaceae, yet such a family arrangement would be paraphyletic. Notably, the topology of the plain neighbour-joining tree does not group the clade with *Microtabella* and *Grammatophora* close to *Striatella*. Whatever character states are shared between *Microtabella* and *Grammatophora* on one hand, and *Striatella* on the other, must be homoplasies. Finally, the Fragilariaceae is not monophyletic, yet it would be premature to rearrange the whole paraphyletic family at this stage. More genera need to be added to get the complete picture.

### 3.4.5 *The Position of* Pseudohimantidium *(Fig. 3B)*

There is a peculiar slit at each end of the valve in the araphid pennate *Pseudohimantidium* (Fig. 2D–F) formed by a short string of labiate processes (Fig. 2F; Simonsen 1970; Gibson 1979; Rivera et al. 1986; Round et al. 1990), and this has been hypothesised to represent an ancestral form of the raphe (Simonsen 1970). For this reason, Simonsen (1970) proposed a separate family Protoraphidaceae, containing *Pseudohimantidium* and *Protoraphis*. In Fig. 3B, *Pseudohimantidium* is located on a single branch in the polytomy of araphid diatoms, thus falsifying Simonsen's conjecture. In a plain neighbour-joining representation of the tree, this taxon groups together with *Cyclophora* and *Licmophora*. *Pseudohimantidium* shares a highly similar organisation of ribs with these taxa and many other araphids. The molecular tree (Fig. 3), therefore suggests that the string of labiate processes in *Pseudohimantidium* evolved from a single one and that this occurred independently from the acquisition of the raphe.

### 3.4.6 *The Raphid Pennates (Fig. 3B)*

Raphid pennate diatoms (Fig. 2I–P) form a single clade in Fig. 3B, though with weak bootstrap support (58%). Apparently, the true raphe has been acquired only once and is believed to have evolved through elongation of labiate processes at the distal ends in the araphid pennate valves (Hasle 1974; Mann 1984; Round et al. 1990, p. 41). *Striatella* appears in Fig. 3B as the nearest sister of the raphid pennates, though with rather weak bootstrap support. Notably, *Striatella* (Fig. 2G, H) is an araphid pennate, adorned with a labiate process at either end of the sternum (Fig. 2H). These processes have an elongated opening on the exterior valve side (Round et al. 1990). One could envisage the evolution of a raphe from such an elongated opening by means of elongation through the midrib towards the centre of the valve.

*Eunotia* has been considered as a primitive raphid pennate. In Fig. 3B this genus is recovered on a single branch collapsing upon the basal topology of the raphid pennates. If this genus is indeed sister to all raphid pennates, then the expanding slit may have reflexed along the side of the sternum, as suggested by Mann (1984). However, the position also allows secondary development of an "incomplete" and off-centre raphe from more normal ancestry. More genera with such raphe systems need to be added.

In most raphid pennate the raphe–sternum runs along the so-called axial area (the keel) of the valve (Fig. 2G, I). In some genera, such as *Amphora*, however, the raphe-sternum is located on one side of the valve allowing the raphes of both valves to be oriented towards the substrate (Fig. 2L–N).

Last, but not least, both *Cocconeis* (Fig. 2P) and *Achnanthes* are monoraphid diatoms; one valve possesses a completely developed raphe, whereas the other does not. Figure 3 shows that such a state of affairs came about through secondary filling in of the raphe in one of the valves. Moreover, this reversal happened at least twice independently because the two genera are not recovered as nearest sisters.

## 4 Phylogenetic Signal in Diatom Chloroplast Structure

The prominent place of frustules in diatom taxonomy does not mean that other characteristics are unimportant and carry no phylogenetic signal. Alternative data sources are simply not as easily accessible and information is often far from complete. Yet, the little that is known often shows distinct patterns across the diatom diversity.

Plastids show great diversity of shape, size and number in diatoms (Mereschkowsky 1901; Cox 1981; Mann 1996). In radial and multipolar centrics including *Toxarium*, and in most araphid pennates, there are many small, disc- or rod-like plastids scattered through the cytoplasm. Virtually all raphids, by contrast, possess one, two or four large plastids, which are often variously lobed and dissected (Round et al. 1990); only a few have many small chloroplasts and these taxa are spread through a number of different lineages, suggesting that these diatoms have acquired polyplastidy independently. Comparison with Fig. 3B indicates, therefore, that there has been a reduction in the number of plastids and an increase in their size relative to the diatom volume during the evolution of raphid pennate diatoms. Increases in the size and complexity of plastids, and reduction in their number, have been accompanied by the development of complex patterns of movements, timed to coincide with critical phases of the cell cycle (mitosis, cell division, valve formation, cell separation), and orchestrated to ensure that plastids are correctly positioned in relation to other cell organelles and the cell wall, and that daughter cells receive equal numbers of plastids from the parent (Mann 1996). The nearest relative of the raphids, *Striatella*, has several elongated plastids that radiate from around the nucleus. A similar arrangement is found in the araphid pennate *Rhabdonema*, which is not related to *Striatella* (Fig. 3B); thus, this plastid morphology must have evolved at least twice independently.

In raphid diatoms, besides orchestrated movements of the plastids, there are often also tightly controlled movements of the nucleus. Some taxa, for example, show "oscillatory" behaviour, the nucleus dividing on opposite sides of the cell at successive divisions; others show "constant" behaviour, always dividing on the same side of the cell (Round et al. 1990). These patterns tend to be constant within lineages. For example, the *Gomphonema–Anomoeoneis–Cocconeis* clade exhibits "constant" behaviour, whereas the *Sellaphora–Pinnularia* and

*Navicula–Pseudogomphonema–Pleurosigma* clades exhibit "oscillatory" behaviour, apart from *Seminavis*, which is not included in Fig. 3B, but is shown to belong to the *Navicula–Pseudogomphonema–Pleurosigma* clade by *rbc*L data (Mann and Simpson, unpubl. data); this has secondarily acquired "constant" behaviour.

Recently, Medlin et al. (1996a, 2000) has brought the location of the Golgi bodies in the cytoplasm into phylogenetic perspective. Unfortunately, this character is not as accessible as cell wall morphology or plastid shape because it can only be examined using transmission electron microscopy. Nonetheless, species examined permit the following tentative generalisation (Medlin et al. 2000). Almost all pennates and all bipolar centrics except *Odontella sinensis* and *Biddulphiopsis titiana* show a perinuclear arrangement of the Golgi bodies, i.e. the Golgi bodies (dictyosomes) form a shell around the nuclear envelope. There are just a few exceptions or modifications, such as in *Synedra*, where the nuclear envelope is extended into a long flat cisterna of ER (a nuclear "tentacle") extending almost the whole length of the cell and to which Golgi bodies are appressed (Schmid 1989) – the so-called Plattenband arrangement. A further slight modification is seen in some *Pinnularia*, *Caloneis* and *Amphora* species (all three are genera of raphid diatoms) and *Striatella*, in which pairs of Golgi bodies are associated with short radiating extensions of the nuclear envelope (for references, see Round et al. 1990). *Biddulphiopsis titiana* shows a distribution of Golgi bodies exclusively along lobes of the ER, not around the nuclear envelope (Medlin et al. 2000). Perinuclear Golgi are also encountered in the radial centric diatoms *Aulacoseira* and its relative *Melosira* (not included in Fig. 3A because of its increased SSU rDNA rate). In the remaining radial centrics, and also in the bipolar *Odontella sinensis*, the Golgi bodies are scattered through the cell, associated with endoplasmic reticulum. Moreover, each Golgi body is also accompanied by a mitochondrion, which is located on the opposite side of the ER (G-ER-M in Medlin et al. 2000).

Irrespective of whether radial centrics are mono- or paraphyletic, the G-ER-M arrangement remains a homoplasy. It apparently evolved initially in the ancestor of the diatoms because the Oomycetes also possess this feature as do the red algae, so this might be a feature of the ancestral endosymbiont and was retained once in the bipolar centrics in *Odontella sinensis*. Moreover, this state changed in the clade with *Aulacoseira*. The perinuclear arrangement is also polyphyletic because it was acquired twice independently, once in the clade with *Aulacoseira* and once in the last common ancestor of the pennates and multi-polar centrics; and the paired Golgi arrangement has apparently evolved at least three times, in *Amphora*, the *Pinnularia–Caloneis* group, and *Striatella* (Fig. 3B). The arrangement in *Biddulphiopsis* and these other pennate genera might be a variant of the perinuclear organisation. Despite the fact that Golgi body arrangements exhibit homoplasy, they do appear to define large clades. Obviously, more data are needed to further extract maximum phylogenetic information from this character.

# 5 Phylogenetic Signal in the Life Cycle and Auxospore Ontogeny

Diatoms are diploid organisms with a life cycle consisting of a prolonged period of mitotic divisions (months to years), a brief stint of gametogenesis (hours) immediately followed by cell fusion and the formation and development of an auxospore (hours to days). Mitotic cell division in diatoms shows several peculiarities. The location of the rigid parental epi- and hypotheca dictates the orientation of the new thecae. The process in which the parental thecae determine the position of the juvenile thecae is called autopoiesis (Günther and Folke 1993). Daughter cells cannot alter their plane of division; they can only be oriented back-to-back forming uniseriate filaments or separating upon division, and so they are unable to form branching multicellular thalli like those encountered in green, red and brown algae (Mann 1999). Another constraint of the rigid silica cell wall is that mitotically reproducing diatoms will almost inevitably decrease in size. The rim of the epitheca fits over the rim of the hypotheca and is, therefore fractionally larger. Consequently, the daughter cell inheriting the parental hypotheca will be smaller than its parent because the parental hypotheca now has become the epitheca (see Round et al. 1990).

In most cases, the only way for a diatom to escape from diminishing to nothing is by means of sexual reproduction and associated size restoration, achieved by the expansion of a specialised cell, the "auxospore". During the life cycle, as cells decrease in size, they are at first unable to undergo sexual reproduction and size restoration via an auxospore. However, when diatoms reach a species-specific critical size window, they gain the potential to become sexual, if other conditions (light, nutrients, internal factors) allow. During the formation of gametes or shortly thereafter, the old cell walls are shed. The zygote formed by fusion of the gametes then becomes surrounded by a special organic or organic + silica wall, which allows expansion. Auxospore expansion is a well-controlled process, during which the shape of the new enlarged cells is generated. Once expansion is complete, new silica cell walls are produced within the auxospore wall. Just a few species scattered throughout the diatoms are known to be able to perform a similar trick also during their vegetative phase. In this case, they shed their cell casing, inflate (often irregularly) and then construct an entirely new pair of thecae. These species go through a sexual phase of enlargement as well.

## 5.1 Gamete Formation

Meiosis and gamete formations vary considerably among diatoms (Drebes 1977; Round et al. 1990; Edlund and Stoermer 1997). All centric diatoms produce motile flagellated micro-gametes (sperm) with one anterior flagellum, and non-motile naked macro-gametes (eggs). In the radial centrics *Melosira*, *Stephanopyxis* and *Actinocyclus* and the multi-polar centric *Pleurosira laevis*, the sperm develop through merogeny, i.e. they separate from the parental cell leaving the plastids behind in a remnant cell without nuclei. In *Coscinodiscus* (a radial centric), the plastids are dumped in a sister cell devoid of nuclei before the remaining cell proceeds to form the gametes. In the multipolar centrics (including Thalassiosirophycidae) *Stephanodiscus*, *Skeletonema*, *Lithodesmium*, *Chaetoceros*, *Odontella*, *Streptotheca* and *Cymatosira*, sperm formed hologenously: the parental spermatogenous cell divides up into equal portions, plastids included. The araphid pennate *Rhabdonema* also produces macro-gametes and micro-gametes, but here the micro-gametes are devoid of a flagellum. As far as is known, all other pennates reproduce sexually by means of isogamy (Mann 1993, 1999). Each cell produces one or two large amoeboid gametes and there is no differentiation into large immobile and small mobile gametes, although there may be differences in behaviour ("physiological anisogamy").

Thus, paraphyletic centrics reproduce by means of flagellated micro-gametes, whereas monophyletic pennates do so through non-flagellated gametes. Merogenous meiosis in *Pleurosira laevis* must be a reversal from the hologenous state in the remainder of the multipolar centrics. Significantly, the fact that *Rhabdonema* produces micro-gametes, even though these are non-flagellated, suggests a position between the centrics and other araphid pennates. In Fig. 1, this araphid species emerges straight from the basal polytomy of the pennates. Therefore, the tree does not falsify an evolutionary sequence in which the micro-gametes of the araphid pennate ancestors first lost the flagellum and subsequently evolved towards isogamy. It is noteworthy that the lobed plastid of *Rhabdonema* hints at something entirely different, a position in between araphids and raphids (cf. *Striatella*: see above). Sexual reproduction in *Toxarium* has never been studied, but given its pennate-like shape and life style, knowledge of its mode of sexual reproduction becomes highly significant.

The method of production of the macro-gametes also varies within (centric) diatoms, but its various modes show little concordance with the phylogenetic tree. Round et al. (1990) list a series of reports on the fate of a diploid nucleus in a maternal cell through the two meiotic divisions. In most centric diatoms, only one of the four haploid nuclei produced by meiosis makes it into a macro-gamete (Round et al. 1990). However, taxa differ in whether there is a cell division at meiosis I, though the authors do not list which species have actually been tested. In the multipolar centrics *Chaetoceros* and *Skeletonema*, and in the radial centrics *Melosira* and *Stephanopyxis*, both meiotic divisions occur in an undivided cell, the superfluous nuclei being resorbed into the cytoplasm. In *Cerataulus* and *Biddulphia*, which presumably belong to the multipolar centrics, there is an unequal cytokinesis after meiosis I, producing a "polar body" that subsequently degenerates. In a third group of centrics, there is an equal cell division after meiosis I and two egg cells are produced. This is known to occur only in multipolar centrics, such as *Lithodesmium* and *Odontella*. The interesting feature is that two eggs are produced per mother cell only in multipolar centrics, and it is this group of centrics that is sister to the pennates, which we can be fairly sure (from the distribution of different types of gamete formation among lineages) primitively produce two isogametes per mother cell. In *Rhabdonema*, which is intermediate between pennates and centrics in having differentiated macro- and micro-gametes, but not flagella, two species produce two macro-gametes per mother cell, whereas the third produces a single macro-gamete and a large polar body (see Drebes 1977).

Pennate diatoms produce large and sedentary or rather slow-moving gametes and fertilisation can only occur if the gamete-producing cells are somehow brought close together. Thus, gametogenesis is closely linked to the pairing up of compatible partners (Round et al. 1990; Mann 1993, 1999; Chepurnov and Mann 1997). The actual fusion involves amoeboid movement of isogametes between the partner cells, or a complete shedding of the walls before fusion. The modes of pairing and gamete behaviour vary wildly among pennates and more species need to be included in the phylogeny before patterns of distribution of these characters can be correlated with molecular phylogeny.

## 5.2 Auxospore Development

After gamete fusion, a specialised zygote, called auxospore is formed (see e.g. Schütt 1889; Mann 1982a,b; von Stosch 1982; Mann and Stickle 1989; Håkansson and Chepurnov 2000). This cell inflates in order to restore the original cell size. Round et al. (1990) in summarising auxospore wall ontogenesis across the diatom diversity discern the following structures and sequences of events.

First, newly established zygotes form an outer polysaccharide layer. In many centric genera and in *Rhabdonema*, numerous silica scales abound in this layer. *Odontella* even possesses scales bearing dichotomously branching spines (Von Stosch 1982). In most radial centrics and possibly in Thalassiosirophycidae, this is all there is to the structure of the auxospore wall and the auxospore itself is a simple expanding ball. When the first mitotic division takes place, the auxospore breaks open along its equator to give rise to the first two sister diatoms with oddly convex epivalves, the heritage of their auxospore stage.

In most multipolar centrics (apparently except Thalassiosirophycidae), a second layer is laid down underneath the primary one. This layer, the properizonium, consists of a complicated system of silica bands and hoops (von Stosch 1982; Round et al. 1990), which force the auxospore to develop anisometrically, producing the characteristic shape of the future vegetative cell.

In pennates, the auxospore is initially either globular or ellipsoidal and is again enveloped in a primary organic wall. This wall ruptures when the auxospore starts to swell, the two more or less equal halves being maintained as caps over the ends of the elongating cell (e.g. Mann and Stickle 1989). As the cell expands, a primary silica ring is formed surrounding the equator of the auxospore. Subsequently, more transverse bands are laid down around the auxospore, one after the other on both sides of the primary band, to force ongoing expansion solely along one axis, like the extension of a telescope (e.g. Mann 1982a; Mann and Stickle 1989). When all that is done, a second series of bands is often added beneath the first. These bands, unlike the first series, are orientated longitudinally. Together, the transverse and longitudinal series constitute the "perizonium". Then, the nucleus divides mitotically twice without division of the cell, the two superfluous nuclei degenerating. After each mitosis a new theca is produced, to create the frustule of the newly enlarged cell.

The properizonium and the perizonium probably represent separate states of the same character, i.e. they are homologous (Kaczmarska et al. 2000 summarises the known distribution of these structures among diatoms). The tree indicates that this form of auxospore encasing came into being along the internode leading to the clade with the multipolar centrics and pennates. The perizonium, which is present in *Rhabdonema* (which has a type of sexual reproduction intermediate between centrics and pennates: see above) was acquired only once, along the internode leading to the pennates, whereas the properizonium appeared at the origin of the multipolar centrics. The Thalassiosiophycidaean auxospores lack a properizonium, but this must be secondary loss, because properizonia are present in other multipolar lineages, including *Lithodesmium*, *Odontella* and *Chaetoceros*, representing three of the lineages of multipolar centrics in Fig. 3A.

# 6
# The Phylogenetic Position of the Diatoms Within Heterokonta

## 6.1
## The Ancestry of the Diatoms

Diatoms (Bacillariophyceae) form the largest lineage (i.e. the lineage containing the most species) within a sprawl of related algal groups, including the Raphidophyceae, Bolidophyceae, Chrysophyceae, Pelagophyceae, Dictyochophyceae, Xanthophyceae and Phaeophyceae (Van den Hoek et al. 1995; Graham and Wilcox 2000). Many of these algae have been discovered only recently by means of molecular phylogenetic methods (Anderson et al. 1993; Guillou et al. 1999). These photosynthetic "algal" lineages, in their turn, belong to a much wider assemblage, which also contains heterotrophic organisms, including various kinds of amoebae and flagellates, as well as organisms that have traditionally been regarded as fungi (e.g. the Oomycota; Cavalier-Smith and Chao 1996).

Although there are few characteristics common to all of these organisms, as a result of secondary loss, in most groups the flagellate cells (some species produce flagellate cells only for reproduction) have two dissimilar flagella, one of which is smooth, whereas the other bears a unique type of complex, tubular, stiff hair; the presence of dissimilar flagella is why this whole group is called the Heterokonta and the algal lineages therein Heterokontophyta.

Molecular phylogenies (Bhattacharya and Medlin 1995; Cavalier-Smith and Chao 1996; Daugbjerg and Andersen 1997; Medlin et al. 1997b) inferred from taxa across the heterokonts reveal monophyly for the photosynthetic heterokontophytes within a paraphyletic assemblage of heterotrophic heterokonts. The autotrophic clade originated from a symbiotic association of a rhodophyte (red alga) or rhodophyte-like endosymbiont (Bhattacharya and Medlin 1995; Delwiche et al. 1995) and a heterotrophic Heterokont host, possibly related to the Oomycota (Cavalier-Smith and Chao 1996). Of the original "rhodophyte" cell only its cell membrane, the two membranes surrounding its plastid, and the plastid internal structures, including the plastid genome, remain. These structures together with the original invagination membrane of the host cell add up to the four membranes around the heterokontophyte plastid.

## 6.2 Origin of Pigmented Heterokontophyta and the End Permian Mass Extinction

Significantly, a large part of the extant eukaryote phytoplankton diversity in the marine realm belongs to four major groups, Haptophyta, Heterokontophyta, Cryptophyta and Dinophyta. All came about through endosymbiosis events (Bhattacharya and Medlin 1995; Medlin et al. 1997a, b), i.e. the endosymbiosis of an eukaryotic autotroph with simple, two-membrane plastid within a eukaryotic host. Molecular phylogenetic studies (e.g. Medlin et al. 2000) and palaeoclimatological reconstructions have suggested that these groups may have acquired their plastids at around the Permian-Triassic boundary (245 Ma), which was a period of intense global change, extreme shortage of dissolved inorganic carbon in ocean surface layers (Lee and Kugrens 2000), and essentially anoxic deep water. The fossil record of that period reveals the demise of almost all phytoplankton species and, as a consequence of that, a catastrophic collapse of marine biodiversity.

During such upheaval, the new chimaeric organisms resulting from secondary endosymbiosis may have possessed a cutting edge above their heterotroph and rhodophyte ancestors (Lee and Kugrens 2000). These new organisms combined the advantage of the autotrophic life-style of the endosymbiont with the opportunity of surviving unfavourable periods through encystment, a trait inherited through the host; red algae lack resting stages. Lee and Kugrens (2000) argue also that the endosymbioses may have been advantageous in relation to carbon acquisition for photosynthesis at a time when $CO_2$ was present in unprecedented low concentrations. In addition, symbiosis permits internal recycling of waste products, an asset in environments where nutrients are in short supply.

## 6.3 Origin of the Silica Cell Wall Within Heterokonta

The diatoms have deployed silica for their cell wall synthesis throughout their evolutionary history (Medlin et al. 2000). The few naked endosymbiotic diatoms have dispensed with their cell wall only secondarily (Chesnick et al. 1997). Hence, the silica cell wall may have evolved before the diatoms came into being and have been shared with heterokontophyte or even heterotrophic heterokont relatives outside the diatoms.

A phylogenetic stroll among heterokontophyte relatives of the diatom lineage may thus shed light on the origin of the silica frustule. Van den Hoek et al. (1995) inferred a phylogeny across Heterokontophyta from morphological, ultrastructural and biochemical characteristics. Significantly, silica metabolism and the precipitation of amorphous silica in membrane-bound silica deposition vesicles seem to appear early in their tree. Most heterokontophyte lineages that lack silica elements in their cell walls seem to have lost this character secondarily.

A small group of marine planktonic heterokontophytes, the Parmales, have been suggested to be the nearest sisters of the diatoms (Round et al. 1990), because of structural similarity between their silica envelope and the diatom frustule (see illustration in Graham and Wilcox 2000, p. 245). Unfortunately, no molecular data are yet available for Parmales and their relationship to diatoms remains a mystery. The Parmales share with the diatoms one unique feature. Alone among all other organisms, these two groups produce a true cell wall (i.e. a complete covering to the cell, capable of resisting cell turgor) made of amorphous silica. A necessity correlated with this is the fact that their morphogenetic machinery must be able to adjust to the size and shape of the cell itself during the formation of the wall. By contrast, in other heterokontophytes, silica is used to form scales of a discrete size and shape, which imbricate to form a coherent envelope, but do not abut to form an integrated cell wall.

Recently, Guillou et al. (1999) have described the Bolidophyceae, a new heterokontophytan class of ocean-dwelling, small and fast-moving flagellates without any scales around their vegetative cells. Molecular sequence data have been obtained for these organisms and they appear as the nearest neighbour of the diatoms (Guillou et al. 1999; Daugbjerg and Guillou 2001). Yet, such a finding does not prove that the ancestors of the diatoms were naked flagellates as well. The Bolidophyceae may just as well represent a morphologically simple remnant genus of a class that originally had silica cell walls. Thus, molecular phylogenies inferred from *rbc*L or nuclear SSU rDNA sequences do not clearly resolve whether silica cell walls represent an ancestral state already present in the last common ancestor of all Heterokontophyta or that they evolved several times independently.

One possible way to find out if silica usage was already in place in the last common ancestor of all Heterokontophyta is to look for the genes actually coding for proteins involved in silica deposition across heterokonts and beyond. If such genes are all related across the diversity, then they almost certainly existed in the last common ancestor and may have been involved in silica cell wall deposition all along. Alternatively, if the genes bear no resemblance among distant relatives that produce silica structures, then silica biomineralisation has been acquired independently several times. The diatom genetic toolbox is currently under scrutiny. Proteins, such as pleuralins and silaffins and putrescine-linked polyamines, are involved with silica deposition in the developing cell wall, and other proteins are responsible for taking in silica across the cell membrane (Kröger and Wetherbee 2000; Kröger et al. 2000; Falciatore and Bowler 2002).

## 7 Historical Ecology

Despite the constraints imposed on their Bauplan by the rigid silica cell wall, diatoms have managed to diversify in a bewildering range of habitats (Round et al. 1990; Hasle and Syvertsen 1996; Günther et al. 1999; Mann 1999). Molecular phylogenies can provide insight into the ways and the order in which the diatoms managed to achieve this.

The use of silica as cell wall material is advantageous apparently because it is sequestered more cheaply than organic material (Raven 1983); it is often, though not always, in ample supply. However, this does not explain why diatoms have encased themselves so rigorously and it is necessary to address the question, why do diatoms have a wall, as opposed to an incomplete scaly envelope, like that present in some Chrysophyceae (Graham and Wilcox 2000). In most species, even the large tubular processes show labiate flattening, as if any trespassers are to be kept outside. Control of cell water content in freshwater diatoms or maintenance of shape, during times of slow or zero growth is one possible reason (e.g. Raven 1982). To keep intruders out is another. Unlike Bolidophyceae, diatoms have no way of escaping by "buzzing off", because their vegetative life stage lacks flagella. Therefore, some alternative mode of defence would be useful. The frustule is believed to provide some protection against an assortment of viruses, bacteria, chytrids, oomycetes and grazers (Baldi et al. 1998; Mann 1999; Holfeld 2000).

The diatoms are apparently extremely well adapted to growth in deeply mixed turbulent water, where cells are only intermittently exposed to high light levels (Mitchell and Holm-Hansen 1991; Falkowski and Raven 1997). Thus, their diversification in the Cretaceous must have had a profound effect on water chemistry in the euphotic zone, through removal of silicate and other nutrients, especially during the early part of the year before other autotrophs are active. This in turn must have affected the invertebrates and fish dependent on the plankton, as well as those organisms more directly affected by silicate availability. There is evidence that sponges became limited by silicate availability in the Late Cretaceous and early Tertiary, probably as a result of the increasing predominance of diatoms in marine phytoplankton (Maldonado et al. 1999).

As pointed out before, the exact shape of the frustule and its ornamentation is of course a function of ontogenetic constraints related to the form of the original auxospore from which the vegetative cell emerged and autopoiesis. However, this is only a short-term explanation of shape. During evolution, the genetic and epigenetic machinery of the cell change in response to a whole battery of selective ecological pressures (Gould and Lewontin 1979). As examples, spines and chain formation, e.g. in *Chaetoceros*, are probably adaptive. Indeed, they may well serve several purposes and their exact form is probably a compromise between effectiveness of grazer defence, adjustment of sedimentation rate and maximisation of nutrient uptake through micro shear, always constrained by the imperative of economical use of resources (energy, material).

Diatom evolution reveals the occurrence of a series of major innovations that allowed the occupation of entirely new habitats. Radial centrics appear now most frequently in the plankton, as single cells or chains. Chains can be viewed as adaptive in relation to sedimentation (Reynolds 1984). Chain-forming genera among the radial centrics maintain the integrity of the chains via interlocking silica projections borne by the valves.

Many multipolar centrics also abound in the plankton. Some among them are connected by extensions of the silica valves, as in radial centric chain-formers. The extensions can be tubular or solid, and may be located at the valve centre or around the periphery of the valve face. Thalassiosirophycidae have acquired a novel way of connection, via chitin threads emanating from the strutted processes in their valve faces. Many among these species also possess interlocking elements, e.g. in *Skeletonema* and *Detonula*; however, these have not evolved from the valve perimeter, but from the strutted processes. Curiously, the spines that are present in some Thalassiosirophycidae genera, e.g. *Stephanodiscus*, appear to be unrelated to chain formation, despite their position and morphology, because the cells generally occur in the plankton as solitary individuals.

Diatoms from among all the major lineages form chains by means of interconnecting mucilage pads. The huge difference between such connections and those effected by siliceous connections is that they are flexible, allowing the chain to bend without stretching or compressing individual diatoms, and that they can be disrupted or jettisoned without damaging the valves. Sometimes the pads are produced more or less centrally, as in *Podosira*, but more often they are produced from the sides or ends of the cells. Lateral mucilage pads often lead to the formation of characteristic zigzag chains or star patterns. Species deploying such a mode of chain or colony formation abound in the plankton, like the centric and pennate relatives with interlocked silica structures.

Yet, the real advance made possible by the evolution of connections via mucilage pads is that these diatoms can establish contact with something else than their adjacent sister cells, e.g. macroalgae, rocks and other objects. Much of the evolution of form in multipolar and araphid pennate diatoms may perhaps be attributed to adaptation to an attached existence. Multipolar centric diatoms like *Pleurosira*, *Lampriscus* and its putative pennate-like sister *Toxarium*, and some *Odontella* (Fig. 3A), as well as others not included in our molecular analysis, such as *Amphitetras*, *Hydrosera*, *Isthmia* and *Biddulphia* are epiphytic or epilithic, and many araphid pennates are found in similar niches. In these groups there appears to be specialisation of different parts of the frustule, with the mucilage secreting regions increasingly localised and demarcated. The primary innovation within the araphid pennate group and in *Toxarium* and its possible associates is the ability to form clusters or tufts, or even complex branched stalks, as a result of the localisation of mucilage secretion to the ends of very narrow, elongated cells. The branched stalks of e.g. *Licmophora* species, which enable some species to project their cells many hundreds of micrometres above the substratum (which must facilitate light and nutrient capture), is impossible in radial or multipolar centrics. Such elongate cells were only possible after the evolution of the sternum (in the pennates) or the elongate bifacial annulus (in *Toxarium* and its possible relatives), which provide a mechanically strong longitudinal element to support the valve (and girdle).

Some araphid pennate diatoms, and even some radial and multipolar centric diatoms, are known to move around actively, although they do so extremely slowly, except in two of the araphid genera with bifacial annuli, viz. *Ardissonea* (Pickett-Heaps et al. 1991) and *Toxarium* (Salma and Mann, unpubl.; Kooistra, unpubl.); in these mucilage is apparently secreted from between valve and girdle at the poles, propelling the cell in the opposite direction. However, even in these motility is very limited and most araphid pennates appear to be entirely sessile.

The most astonishing revolutionary innovation is the acquisition of the raphe. This structure appeared only once from araphid ancestry and is of great ecological significance. It permits the diatom to move around actively, and to migrate directionally in relation to environmental stimuli, such as light, gravity and chemicals, apparently including sex hormones (Bates and Davidovich 2002). These species do not exclusively rely on passive migration by current, but can actively search for a more desirable spot to be. Thus, in addition to environments where their sessile ancestors live, these species also have access to unstable environments, such as intertidal or subtidal sand and mudflats, where they play a very important role in stabilising sediments (e.g. Paterson and Hagerthey 2001). The evolution of raphid diatoms must thus have had profound effects on sedimentary processes and food webs in near-shore marine environments and in lakes and rivers, paralleling the changes made earlier when diatoms diversified into the plankton.

Of course, the evolution of the diatoms is full of reversals, some of these quite astounding in their own right. The raphid pennate *Pseudo-nitzschia* spp. abounds in the phytoplankton, where it forms chains appearing superficially like those of the radial centric diatom *Rhizosolenia*. In *Pseudo-nitzschia*, however, individuals form chains because, following cell division, the daughter cells move along one another, so that they remain connected only by their tips. *Bacillaria paxillifer*, also a raphid pennate, performs the same trick, except that the movement is reversible; neighbouring cells possess interlocking extensions along their valve face along which they can slide rapidly, thus quickly adjusting the form of their colony from sheet to string and back. Another example of reversal can be observed in *Achnanthes* and in *Cocconeis* and its relatives. These monoraphid pennate genera show a rather sedentary lifestyle. At least *Cocconeis* exposes one valve constantly to the environment. The raphe in that exposed valve has no function and has become a liability, a mechanically and biologically weak spot. Not surprisingly, the raphe on the exposed valve is gone.

Because many of the most ancestral lineages are found in the plankton, this environment appears to be the ancestral habitat; that is, the environment in which the diatoms primarily evolved. The appearance of an alternative form of connection, by means of distal mucilage pads and the elongation of the diatom valve made possible by the evolution of the sternum opened up a new suite of ecological opportunities associated with living on or attached to surfaces. Once there, acquisition of the raphe provided the means of actively moving around over that surface. However, a caution should be added. Although many of the best-known radial centric diatoms are planktonic, there are several that are benthic. Examples include the massively silicified *Ellerbeckia*, *Orthoseira* and *Podosira*, but also uncommon genera like *Druridgia*, as well as most species of *Melosira*. It has also been suggested that the primary habitat of *Paralia*, which is possibly the earliest lineage yet known among the diatoms (see above) and also has very heavily silicified frustules is not the plankton, but in an unattached or entangled bottom community (e.g. Cooper 1995, p. 55). Hence, it is very possible that, as a group, the diatoms are ancestrally benthic and that there has been repeated invasion of the plankton by a variety of unrelated lineages.

## 8 Palaeontology and Phylogeny

Apart from reconstructing the past from information available in the extant diversity, one might also look into the past itself. Diatoms possess a marvellous fossil record, from which hypotheses can be gleaned concerning their evolutionary history (Gersonde and Harwood 1990; Barron and Mahood 1993). Nevertheless, information from the present is still needed because similarities between fossil and extant forms may reflect shared evolutionary history, but could also signify convergence (e.g. Fryxell and Medlin 1981). If morphological character states show homoplasy already in the extant diversity, then they are of little value when comparing extant and fossil taxa.

Fortunately, many fossil diatom taxa are available in large quantities from facies all over the world, especially during the Tertiary and to a lesser extent the Upper Cretaceous, permitting fairly precise and accurate timing of their beginning and demise. First appearances of morphological synapomorphies or genera as deduced from molecular phylogenies can thus be dated using the fossil record (e.g. Kooistra and Medlin 1996; Medlin et al. 1996b, 1997a, 2000).

Although there is an earlier record from the Upper Jurassic at 180 Ma (e.g. see Harwood and Nikolaev 1995), the most ancient deposits in which well preserved diatoms have been recovered are from ca. 120 Ma sediment layers around Antarctica (Gersonde and Harwood 1990; Harwood and Gersonde 1990). The diatom fossil layers on oceanic bottoms there are among the few Lower Cretaceous sediments that have escaped subduction or destruction by volcanic deposits. All these fossils resemble modern centrics in their overall organisation (i.e. they have valves and girdle) and the annulus is visible in several cases (e.g. Gersonde and Harwood 1990, p. 14, Fig. 3). Some possess interlocking ornamentations emanating from the valve perimeter, whereas a few show centrally located simple tubes. These fossils could be direct ancestors of modern radial centric and multipolar centric genera. If so, it implies that the primary radiation that produced the various lineages of radial centric diatoms and the first multipolar diatoms occurred in or before the Lower Cretaceous.

What is suspicious, however, is that none of these ancient fossil radial centrics possess peripheral labiate processes, whereas most extant radial centric ones do. The ones with a central tube-like structure are like modern multipolar centrics in that the tube lies within the annulus. However, the tube-like processes are unlike the labiate process found in a similar position in e.g. *Ditylum*, because they open inwardly by a round pore or a cluster of small pores, rather than via a pair of tightly appressed lips (e.g. Fig. 7 in Harwood and Nikolaev 1995). Furthermore, there is no sign of any specialised pore fields for mucilage secretion. Of course, one could invoke independent and sudden acquisition of labiate processes and pore fields in all existing diatom lineages during the mid or late Cretaceous, but such an assumption is not very parsimonious.

Instead, one has to ask how reliable overall shape is. Perhaps particular shapes have been generated more than once. The molecular phylogeny already reveals many independent acquisitions among valve structure elements and fine ornamentations and, conversely, very differently shaped cell walls are encountered among close relatives (e.g. in the Naviculaceae, which is represented by the boat shaped *Navicula* and the wedge-shaped *Pseudogom-*

*phonema* in Fig. 3B). Such character states may have appeared and then been lost over and over again. Consequently, fossil genera and their modern counterparts do not need to be relatives; instead they might be simply cognates or look-alikes in time. Such patterns are encountered in fossil species groups, such as ammonites (Newell 1967), foraminifera (Cifelli 1969) and calcified macroalgae (Kooistra, in prep).

Heterokontophyte phylogenies do not reveal long branches (= extended time) between the separation of the bolidophycean lineage and the basal ramification of the extant diatom diversity (Guillou et al. 1999). Nevertheless, these trees permit time for an older diversification, unrelated to the one that gave rise to the diatoms alive today. All extant diversity may have come about from a radial centric diatom with a central simple process. In modern radial centrics, this tube was duplicated many times and moved to the periphery, whereas in the other clade mucilage pads were acquired and the tube remained initially where it was. If this is correct, the fossil radial centrics became completely extinct and were replaced by the modern radial centrics with labiate processes.

Instead of tracking genera and families in the fossil record, it may be better to track the appearance of the synapomorphies as defined by molecular phylogenies. Unfortunately, large gaps abound in the diatom fossil record. These gaps could indicate just a temporal lack of fossilised material of a continuous lineage, but they could also represent extinction events and a subsequent appearance of cognates. For example, the exceptionally well-preserved diatoms from the Lower Cretaceous are followed by a period of diatom fossils that are poorly preserved and often pyritised. Well-preserved fossil diatoms then reappear at ca. 90 Ma, in the Mid Cretaceous. By the Upper Cretaceous and Palaeocene, many of the diatoms present are so similar to extant genera (e.g. Strel'nikova 1974, 1992) that there seems no question that they belong to the same clades. There is apparently no major extinction of diatom lineages at the K/T boundary (65 Ma), but perhaps it is no coincidence that the explosive radiation of pennates, maybe including the evolution of raphid diatoms, seems to have occurred at around this time.

Pennate diatoms have never been observed in the Lower Cretaceous. They appear only after 90 Ma in the fossil record. Raphid diatoms appear only at ca. 45 Ma. Therefore, the sequence of centrics first – then araphid pennates – then raphids, as inferred from the molecular phylogenies, is confirmed by stratigraphic evidence in the fossil record, because *Toxarium* and its allies already show that pennate like morphology and behaviour have evolved twice independently. The fossil record may be riddled with many more such experimental designs. However, the earliest pennates known so far have a simple sternum (e.g. Strel'nikova 1974), not a bifacial annulus.

# 9 Conclusions

Phylogenies derived from nuclear SSU rDNA sequences show that

- Radial centric diatoms begot multipolar centrics, multipolar centrics begot pennates and the araphid pennates begot the raphid pennates.
- Convergence does abound. Notably *Toxarium* is a diatom with a typical pennate shape and lifestyle, but it lacks the characters typical for pennates. It is in fact a multipolar centric diatom. Another example, Thalassiophycidae have regained the typical radial centric form and lifestyle within the multipolar centrics.
- Although the fossil record supports the above evolutionary sequence, convergence in the extant diversity challenges interpretation of fossil genera that appear similar to extant ones.
- Ultrastructural and life-history traits show clear evolutionary patterns in accordance with molecular phylogenies.
- Silica deposition may be a feature inherited from heterokont ancestors; it likely did not evolve at the root of the diatoms.
- A tentative historical ecological interpretation suggests that radial centrics evolved in the benthos and radiated into the plankton and then gave rise to multipolar centrics that occurred attached to the substrate. Pennates came about as a special form of a bipolar centric. From the (araphid) pennates, the raphid pennates invaded yet a new series of environments by being able to move actively.

*Acknowledgements.* The authors wish to thank Donato Marino and Gandi Forlani for electron micrographs and Carmen Minucchi for keeping unruly diatom cultures happy. LKM's research was supported in part by the DFG (Sm22/7-1 and Me1480/1-2).

## References

Anderson RA, Saunders GW, Paskind MP, Sexton JP (1993) Ultrastructure and 18S ribosomal-RNA gene sequence for *Pelagomonas calceolata* gen. et sp. – nov. and the description of a new algal class, the Pelagophyceae *classis nov.* J Phycol 29:701–715

Baldauf JG (1992) Middle Eocene through early Miocene diatom floral turnover. In: Prothero DR, Berggren WA (eds) Eocene–Oligocene climatic and biotic evolution. Princeton Univ Press, Princeton, pp 310–326

Baldi F, Malej A, Minacci A, Milanesi C, Vignani R (1998) Is diatom lysis in extensive mucilage of Adriatic Sea due to a viral infection? Abstracts, Algal Virus Workshop, 14–18 June 1998, Department of Microbiology, University of Bergen, Bergen, Norway

Bates SS, Davidovich NA (2002) Factors affecting the sexual reproduction of diatoms, with emphasis on *Pseudo-nitzschia* spp. In: Proceedings of the LIFEHAB Workshop, Majorca, October 2001 (in press)

Barron JA, Mahood AD (1993) Exceptionally well-preserved early Oligocene diatoms from glacial sediments of Prydz Bay, East Antarctica. Micropaleontology 39:29–45

Bhattacharya D, Medlin L (1995) The phylogeny of plastids: a review based on comparisons of small-subunit ribosomal RNA coding regions. J Phycol 31:489–498
Brott LL, Naik RR, Pikas DJ, Kirkpatrick SM, Tomlin DW, Whitlock PW, Clarson SJ, Stone MO (2001) Ultrafast holographic nanopatterning of biocatalytically formed silica. Nature 413: 291–293
Cavalier-Smith T, Chao EE (1996) 18S rRNA sequence of *Heterosigma carterae* (Raphidophyceae) and the phylogeny of the heterokont algae (Ochrophyta). Phycologia 35:500–510
Chai JY, Lee JJ (1999) Initial recognition of endosymbiotic diatom surface antigens by the larger foraminifer *Amphistegina lobifera*. Symbiosis 26:39–53
Chai JY, Lee JJ (2000) Recognition, establishment and maintenance of diatom endosymbiosis in foraminifera. Micropaleontology 46:182–195
Chepurnov VA, Mann DG (1997) Variation in the sexual behaviour of natural clones of *Achnanthes longipes*. Eur J Phycol 32:147–154
Chesnick JM, Kooistra WHCF, Wellbrock U, Medlin LK (1997) Ribosomal RNA analysis indicates a benthic pennate diatom ancestry for the endosymbionts of the dinoflagellates *Peridinium foliaceum* and *Peridinium balticum* (Pyrrhophyta). J Eukaryotic Microbiol 44:314–320
Cifelli R (1969) Radiation of Cenozoic planktonic Foraminifera. Syst Zool 18:154–168
Cooper SR (1995) Diatoms in sediment cores from the mesohaline Chesapeake Bay, USA. Diat Res 10:39–89
Cox EJ (1981) The use of chloroplasts and other features of the living cell in the taxonomy of naviculoid diatoms. In: Ross R (ed) Proceedings of the 6th symposium on recent and fossil diatoms. Koeltz, Koenigstein, pp 115–133
Cox EJ (1993) Diatom systematics – a review of past and present practice and a personal vision for future development. Nov Hedw Beih 106:1–20
Crawford RM, Gardner C, Medlin LK (1994). The genus *Attheya*. I. A description of four new taxa, and the transfer of *Gonioceros septentrionalis* and *G. armata*. Diat Res 9:27–51
Daugbjerg, N, Andersen RA (1997) A molecular phylogeny of the heterokont algae based on analyses of chloroplast-encoded rbcL sequence data. J Phycol 33:1031–1041
Daugbjerg N, Guillou L (2001) Phylogenetic analyses of Bolidophyceae (Heterokontophyta) using rbcL gene sequences support their sister group relationship to diatoms. Phycologia 40:153–161
Delwiche CF, Kuhsel M, Palmer JD (1995) Phylogenetic analysis of tufA sequences indicates a cyanobacterial origin of all plastids. Mol Phylogenet Evol 4:110–128
Drebes G (1977) Sexuality. In: Werner D (ed) The biology of diatoms. Botanical monographs. Blackwell, Oxford, pp 250–283
Droop SJM, Sims PA, Mann DG, Pankhurst RJ (1993) A taxonomic database and linked iconograph for diatoms. Hydrobiologia 269/270:503–508
Edlund MB, Stoermer EF (1997) Ecological, evolutionary, and systematic significance of diatom life histories. J Phycol 33:897–918
Ehara M, Inagaki Y, Watanabe KI, Ohama T (2000) Phylogenetic analysis of diatom coxI genes and implications of a fluctuating GC content on mitochondrial genetic code evolution. Curr Genet 37:29–33
Falciatore A, Bowler C (2002) Revealing the secrets of marine diatoms. Annu Rev Plant Physiol 53:109–130
Falkowski PG, Raven JA (1997) Aquatic photosynthesis. Blackwell, Oxford, 375 pp
Falkowski PG, Barber RT, Smetacek V (1998) Biochemical controls and feedbacks on ocean primary production. Science 281:200–206
Fryxell GA, Medlin LK (1981) Chain forming diatoms: evidence of parallel evolution in *Chaetoceros*. Cryptogamie Algol 2:3–29
Gersonde R, Bárcena MA (1998) Revision of the late Pliocene-Pleistocene diatom biostratigraphy for the northern belt of the Southern Ocean. Micropaleontology 44:1–15
Gersonde R, Harwood DM (1990) Lower Cretaceous diatoms from ODP leg 113 site 693 (Wedell Sea) II. Vegetative cells. In: Barker PF, Kennett JP (eds) Proceedings of the ocean drilling program scientific results, vol 113. Publications Distribution Center, Ocean Drilling Program, 1000 Discovery Drive, College Station, Texas, TX 77845-9547, pp 365–403

Gibson R (1979) An ultrastructure study of *Pseudohimantidium pacificum* Hust. & Krasske (Bacillariophyceae: Protoraphidaceae) with special reference to the labiate processes. Nov Hedw 64:147–60

Goldman JC (1993) Potential role of large oceanic diatoms in new primary production. Deep Sea Res I/40:159–168

Gould SJ, Lewontin RC (1979) The spandrels of San Marco and the Panglossian paradigm: a critique of the adaptationist programme. Proc R Soc Lond B 205:581–598

Graham LE, Wilcox LW (2000) Algae. Prentice-Hall, London

Guillard RRL, Kilham P (1978) The ecology of marine planktonic diatoms. In: Werner D (ed) The biology of diatoms. Botanical monographs 13. Univ California Press, Berkeley, pp 372–469

Guillou L, Chrétiennot-Dinet M-J, Medlin LK, Claustre H, Loiseaux-de Goër S, Vaulot D (1999) Bolidomonas: a new genus with two species belonging to a new algal class, the Bolidophyceae class. nov. (Heterokonta). J Phycol 35:368–381

Günther F, Folke C (1993) Characteristics of nested living systems. J Biol Syst 1:257–274

Günther S, Gleitz M, Dieckmann GS (1999) Biogeochemistry of Antarctic Sea ice: a case study on platelet layers at Drescher Inlet, Weddell Sea. Mar Ecol Progr Ser 177:1–13

Håkansson H, Chepurnov V (2000) A study of variation in valve morphology of the diatom *Cyclotella meneghiniana* in monoclonal cultures: effect of auxospore formation and different salinity conditions. Diat Res 14:251–272

Harwood DM, Gersonde R (1990) Lower Cretaceous diatoms from ODP leg 113 site 693 (Wedell Sea) II. Resting spores chrysophycean cysts an endoskeletal dinoflagellate and notes on the origin of diatoms. In: Barker PF, Kennett JP (eds) Proceedings of the ocean drilling program scientific results, vol 113. Publications Distribution Center, Ocean Drilling Program, 1000 Discovery Drive, College Station, Texas, TX 77845-9547, pp 403–425

Harwood DM, Nikolaev VA (1995) Cretaceous diatoms: morphology, taxonomy, biostratigraphy. In: Blome CD (eds) Siliceous microfossils. Paleontological society short courses in paleontology, vol 8. Publications Distribution Center, Ocean Drilling Program, 1000 Discovery Drive, College Station, Texas, TX 77845-9547, pp 81–106

Hasle GR (1974) The "mucilage pore of pennate diatoms. Nov Hedw Beih 45:167–194

Hasle G, Syvertsen E (1996) Marine Diatoms. In: Tomas CR (ed) Identifying marine diatoms and dinoflagellates. Academic Press, San Diego, pp 5–385

Hillis DM, Moritz C, Mable BK (eds) (1996) Molecular systematics, 2nd edn. Sinauer Associates, Sunderland, MA, 655 pp

Holfeld H (2000) Infection of the single-celled diatom *Stephanodiscus alpinus* by the chytrid *Zygorhizidium*: parasite distribution within host population, changes in host cell size, and host-parasite size relationship. Limnol Oceanogr 45:1440–1444

Kaczmarska I, Bates SS, Ehrman JM, Léger C (2000) Fine structure of the gamete, auxospore and initial cell in the pennate diatom *Pseudo-nitzschia multiseries* (Bacillariophyta). Nov Hedw 71: 337–357

Kilham SS, Theriot EC, Fritz SC (1996) Linking planktonic diatoms and climate change using resource theory in the large lakes of the Yellowstone ecosystem. Limnol Oceanogr 41: 1052–1062

Kociolek JP (1986) Determining character polarity in diatoms: an alternative approach. Diat Res 1:189–192

Kociolek JP, Stoermer EF (1986) Phylogenetic relationships and classification of monoraphid diatoms based on phenetic and cladistic methodologies. Phycologia 25:297–303

Kociolek JP, Stoermer EF (1988) A preliminary investigation of the phylogenetic relationships among the freshwater, apical pore field-bearing cymbelloid and gomphonemoid diatoms (Bacillariophyceae). J Phycol 24:377–385

Kociolek JP, Stoermer EF (1989) Phylogenetic relationships and evolutionary history of the diatom genus *Gomphoneis*. Phycologia 28:438–454

Kociolek JP, Stoermer EF (1993) The diatom genus Gomphocymbella O. Müller: taxonomy, ultrastructure and phylogenetic relationships. Nov Hedw Beih 106:71–91

Kociolek JP, Williams DM (1987) Unicell ontogeny and phylogeny: examples from the diatoms. Cladistics 3:274–284
Kociolek JP, Theriot EC, Williams DM (1989) Inferring diatom phylogeny: a cladistic perspective. Diat Res 4:289–300
Kooistra WHCF, Medlin LK (1996) The evolution of the diatoms (Bacillariophyta) IV. A reconstruction of their age from small subunit rRNA coding regions and the fossil record. Mol Phylogenet Evol 6:391–407
Kröger N, Wetherbee R (2000) Pleuralins are involved in theca differentiation in the diatom *Cylindrotheca fusiformis*. Protist 151:263–273
Kröger N, Deutzmann R, Sumper M (2000) Polycationic peptides from diatom biosilica that direct silica nanosphere formation. Science 286:1129–1132
Lange-Bertalot H (1980) Zur systematischen Bewertung der Bandförmigen Kolonien bei *Navicula* und *Fragilaria*. Nov Hedw 33:723–787
Lee JJ, Reimer CW (1984) Isolation and identification of endosymbiotic diatoms. In: Mann DG (ed) Proceedings of the 7th Int Diatom Symp. Koeltz, Koenigstein, pp 327–343
Lee RE, Kugrens P (2000) Commentary: ancient atmospheric $CO_2$ and the timing of evolution of secondary endosymbioses. Phycologia 39:167–172
Li C-W, Volcani BE (1987) Four new apochlorotic diatoms. Br Phycol J 22:375–382
Lundholm N, Daugbjerg N, Moestrup → (2001) Phylogeny of the Bacillariaceae with emphasis on the genus *Pseudo-nitzschia* (Bacillariophyceae) based on partial LSU rDNA. Eur J Phycol 37:115–134
Maldonado M, Carmona MC, Uriz MJ, Cruzado A (1999) Decline in Mesozoic reef-building sponges explained by silicon limitation. Nature 401:785–788.
Mann DG (1982a) Structure, life history and systematics of *Rhoicosphenia* (Bacillariophyta) II. Auxospore formation and perizonium structure of *Rh. curvata*. J Phycol 18:264–274
Mann DG (1982b) Auxospore formation in *Licmophora* (Bacillariophyta). Plant Syst Evol 139: 289 294
Mann DG (1984) An ontogenetic approach to diatom systematics. In: Mann DG (ed) Proceedings of the 7th Int Diatom Symp. Koeltz, Koenigstein, pp 113–144
Mann DG (1993) Patterns of sexual reproduction in diatoms. Hydrobiologia 269/270:11–20
Mann DG (1996) Chloroplast morphology, movements and inheritance in diatoms. In: Chaudhary BR, Agrawal SB (eds) Cytology, genetics and molecular biology of algae. SPB Academic Publ, Amsterdam, pp 249–274
Mann DG (1998) Shifting sands: the use of the lower taxonomic ranks in diatoms. Diatom 13: 13–17
Mann DG (1999) The species concept in diatoms. Phycological reviews 18. Phycologia 38:437–495
Mann DG, Droop SJM (1996) Biodiversity, biogeography and conservation of diatoms. Hydrobiologia 336:19–32
Mann DG, Stickle AJ (1989) Meiosis, nuclear cyclosis, and auxospore formation in Navicula sensu stricto (Bacillariophyta). Br Phycol J 24:167–181
Mann DG, Simpson GE, Sluiman HJ, Möller M (2001) *rbc*L gene tree of diatoms: a second large data-set for phylogenetic reconstruction. Phycologia 40(4) [Suppl]:1–2
McLaughlin RB (1992) Diatom microscopy: harmonic analysis of the diatom valve outline. Microscope 40:265–298
Medlin L, Elwood HJ, Stickel S, Sogin ML (1988) The characterization of enzymatically amplified eukaryotic 16S-like rRNA coding regions. Gene 71:491–499
Medlin, LK, Elwood HJ, Stickel S, Sogin ML (1991) Morphological and genetic variation within the diatom *Skeletonema costatum* (Bacillariophyta): evidence for a new species, *Skeletonema pseudocostatum*. J Phycol 27:514–524
Medlin LK, Williams DM, Sims PA (1993) The evolution of the diatoms (Bacillariophyta). I. Origin of the group and assessment of the monophyly of its major divisions. Eur J Phycol 28:261–275
Medlin LK, Gersonde R, Kooistra WHCF, Wellbrock U (1996a) Evolution of the Diatoms (Bacillariophyta). II. Nuclear-encoded small-subunit rRNA sequence comparisons confirm a paraphyletic origin for the centric diatoms. Mol Biol Evol 13:67–75

Medlin LK, Kooistra WHCF, Gersonde R, Wellbrock U (1996b) Evolution of the diatoms (Bacillariophyta): molecular evidence for the origin of the Thalassiosirales. Nov Hedw 112:221–234

Medlin LK, Kooistra WHCF, Gersonde R, Sims PA, Wellbrock U (1997a) Mini-review: is the origin of the diatoms related to the end-Permian mass extinction? Nov Hedw 65:1–11

Medlin LK, Kooistra WHCF, Potter D, Saunders GW, Andersen RA (1997b) Phylogenetic relationships of the "golden algae" (haptophytes, heterokont chromophytes) and their plastids. In: Bhattacharya D (ed) Origins of algae and their plastids. Springer, Berlin Heidelberg New York, pp 187–219

Medlin LK, Kooistra WHCF, Schmid A-MM (2000) A review of the evolution of the diatoms – a total approach using molecules, morphology and geology. In: Witkowski A, Sieminska J (eds) The origin and early evolution of the diatoms: fossil, molecular and biogeographical approaches. W Szafer Institute of Botany, Polish Academy of Sciences, Cracow, pp 13–35

Mereschkowsky C (1901) Études sur l'endochrome des diatomées. Mem Acad Imp Sci St Petersb Ser 8/11:1–40

Mitchell BG, Holm-Hansen O (1991) Observations and modeling of the Antarctic phytoplankton crop in relation to mixing depth. Deep-Sea Res 38:981–1007

Newell ND (1967) Revolutions in the history of life. Geol Soc Am Spec Pap 89:63–91

Page RDM, Holmes EC (1998) Molecular evolution: a phylogenetic approach. Blackwell, Malden, MA

Pappas JL, Stoermer EF (1995) Multidimensional analysis of diatom morphological phenotypic variation and relation to niche. Ecoscience 2:357–367

Paterson DM, Hagerthey SE (2001) Microphytobenthos in contrasting coastal ecosystems: biology and dynamics. In: Reise K (ed) Ecological comparisons of sedimentary shores. Springer, Berlin Heidelberg New York, pp 105–125

Pickett-Heaps J, Schmid A-MM, Edgar LA (1990) The cell biology of diatom valve formation. Progr Phycol Res 7:1–168

Pickett-Heaps JD, Hill DRA, Blaze KL (1991) Active gliding motility in an araphid marine diatom, *Ardissonea* (formerly *Synedra*) *crystallina*. J Phycol 27:718–725

Popper K (1959) The logic of scientific discovery. Hutchinson, London

Posada D, Crandall KA (1998) Modeltest: testing the model of DNA substitution. Bioinformatics 14:817–818

Raven JA (1982) The energetics of freshwater algae; energy requirements for biosynthesis and volume regulation. New Phytol 92:1–20

Raven JA (1983) The transport and function of silicon in plants. Biol Rev 58:179–207

Reynolds CS (1984) The ecology of freshwater phytoplankton. Cambridge Univ Press, Cambridge

Rivera P, González U, Barrales H (1986) Cingulum and valve morphology of *Pseudohimantidium* Husted & Krasske (Bacillariophyceae). Phycologia 25:19–27

Round FE (1981) The ecology of algae. Cambridge Univ Press, Cambridge

Round FE, Crawford RM (1990) Phylum Bacillariophyta. In: Margulis L, Corliss JO, Melkonian M, Chapman DJ (eds) Handbook of protoctista. Jones and Barlett, Boston, pp 574–596

Round FE, Crawford RM, Mann DG (1990) The diatoms. Biology and morphology of the genera. Cambridge Univ Press, Cambridge

Schmid A-MM (1989) Geitler's "Plattenband" in the diatom *Synedra* cf. *ulna* in the light of TEM investigations. Plant Syst Evol 164:239–252

Schmidt A, Schmidt M, Fricke F, Heiden H, Müller O, Hustedt F (1874–1959) Atlas der Diatomeen-Kunde. Reisland, Leipzig, Heft 1–120 Tafeln 1–480

Schütt F (1889) Über Auxosporenbildung der Gattung *Chaetoceros*. Ber Dtsch Bot Ges 7:361–363

Simonsen R (1970) Protoraphidaceae, eine neue Familie der Diatomeen. Nov Hedw 31:377–394

Simonsen R (1979) The diatom system: ideas on phylogeny. Bacillaria 2:9–71

Smetacek V (1998) Biological oceanography: diatoms and the silicate factor. Nature 391:224–225

Smith W (1853) A synopsis of the British Diatomaceae, vol 1. Van Voorst, London

Sober E (1990) Let's razor Ockham's razor. In: Knowles D (ed) Explanation and its limits. Royal Institute of Philosophy [Suppl] 27:73–94. Cambridge Univ Press, Cambridge

Sörhannus U, Gasse F, Perasso R, Baroin-Tourancheau A (1995) A preliminary phylogeny of diatoms based on 28S ribosomal RNA sequence data. Phycologia 34:65–73
Stoermer EF, Smol JP (1999) The diatoms. Applications for the environmental and earth sciences. Cambridge Univ Press, Cambridge
Strel'nikova NI (1974) Diatomei pozdego mela (Zapadnaya Sibir'). Nauka, Moscow
Strel'nikova NI (1992) Paleogenovye diatomovye vodorosli. Saint Petersburg Univ, St Petersburg
Swofford DL (2000) "PAUP* Phylogenetic Analysis Using Parsimony (* and other methods) Version 4.0b8". Sinauer Associates, Sunderland, MA
Syvertsen EE (1977) *Thalassiosira rotula* and *T. gravida*: ecology and morphology. Nov Hedw Beih 54:99–112
Theriot EC (1992) Clusters, species concepts and morphological evolution of diatoms. Syst Biol 41:141–157
Theriot EC, Fritz SC, Gresswell RE (1997) Long-term limnological data from the larger lakes of Yellowstone National Park. J Arct Alp Res 29:304–331
Van den Hoek C, Mann DG, Jahns HM (1995) Algae, an introduction to phycology. Cambridge Univ Press, Cambridge
VanLandingham SL (1967–1979) Catalog of the fossil and recent genera and species of diatoms and their synonyms, vols 1–8. Cramer, Lehre, 4654 pp
Von Stosch HA (1982) On auxospore envelopes in diatoms. Bacillaria 5:127–156
Vrieling EG, Beelen TPM, van Santen, RA, Gieskes WWC (2000) Nanoscale uniformity of pore architecture in diatomaceous silica: a combined small and wide angle X-ray scattering study. J Phycol 36:146–159
Vyverman W, Sabbe K, Mann DG, Kilroy C, Vyverman R, Vanhoutte K, Hodgson D (1998) *Eunophora* gen. nov. (Bacillariophyta) from Tasmanian highland lakes: description and comparison with *Eunotia* and amphoroid diatoms. Eur J Phycol 33:95–111
Wiley EO, Siegel-Causey D, Brooks DR, Funk VA (1991) The compleat cladist. Univ Kans Mus Nat Hist Spec Publ 19:1–158
Williams DM (1993) Diatom nomenclature and the future of taxonomic database studies. Nov Hedw Beih 106:21–31
Zechman FW, Zimmer EA, Theriot EC (1994) Use of ribosomal DNA internal transcribed spacers for phylogenetic studies in diatoms. J Phycol 30:507–512

# Silicon – a Central Metabolite for Diatom Growth and Morphogenesis

Véronique Martin-Jézéquel[1] and Pascal J. Lopez[2]

## 1 Introduction

Diatoms consist of a large group of micro-algae which exhibit a very large size range from a few to several hundreds microns. The number of extinct and extant species has been estimated at about 60,000. The hallmark of the diatom is its cell wall, which is highly differentiated and is a remarkable example of the artistry of nature; for an illustration of the aesthetic sense of diatom "glass-box" see Fig. 1. Moreover, the precise replication, during each cell division, of the frustule pattern argues for a genetic control of this biomineralization process. Abundant throughout the world, they are found in a large variety of habitats including aquatic and moist terrestrial ones; some can cope with harsh conditions, polluted or physically disturbed biotopes. They are found in fresh and marine waters either as planktonic or benthic forms. This wide distribution reveals their adaptive capacity and indicates that they have evolved robust biochemical and cellular processes to maintain growth under these various conditions. Nevertheless, almost all diatoms show a strict dependence on silica for growth, which suggests that this metabolite occupies a central position in diatom metabolism.

## 2 Silicon Uptake and Transport: Regulation and Influencing Factors

### 2.1 Uptake, Transport and Soluble Pools

Silicon concentrations in the sea are generally lower than in many freshwater environments (Paasche 1980), and this difference is reflected in the ability of marine diatoms to grow at lower concentrations as compared to freshwater

[1] UMR CNRS 6539, Université de Bretagne Occidentale, Institut Universitaire Européen de la Mer, Technopole Brest-Iroise, Place Nicolas Copernic, 29280 Plouzané, France
[2] UMR CNRS 8543, Ecole Normale Supérieure, 46 rue d'Ulm, 75005 Paris, France

Progress in Molecular and Subcellular Biology, Vol. 33
W. E. G. Müller (Ed.)

**Fig. 1.** Example of shape and ornamentation of different frustules of vegetative cells from pennate or centric diatoms and of resting spore. From *left* to *right* and *top* to *bottom*: resting spore supposed from *Chaetoceros*, *Paralia marina*, *Campylodiscus clypeus*, *Diploneis didyma*, *Kittonia elaborata*, *Thalassiosira nordenskioeldii*, *Sheshukovia kolbei* var. *uralensis*, *Strangulonema barbadense*, *Cymatoneis sulcata*. (Original photographs are from Round 1990)

diatoms (cf. Olsen and Paasche 1986). Half the maximum growth rate for marine species is usually attained at about 0.1 μM monosilicic acid, whereas freshwater species reach their half-maximum growth rate at about 1 μM. The chemical forms of the dissolved silicon available to marine diatoms in seawater at pH 8.0 are undissociated silicic acid, $Si(OH)_4$, accounting for about

97%, and $SiO(OH)_3^-$, accounting for the remaining 3%. Diatoms use undissociated silicic acid for their growth, with the exception of *Phaeodactylum tricornutum* which may also take up $SiO(OH)_3^-$ (Del Amo and Brzezinski 1999). Silicon uptake is carrier-mediated (Sullivan 1986; Sullivan and Volcani 1981 and references therein). In marine diatoms, this transport is sodium-dependent (Bhattacharyya and Volcani 1980), and may be coupled to sodium or perhaps potassium in freshwater species (Sullivan 1976). The transporters in marine species have characteristics of a sodium/silicic acid symporter (Bhattacharyya and Volcani 1980; Hildebrand et al. 1997) and an electrogenic transport was hypothesized with a $Si(OH)_4:Na^+$ ratio of 1:1 (Bhattacharyya and Volcani 1980).

The specific rate of silicon uptake (V) and the specific Si-dependent cell division rate ($\mu$) follow Michaelis-Menten (Eq. 1) or Monod (Eq. 2) saturation functions:

$$V = \frac{V_{max} \cdot [Si(OH)_4]}{K_s + [Si(OH)_4]} \quad (1)$$

$$\mu = \frac{\mu_{max} \cdot [Si(OH)_4]}{K_\mu + [Si(OH)_4]} \quad (2)$$

where $V_{max}$, and $\mu_{max}$ represent the maximum rates of uptake and division, respectively, at infinite substrate concentration; and $K_s$ is the $Si(OH)_4$ concentration at half $V_{max}$; $K_\mu$ is the $Si(OH)_4$ concentration that limits $\mu$ to $0.5\,\mu_{max}$ These kinetic parameters $K_s$, $K_\mu$, $V_{max}$, and $\mu_{max}$ for silicic acid uptake and Si-dependent growth have been obtained for many diatom species in culture, in a range from 0.2 to 8.32 µmol for the $K_s$ and 0.022–8.6 µmol for the $K_\mu$ obtained under autotrophic growth of the diatoms (Martin-Jézéquel et al. 2000; Paasche 1980). The half-saturation constant for growth, $K_\mu$, is less than that for uptake, $K_s$, which is common for nutrients. As a consequence, diatoms can maintain division rates very close to $\mu_{max}$ at extracellular $Si(OH)_4$ concentrations that limit silicon uptake. Moreover, the coupling between the different steps of silicon incorporation and the cell cycle (Schmid 1994), have important implications for assessing kinetic parameters for uptake (Martin-Jézéquel et al. 2000). Indeed, calculations by Brzezinski (1992) have shown that the silicon uptake systems of diatoms could have lower affinity and greater maximum capacity than the values obtained (cf. values in Martin-Jézéquel et al. 2000; Paasche 1980), because cultures were asynchronous and not all cells were in a state to take up silicon during these experiments.

In general, marine diatoms have less Si/cell than freshwater species; this may have a genetic basis or result from other factors such as salinity (Conley et al. 1989). Among the different species, the total silicon content of diatoms covers a large range, from 0.14 to 225 pmol/cell, and within a single species, the cellular amount may vary by almost a factor of 10 (Table 1). The intracellular

**Table 1.** Silica content of different pennate and centric diatoms, from marine or freshwater habitats

| Species | Silica content[a] | Cell volume[b] | Reference[c] |
|---|---|---|---|
| *Asterionella formosa* | 0.38–6.83 | 260–700 | Conley et al. (1989) |
| *Asterionella glacialis* | 0.68–1.1 | 822–1100 | Brzezinski (1985) |
| *Bacteriastrum furcatum* | 0.91 | 2000 | Brzezinski (1985) |
| *Coscinodiscus granii* | 24–225 | – | Martin-Jézéquel et al. (2000) |
| *Chaetoceros affinis* | 1.1–1.39 | – | Martin-Jézéquel et al. (2000) |
| *Chaetoceros constrictus* | 1 | 3350 | Brzezinski (1985) |
| *Chaetoceros convolutus* | 7.3 | 15,100 | Brzezinski (1985) |
| *Chaetoceros curvisetum* | 0.26–2.80 | 100–2200 | Furnas (1978) |
| *Chaetoceros debilis* | 0.11–0.30 | 229–380 | Conley et al. (1989) |
| *Cerataulina pelagica* | 3.15–3.45 | 21,400 | Conley et al. (1989) |
| *Corethron criophilum* | 11 | 30,160 | Brzezinski (1985) |
| *Cyclotella bodanica* | 533 | 40,000 | Conley et al. (1989) |
| *Cyclotella comta* | 31.7–39.8 | 1900 | Conley et al. (1989) |
| *Cyclotella cryptica* | 0.28–0.43 | – | Werner (1977) |
| *Cyclotella glomerata* | 3.67 | 190 | Conley et al. (1989) |
| *Cyclotella melosiroides* | 12.5 | 650 | Conley et al. (1989) |
| *Cyclotella meneghiniana* | 2.43 | 1270 | Conley et al. (1989) |
| *Cyclotella speudostelligera* | 0.52–0.70 | 200 | Conley et al. (1989) |
| *Cyclotella socialis* | 14.2–19.2 | 1600–1900 | Conley et al. (1989) |
| *Cylindrotheca fusiformis* | 0.025–0.110 | 190–420 | Brzezinski et al. (1990) |
| *Diatoma elongatum* | 4.17 | 1000 | Conley et al. (1989) |
| *Ditylum brightwellii* | 6.17–24 | 23,900–90,500 | Conley et al. (1989) |
| *Fragilaria capucina* | 1.48–1.77 | 400 | Conley et al. (1989) |
| *Fragilaria crotonensis* | 2.2–5.07 | 800–623 | Conley et al. (1989) |
| *Fragilaria intermedia* var. *Fallax* | 1.63 | 300 | Conley et al. (1989) |
| *Hemiaulus sinensis* | 3.4–5.1 | 7550–1920 | Brzezinski (1985) |
| *Lauderia borealis* | 1.5 | 5950 | Brzezinski (1985) |
| *Leptocylindrus danicus* | 0.19–7.8 | 1660–2130 | Brzezinski (1985) |
| *Licmophora hyalina* | 2.67–7.00 | – | Paasche (1973) |
| *Melosira granulata* | 1.4–10.1 | 283–1030 | Conley et al. (1989) |
| *Melosira islandica* | 4.67–17.7 | 1825–3633 | Conley et al. (1989) |
| *Melosira italica* | 3.22–5.33 | 800 | Conley et al. (1989) |
| *Navicula salinarum* | 0.14–1.24 | – | Vrieling et al. (1999b) |
| *Rhizosolenia alata* | 39 | 482,000 | Brzezinski (1985) |
| *Rhizolenia fragilissima* | 0.92–1.7 | – | Martin-Jézéquel et al. (2000) |
| *Skeletonema costatum* | 0.048–7.9 | 131 | Davis (1976); Martin-Jézéquel et al. (2000) |
| *Stephanodiscus alpinus* | 2.62–13.2 | 649–5192 | Conley et al. (1989) |
| *Stephanodiscus astraea* | 26.7–142 | 5930–25,000 | Conley et al. (1989) |
| *Stephanodiscus binderanus* | 0.78–1.03 | 200–830 | Conley et al. (1989) |
| *Stephanodiscus hantzschii* | 0.18–0.80 | 50–828 | Conley et al. (1989) |
| *Stephanodiscus minutus* | 0.27–0.45 | 128–158 | Conley et al. (1989) |
| *Stephanodiscus niagarae* | 51 | 21,360 | Conley et al. (1989) |
| *Stephanodiscus rotula* | 2.12–12.2 | 1150–7240 | Conley et al. (1989) |
| *Stephanodiscus subtilis* | 0.27 | 185 | Conley et al. (1989) |

**Table 1.** *Continued*

| Species | Silica content[a] | Cell volume[b] | Reference[c] |
|---|---|---|---|
| *Stephanodiscus tenuis* | 4.5 | 3540 | Conley et al. (1989) |
| *Stephanopyxis palmeriana* | 28–95 | 332,000–818,000 | Brzezinski (1985) |
| *Surirella* spp. | 1.5–2.2 | 2010–1660 | Brzezinski (1985) |
| *Synedra acus* | 18.3–25.7 | 3200–2000 | Conley et al. (1989) |
| *Synedra delicatissima* | 26.2 | 4787 | Conley et al. (1989) |
| *Synedra ostenfeldii* | 12.2 | 2985 | Conley et al. (1989) |
| *Synedra ulna* var. *danica* | 35.0 | 8500 | Conley et al. (1989) |
| *Tabellaria fenestrata* | 6.17–7.00 | 1150–1400 | Conley et al. (1989) |
| *Tabellaria flocculosa* | 13.4–29.3 | 2294–2419 | Conley et al. (1989) |
| *Thalassionema nitzschoides* | 0.21 | 501 | Brzezinski (1985) |
| *Thalassiosira aestivalis* | 13 | 16,300 | Brzezinski (1985) |
| *Thalassiosira decipiens* | 5.34–11.75 | – | Paasche (1973) |
| *Thalassiosira gravida* | 1.08–2 | 3710–6030 | Conley et al. (1989) |
| *Thalassiosira nordenskioeldii* | 1.05–2.66 | 1384–24,680 | Conley et al. (1989) |
| *Thalassiosira oceanica* | 0.093–0.11 | 184–153 | Brzezinski (1985) |
| *Thalassiosira partheneia* | 0.31 | 893 | Brzezinski (1985) |
| *Thalassiosira pseudonana* | 0.02–0.96 | – | Martin-Jézéquel et al. (2000) |
| *Thalassiosira rotula* | 2.8–5.3 | 11,500–10,400 | Brzezinski (1985) |
| *Thalassiosira weissflogii* | 0.4–1.6 | – | Martin-Jézéquel et al. (2000) |

[a] Single or minimum and maximum values measured of silica content (pmol $cell^{-1}$).
[b] Cell volume ($\mu m^3$) corresponding to the silica content, single or minimum and maximum value, respectively.
[c] Main references, for specific information see therein.

soluble pools can account for a sizable fraction of the cellular silicon, 0.5–56% of the total pool, depending on the conditions, with values between 0.9 and $6 \times 10^3$ fmol/cell (reviewed in Martin-Jézéquel et al. 2000). Measurements in sonicated cells by the molybdate method suggested that the predominant form of the soluble pool was monosilicic acid (Binder and Chisholm 1980). This chemical form was found during surge uptake (Binder and Chisholm 1980; Martin-Jézéquel et al. 2000), and its concentration decreased 6h after replenishment (Blank et al. 1986). There is also some evidence that the size of the internal soluble pool after surge uptake was related and positively correlated to the amount of the external silicon provided after silicon deficiency (V. Martin-Jézéquel, unpubl. data). In *Cylindrotheca fusiformis*, soluble pools did not have to increase during surge uptake to maximum levels before uptake was controlled, and these pools changed only gradually over a long period (Hildebrand, pers. comm.). This soluble form was also measured as a few percent of the total silicon content under limited-growth rate regulated by light, nitrogen or phosphorus (Martin-Jézéquel and Claquin, unpubl. data).

The internal molarity obtained by direct measurement of cell water was rarely measured. M. Hildebrand (pers. comm.) determined that the soluble form ranged from 19 mM in *C. fusiformis* to 340 mM in *Stephanopyxis turris*. These values correspond to supersaturated concentrations, well above saturation for silica solubility, which is around 2 mM at pH < 9 (Iler 1979). This large difference between external and internal silicon concentration suggests that different chemical forms exist and these forms could be monomeric silicic acid, polymerized or colloidal silica, or organosilicon compounds. Several studies suggest an association of silicon with organic material. For example, 80% of the intracellular pool of dissolved silicon in *Nitzschia alba* is precipitable with trichloroacetic acid (Azam et al. 1974). A possibility to maintain a soluble silicon form is by an interaction with organic silicon-binding components. These later components presumably composed of mono- or poly-silicic acid complexed with organic material (Azam et al. 1974; Bhattacharya and Volcani 1983; Sullivan 1977), which is consistent with the various intra- and extracellular chemical forms found. Hildebrand (pers. comm.) proposed that soluble-pool concentrations are determined by the capacity of intracellular silicon-binding components and that control of the transport is modulated by the relative amounts of bound and unbound components. This hypothesis could also explain the lack of measured silicon efflux in the absence of external silicate (Sullivan 1976) and the increase with increasing external silicate concentrations up to saturation. Another possibility to maintain the soluble pool is that specialized vesicles exist with intralumenal conditions able to maintain silicon solubility (Schmid and Schulz 1979; Sullivan 1986), even if there is no evidence that these vesicles contain silicon in either soluble or polymerized form.

Cloning and functional characterization of a cDNA encoding a silicic acid transporter was made by Hildebrand et al. (1997) for *Cylindrotheca fusiformis*, and further investigation revealed five silicic acid transporter (*sit*) genes in the species (Hildebrand et al. 1998). Heterologous hybridization of genomic DNA from pennate or centric species with a SIT-specific probe indicated that SIT genes exist in marine and freshwater diatoms. The molecular recognition of multiple transport proteins in *C. fusiformis* and their different levels of expression (Hildebrand et al. 1998), are consistent with earlier work on the kinetics of Si-uptake in diatoms, which was shown to vary during the course of cell wall synthesis (Hildebrand et al. 1998; Hildebrand and Wetherbee, this Vol.). A low affinity and low capacity for silicon uptake were measured prior to frustule deposition and a high affinity, high capacity during deposition (Volcani 1981; Sullivan 1986). The overall regulation of this system would allow the cell, under conditions of maximum uptake, to transport enough silicic acid to construct the wall in a very short time. However, the silicon pool sizes differ in different species, and measurements in *Thalassiosira weissflogii* showed that the pool could be sufficient for the synthesis of an entire cell wall (Binder and Chisholm 1980; Brzezinski and Conley 1994). When uptake and deposition are nearly simultaneous, diatoms such as *Chaetoceros gracilis* and other species have only a small soluble pool of Si, but others, such as *Ditylum brightwellii* in which uptake and deposition are temporally uncoupled, accumulate substantial soluble pools (Chisholm et al. 1978). Moreover, the transport of silicon seems to some extent to be independent of the cell division. Finally, results of Blank and Sullivan (1983b) have shown that the microtubule inhibitors do not affect the silicic acid transport, the total uptake and the pool size of silicon, resulting in the same qualitative and quantitative incorporation of silicon in treated *Navicula saprophila* as in untreated cells. These results suggest that, at least in this species, the total transport of silicon is to some extent independent of the cytoskeleton network and/or trafficking.

To summarize, three modes of silicic acid uptake are known: (1) surge uptake, during which maximum uptake rates were observed; (2) internally controlled uptake, regulated by the intracellular utilization of silicon for the building up of the cell wall; and (3) controlled uptake, when silicon is at very low levels in the external medium (see Martin-Jézéquel et al. 2000 and references therein). This flexibility could allow the cells to react to their environment by the regulation of both the uptake and the transport of silicon.

## 2.2 Energy Requirement

Diatoms have low energetic requirements for silicon-uptake (Raven 1983). Low-energy pathways may help to explain why division of diatoms is not strictly light-dependent and why cells can take up and deposit silicon in the dark with carbon provided by reserves built up in the light (Coombs et al. 1967a). The energy for transport and biomineralization seems to be provided by aerobic respiration and oxidative phosphorylation (Sullivan 1976, 1980) rather than by ongoing photosynthesis (Sullivan 1986). Nevertheless, measured silicon uptake rates (with the analog $^{68}Ge$) in darkness were 44% of those obtained in the light (Azam and Chisholm 1976) and increases in respiration rate were measured during the transport period (Coombs et al. 1967b; Coombs and Volcani 1968). It is noteworthy that some coupling exists between silicon metabolism and the photocycle, but this coupling which seems species-specific probably reflects the differences in adaptation. *Ditylum brightwellii* can take up silicon during the light or dark period of a diel photocycle, but deposition only occurs in the light and in *Skeletonema costatum* all these processes take place within the dark period (Chisholm et al. 1978).

## 2.3 Factors Affecting the Uptake and Transport Processes

Environmental factors affect uptake or incorporation processes differently, thus inducing variation in the soluble silicon pool. Temperature affects the rate of silicon polymerization more than it does the uptake system, leading to a higher soluble pool at low temperatures and vice versa. This effect is balanced by the high uptake of silicon under high temperatures compared to low temperatures which decreases the uptake rate (Sullivan 1980; Blank et al. 1986). Light and nutrients affect also the balance between uptake and incorporation of silicon, but the regulations have not yet been explained. Low light increases soluble pool size in *Coscinodiscus granii* (Taylor 1985) and *Thalassiosira pseudonana* (Martin-Jézéquel and Claquin, unpubl. data). Conversely, nitrogen or phosphorus limitation results in a smaller soluble pool size in *Thalassiosira pseudonana* (Martin-Jézéquel and Claquin, unpubl. data).

Metal ions can also affect silicon uptake and incorporation, but it is unclear whether they affect silicon metabolism directly or indirectly. In their study of *Thalassiosira weissflogii*, De La Rocha et al. (2000) proposed that $Zn^{2+}$ has a more direct action on the uptake system than $Fe^{2+/3+}$. Low $Fe^{2+/3+}$ did not significantly change the half-saturation constant for silicon uptake, while the $K_s$ increases significantly under low $Zn^{2+}$. This result suggests that while the number of silicon transporters decreased, their affinity for silicic acid is not Fe-dependent. Diatoms are also sensitive to copper concentration and it was shown that this toxicity is linked to the silicon metabolism in several marine diatoms. *Skeletonema costatum* and *T. pseudonana* exhibit slower growth rates in the presence of copper, but growth inhibition by copper was alleviated by increasing the silicic acid concentration in the media (Morel et al. 1978; Rueter and Morel 1981). It is likely that copper interferes with the silicic acid transport across the plasma membrane and, hence slows down the uptake of silicic acid (Rueter and Morel 1981; Rueter 1983). When exposed to elevated concentrations of copper or zinc, *Asterionnella japonica* showed a reduced cell-division rate and a marked increase in cell size, with two protoplasts often contained within one frustule (Fischer et al. 1981). However, the cellular silicon concentration did not increase in these cells, confirming the reduced uptake of silicic acid. Studies with other metals, such as aluminium, suggested that the silicon transporters were not affected and that the impact on the cells was essentially due to other metabolic perturbations (Thomas et al. 1980; Gensemer 1990). However, the absence of perturbation in the silica uptake would result in quasi-constant silicon content per cell independently of cell size, but could affect the degree of the silicification process (see below).

# 3 Link Between Silicon Metabolism, Growth and Cell Division

## 3.1 Coupling Between Silicon Metabolism and Cell Growth

Diatoms maintain pools of internal silicon that may be expected to expand or contract depending on the cell-cycle stage, since more silicon would be required during cell wall synthesis than at other stages. The deposition of new siliceous valves occurs between mitosis and daughter cell separation and differences exist between pennate and centric diatoms in the duration of these events (Schmid 1994). Deposition of the girdle bands and other silica structures is less well documented; however, it is known that the girdle bands are formed sequentially following deposition of the new valves and that they can contribute to a large fraction of the biogenic silica. In *Navicula pelliculosa*, new valves are deposited after cytokinesis, followed by the deposition of the girdle bands before mitosis (Chiappino and Volcani 1977). Thus, in this species, silicic acid uptake and deposition of the entire frustule occur during one continuous stage of the cell cycle, beginning just before division and ending just prior to daughter cell separation (Sullivan 1977).

Because of these sequential events, both cell division and growth rate seem to have a direct effect on the degree of silicification of the cells. The time a given diatom species spends at the different stages of the cell cycle can vary depending on the growth rate (Brzezinski et al. 1990; Martin-Jézéquel et al. 2000; Claquin et al. 2002). Under nonlimiting silicon conditions, the incorporation of silicon is mainly linked to the duration of the cell wall synthesis phase. An increase in the length of this time period at low growth rates allows maximum incorporation of silicon into the frustule, and vice versa. Thus, under nonlimiting silicon conditions, mineralization is generally inversely correlated to growth rates. This general effect was shown when growth is regulated by nutrients (nitrogen, phosphorus), trace metals, light and temperature, for different species: *Asterionella ralfsii*, *Chaetoceros affinis*, *C. debilis*, *C. curvisetum*, *Coscinodiscus granii*, *Skeletonema costatum*, *Thalassiosira gravida*, *T. pseudonana*, *T. nordenskioldii* and *Rhizosolenia fragilissima* (Martin-Jézéquel et al. 2000 and references therein; Claquin et al. 2002). Indeed, when the growth rate is triggered by silicon, diatoms growing with higher extracellular silicon have increased cellular silicon contents (Harrison et al. 1977; Paasche 1980). In this case, the mass action effect that drives more silica incorporation per time unit and the level of silicification directly varies with growth rates.

## 3.2 Cell-Cycle Regulation

The link between silicon availability and cell cycle has been studied in synchronized cultures, which can be obtained by manipulating light–dark regimes or silicon availability. Darley and Volcani (1969) have shown with synchronized *Cylindrotheca fusiformis* cultures that in silica-free media the DNA cell content increases only by 4–10%, compared to the other major constituents - RNA and proteins - that increase by at least 75%. Correspondingly, the predominant arrest point was at the G1/S boundary (prior to DNA replication), the other one being at G2/M. Upon repletion of silica, the cellular DNA content increased within the first 10 min and cell numbers began increasing after ca. 2 h (Darley and Volcani 1969). Cell recovery was, however, different depending on the point at which the cell cycle was arrested. Cells arrested in G2 had a much lower recovery than those arrested at G1/S. The latter were able to complete the cell cycle (S + G2 + M) in only 3 h. Flow cytometry was also used to determine the cell-cycle stage arrest in asynchronous cultures by analyzing the relative DNA content per cell. This technique revealed that *Thalassosiria weissflogii* cells arrest at both G1 and G2/M in the dark or in the absence of silicon (Vaulot et al. 1987; Brzezinski et al. 1990). *Chaetoceros* spp. show, in addition to the G1/S growth, an arrest in early G1 at the time of siliceous setae synthesis (Brzezinski et al. 1990). Analysis of several other centric or pennate diatoms proved that in the absence of silica the main arrest point was at the G1/S stage prior to DNA synthesis (Brzezinski et al. 1990). An exception to the dependence on silicate for growth among diatoms is *Phaeodactylum tricornutum*, a polymorphic pennate that usually grows as a fusiform morphotype lacking siliceous structures (Lewin et al. 1958). Correlated with the absence of silica biomineralization in these cells is the lack of dependence on silica during any part of the cell cycle (Brzezinski et al. 1990). However, Borowitzka and Volcani (1978) have found silica bands embedded in the surface of the grid in all cell types. Interestingly, the absence of silica dependency for cell-cycle progression has also been observed for a different algae group, the freshwater colonial flagellate *Synura*. In this cell type, which is ornamented by a layer of siliceous scales, variation in silica availability results in various patterns and rigidity of the scales; they are less robust and more aberrant under high silica stressed conditions. The process may continue until cells become scale-free. However, in *Synura* and in contrast to diatoms, the silica deposition vesicles (SDVs) are still produced even in the absence of silica and the cells still divide, although at a lower rate (Sandgren et al. 1996).

Experiments have also tended to correlate the cell-cycle arrest to a specific cause. After silicon repletion, the renewal of DNA synthesis, of cell arrest at G1/S have been correlated with the induction of specific DNA binding proteins: two DNA polymerases and a single-stranded binding protein (Okita and Volcani 1978). However, tests for activity of these DNA polymerases did not show any significant dependency on silicon in vitro (Okita and Volcani 1978). Based on these experiments, the G1/S arrest-point was suggested as a silicon dependency for the synthesis of proteins involved in DNA replication. Variation in the pattern of expression of DNA polymerases has now been observed for the α-DNA polymerase in the late exponential and the early stationary phases in two *Skeletonema costatum* strains (Hwang et al. 1999). It would be interesting to study the expression pattern of these α-DNA polymerases during the cell cycle and test if this enzyme is homologous to those identified in *C. fusiformis*. However, experiments performed in a different organism, the yeast *Saccharomyces cerevisiae*, have shown that the transcription of α-DNA polymerase, which is involved in DNA replication at the G1/S boundary, is not required for entrance into S phase (Falconi et al. 1993). Therefore, even if regulation of the expression of enzymes involved in the replication of DNA exists, it might not be sufficient to explain the arrest point observed before entering the S phase. Moreover, when 15% of the silicon needed for cell wall formation was given to *Cylindrotheca fusiformis*-starved cells, no effect on DNA synthesis was observed, whereas 60% satisfied most of the silicon requirement and allowed cell-cycle progression through the G1/S arrest point (Darley and Volcani 1969). To explain how the concentration of silica regulates cell-cycle progression, we can envision a threshold-type mechanism, in which DNA synthesis is blocked below a certain concentration. Accumulation (and/or activation) of key components during the cell cycle at critical threshold levels is reminiscent of the situation encountered for the cyclins and kinases of the *cdc* family. Treatment with olomoucine, a competitive inhibitor of the cell cycle regulating p34cdc2/cyclin B, p34cdc2/cyclin A, p34cdc2/cyclin E kinases and the ERK1/MAP-kinase, was shown to inhibit the G1/S transition in diatoms (Abraham et al. 1995). We can hypothesize that silica directly regulates the activity of proteins involved in cell-cycle progression, and that the inhibition of this (these) activity(ies) induces blockage at the G1/S and/or G2 boundaries. According to this hypothesis, the dependency on silica for diatoms might be explained by the involvement of silica as a sensor/messenger, controlling cell-cycle progression.

# 4 Diatom Morphogenesis

## 4.1 Overview of the Morphogenesis Process

The initial silica deposition events occur inside a primordial SDV, at a specific site within the cell, which generally corresponds to the center of the cleavage furrow. In this specialized vesicle, silica nanoparticles in the range of 3–200 nm (Volcani 1981) with a usual size of 20–50 nm (Schmid and Schulz 1979; Crawford et al. 2001) are formed and aggregate. The precipitation/aggregation processes are likely to involve silicate-associated proteins and polyamines (see Hildebrand and Wetherbee, this Vol.) and to be favored by an acidic environment (Vrieling et al. 1999a). The silica particles are themselves supposed to be built outside the SDV and transported into it by silica transport vesicles (STV; Schmid and Schulz 1979) or to be directly produced inside the SDV itself. The aggregation mode of silica particles could involve a process termed sintering, which enables a molecule (or particles) to diffuse at the surface of the aggregate and re-organize towards a more favorable thermodynamic equilibrium. From the embryonic aggregate a network of silica ribs grows, making an initial comb-like structure along the raphe for the pennate diatoms, and a radial branching pattern for the centric diatoms. Later, this pattern is complicated by the superimposition of more silica in a parallel or right-angle fashion. The space between ribs is then filled with some regulated empty territories (the areola). For the pennates this later silica deposition occurs, depending on the species, either in a proximal to distal direction (from the inner to the external valve surface) or in the reverse mode (Cox 1999).

Experiments using electron microscopy support the idea that silica deposition and SDV growth result from a permanent interplay between micro- and macromorphogenesis and require the contribution of several organelles (for reviews see Gordon and Drum 1994; Pickett-Heaps et al. 1990; Round et al. 1990). These observations can be summarized as follow: (1) the position of the nucleus changes during the cell cycle, being generally associated with the girdle bands during interphase, it migrates close to the position of the future hypotheca from the formation of the initial central module until the end of the valve formation; (2) microtubule center (MC) and microtubules (MT) have been seen in the vicinity of the growing SDV, and it has been proposed that they could direct SDV elongation and so dictate valve morphology. Motor proteins associated with microtubules could be involved in the trafficking of silica deposition vesicles towards a specific SDV location. Beside the observations by electron microscopy, the observations that favor the involvement of the cytoskeleton are the description of frustule malformation when cells are treated with actin depolymerizers (Pickett-Heaps et al. 1990). Malformations in different parts of the frustule appeared after treatment with the anti-microtubular drugs colchicine, podophylloxin and aminoprophos-methyl. These drugs, which cause the rapid disappearance of the mitotic spindle and affect migration of the nucleus, led to a complete disorganization of the valve and raphe structure (Schmid 1980; Blank and Sullivan 1983a; Pickett-Heaps et al. 1990). Moreover, luminocolchicine, the analog of colchicines which does not affect MT formation, produces no disturbance in valve formation in *Thalassiosira eccentrica* and *Coscinodiscus wailesii* (Volcani 1981), and very few aberrant valves in *Navicula saprophila* (Blank and Sullivan 1983a); (3) Golgi vesicles have been observed arranged at the margin of the SDV or in close vicinity to it, supporting their role in providing silica precursors and/or membrane components (Sullivan 1986; Schmid 1994). Moreover, participation of the endoplasmic reticulum cannot be excluded (Pickett-Heaps et al. 1990); (4) mitochondria are often found associated along the raphe during its formation, suggesting a requirement for energy. They might also physically restrict the targeting of silica and SDV expansion. They are also associated in the region that will become the areola; (5) the SDV is closely associated to the plasmalemma and follows its curvature, and cell turgor influences the final frustule shape; and (6), synthesis and secretion of polymers, which could influence the plasmalemma characteristics, could have repercussions on morphogenesis. Many of these cellular and subcellular structures could be involved in the regulation of the micro-and macrophogenesis interplay, necessary for the evolution of the exact three-dimensional species-specific pattern.

Switching of valve ontogeny between alternative pathways, illustrates the involvement of cellular organelles and the plasticity in diatom morphogenesis. In *Hantzschia*, the raphe system of each valve lies on the same side, while in *Nitzschia* the raphe system lies diagonally opposed. Species of *Hantzschia* always produce hantzschioid valves, with the two daughter frustules being positioned such that all four raphe systems are located on the same side. The valve is initiated toward the center of the cleavage furrow (at the center in an apical or transapical view), but then the SDVs, microtubule centers and associated organelles move laterally to the same side of the cells adjacent to the keels in the parent valves. The valve morphogenesis is completed at these positions. In *Nitzschia sigma*, the SDVs also appear centrally, but the microtubule center is not closely located. The new primordial SDVs will then move in opposite directions to each other and in opposite directions of their parents to maintain nitzschioid symmetry. However, some species of *Nitzschia* produce some hantzschioid valves. In *N. sigmoidea* the two primordial SDVs still migrate in opposite directions, but can move in the same or opposite direction as their parents, resulting in the formation of daughter cells with either hantzschioid or nitzschioid symmetry; in *N. tryblionella* the SDVs are initiated near girdle bands and SDVs move in close attendance. Then only one SDV-microtubule complex moves across the cell to create stable nitzschioid symmetry in both daughter cells (Kociolek and Williams 1987; Pickett-Heaps et al. 1990).

## 4.2 Differentiation Programs Involving Silicon Morphogenesis

The vegetative state is not the only one in the diatom life cycle that exhibits a silica shell and at least three differentiation programs requiring silica morphogenesis exist: (1) after sexual reproduction the diploid zygote is encased within an organic wall, named the auxospore that exhibits a different morphology from that of the parental cells. Once its extension is complete, mitosis will lead to the formation of the first cell of the new generation, the frustule of which again has a different morphology to the parents. Finally, at the next vegetative reproduction, the cell will recover the original size and pattern; (2) some diatoms, mainly centric, can respond to environmental conditions by the formation of a resting stage. The resting spore presents a different thecae morphology compared to the vegetative cell (Fig. 1), germination, after mitosis, gives rise to a vegetative cell; and (3) under certain circumstances, internal valves are produced closely molded to the normal valve. Internal valves are not associated with division of the cell. They appear to be formed in many cases as a reaction to an osmotic stress or associated with dormancy, but their function and the reason for their formation are still not known (Werner 1977; Round et al. 1990). Even if little information exists concerning the formation of these specific silica-structures, it is likely that the morphogenesis pathway uses some biochemical or subcellular components already available for frustule formation.

# 5 Morphological Plasticity and Variation

Taxonomic recognition of diatom species is complicated by the plasticity of the cells and the number of causes that can affect patterns (for an exhaustive review see Mann 1999). Barber and Carter (1981) suggested seven categories of possible causes for abnormal diatom cells: (1) presence of chemicals in the habitat, including pollutants, high or low pH, etc.; (2) mechanical causes, such as crowding; (3) viral or bacterial pathogens; (4) perturbations occurring during sexual activities; (5) size reduction; (6) cells produced under extreme conditions of light or temperature; and (7) genetic change, including apparent reversion to other species or even genera.

Moreover, morphological data and studies on phenotypic plasticity have revealed heterogeneity among sympatric (populations growing in the same place), but also among allopatric (populations growing in different places) demes. Examples of two sympatric species of *Fragilaria* and of different morphotypes induced by environmental conditions and size reduction are shown in Fig. 2. In most cases, the difficulty is to quantify subtle morphological differences between subpopulations, because the different morphotypes found under environmental stresses might be different phenotypes within the genetically defined range of a polymorphic species (Schmid 1979). It is often unclear whether detectable differences are indicative of morphologically distinct groups or are just individual points along a continuum of morphological variability within a single species. Taxonomic recognition is complicated by the fact that abnormalities can induce unrecognizable and unidentifiable forms, even at the genus level (McLaughlin 1988). The problem then is to recognize a teratologic form in nature, as distinct from a new species or variety. Therefore, descriptions from field samples have been essential to describe variation patterns and to assign them to size reduction through the life cycle, to differences of the same clone in different ecotypes or to genetic variants of the same species.

## 5.1 Size Reduction and Polymorphism

As each daughter cell inherits one theca from the mother and secretes a new one (by formation of two new hypothecae), the cell size of the population

→

**Fig. 2A–P.** Different morphotypes of *Fragilaria* (= *Staurosira*) from Avery Pond, Connecticut, USA, as seen under scanning electron microscopy. **A, B, C, D** Typical *F. construens* var. *venter* form. **E, F, G, H** *F. construens* var. *venter* form close to the *F.* (= *Pseudostaurosira*) *brevistriata* type. **I, J, K, L** *F. construens* var. *venter* form close to the *F.* (= *Staurosirella*) *pinnata* type. **M, N, O, P** Typical morphology of *F. pinnata* (= *Staurosirella pinnata*) form found in the same sample from Avery Pond. *Scale bars* 2 μm. (Photographs are courtesy of Dr. E.A. Morales)

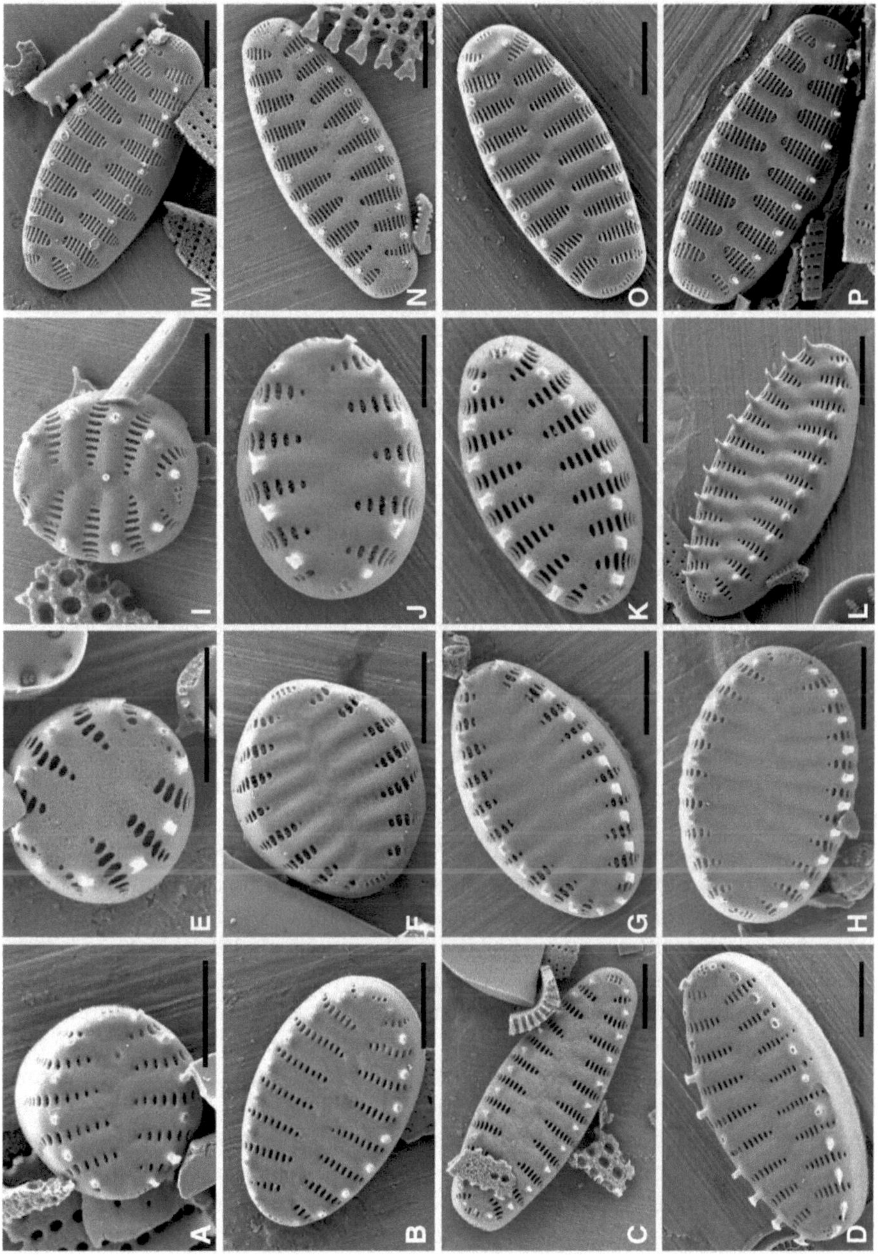
A
B
C
D
E
F
G
H
I
J
K
L
M
N
O
P

necessarily decreases. In general, the size distribution series may be accompanied by allometric changes in structure and shape, which means a general uniform scale transformation allowing the preservation of shape at all sizes. However, size reduction during the life cycle can be accompanied by changes in morphology (i.e., the production of aberrant frustules), either in the valve pattern alone or in both pattern and shape. Indeed, during this process, pennate diatom valves become more rounded, as a consequence of the greater reduction in length, compared to width, and present different morphological patterns, altered in the striae, central area or raphe (Round et al. 1990). This was shown, for example, in *Pinnularia brebissonii* (Hostetter and Rutherford 1976) and in *Stauroneis anceps* (Hostetter and Hoshaw 1972). During frustule formation, some steps in silicification are disturbed and induce morphological aberrations, as found in *Achnantes brevipes*, *Nitzschia palea* and *Synedra ulna*, including fragmented and duplicated raphe systems, disrupted orientation of areolae and fibulae, and obscure areolae. In addition, some aberrant valves have pore-like structures, siliceous nodules on the inner face, or incomplete silica deposition (Estes and Dute 1994). The morphological pattern of the frustule during size reduction is often very close to that obtained after an osmotic stress, and an explanation for the abnormalities is that the surface area/unit volume increases with decreasing size and may alter the balance of ion transport.

Some species are polymorphic and are capable of producing different valve patterns (cf. Mann 1999). Tripolarity is a case of polymorphism which can be triggered by environmental conditions in numerous genera such as *Fragilaria*, *Biddulphia*, *Biddulphiopsis*, *Achnantes*, *Denticula* and *Phaeodactylum*. Polarity polymorphism in diatoms can be reversible, and might be the manifestation of an epigenetically controlled process, environmental conditions serving as a sensor/inducer device (Schmid 1997). As indicated above, *P. tricornutum* has some morphological plasticity: it is cultured has a fusiform morphotype, but can also be oval or triradiate. Even if the amount of biogenic silica is the same for all the different morphotypes (Lewin et al. 1958), the oval state presents a rudimentary frustule. The ratio of fusiform to oval cells can be reversibly increased by repeatedly transferring the cells onto solid medium and differences in the total protein patterns have been observed between these two morphotypes (Gutenbrunner et al. 1994). Polymorphism could even allow, in some rare genera, thecae-bearing morphology of different species (heterovalvy), which can be observed in natural populations, and are termed "Janus cells" (cf. Mann 1999). This kind of polymorphism is well known in the genus *Mastogloia* that may form frustules of two strikingly morphologic types. This occurs during a single division and no intergradations between the two forms are evident. Stoermer (1967) analyzed this phenomenon in a *M. grevillei* population possessing a morphotype referred to as *M. elliptica* var. *dansei*. In this population, the two forms were nearly identical in the shape of the valves and the structure of the raphe and septae. However, they differed in the number and apparent structure of the striae. Rare Janus cells have also been observed in this population (Stoermer 1967).

## 5.2
## Impact of Growth Conditions and Environment

### *5.2.1*
### *Light, Major Nutrients and Temperature*

A complex relationship exists between the need of silica for diatoms to divide and their life cycle, which depends on ambient silicic acid concentrations and the overall metabolism of the cells. In silicate-depleted conditions, the morphology changes, cells have thinner frustules (Paasche 1980; Brzezinski et al. 1990), and other siliceous parts, such as setae in *Chaetoceros* species, are diminished or absent (Brzezinski et al. 1990; Harrison et al. 1977). Under more severe limitations, diatoms may remain viable if kept under dim light or darkness, and exhibit morphological changes in their cell walls (Lewin and Reimann 1969). For example, some *Cylindrotheca fusiformis* cells exhibited normal siliceous skeletons, whereas others had one theca normal and one thin, and some cells had a significant reduction of the siliceous structures of both thecae. Conversely, when the same species was kept in the absence of silicic acid, but under high light intensities, various metabolic syntheses still occurred, even though the cells did not divide; the siliceous parts of the cell wall were greatly reduced, the cell increased in volume six times over that of a normal cell, and showed a total disorganization, rather than a reduction of the various cellular components (Lewin and Reimann 1969). In addition, when silicon-deficient diatoms were placed in the dark in a medium containing silicon, but lacking an assimilable nitrogen or phosphorus source, they continued to take up silicic acid, but did not divide. They absorbed silicic acid until the silicon concentration per cell had doubled. This silica was firmly bound inside the cell and it was not released when the cells were killed (Lewin and Reimann 1969).

Changes in the outline of the valve were also found in nutrient-depleted conditions, as observed in *Nitzschia pungens*, in which normal cells have linear-lanceolate, acute and symmetrical forms, but in depleted cultures, sometimes became sickle-shaped, beaked or lobed (Subba Rao and Wohlgeschaffen 1990). Physical factors, such as temperature, can induce morphological variations of the valves and extreme conditions such as UV radiation, high light and high temperature elicit teratological forms. For example, UV radiation was shown to cause an increase in cell size and abundance of stalked diatoms (Bothwell et al. 1993). High light intensity and high temperature produced abnormal patterns of silica deposition, including deformities of outline, interrupted striations and raphe, and bifurcated or eccentric raphes (see Schmid 1979; Mann 1999 and references therein).

### 5.2.2
### Salinity and Osmotic Stress

The morphometric effects of salinity and osmotic stress are apparent in centric as well as pennate diatoms, as shown for *Navicula salinarum* and *Thalassiosira weissflogii* (Vrieling et al. 1999b, 2000). At low salinities (15 and 20 psu), the size of the pores became smaller, and the transapical ribs and pores in the costae of *N. salinarum* were more clearly visible compared with the higher salinities. The branched network at the outside of the valves of *T. weissflogii* showed a nearly completed differentiation in the periphery over the whole surface, whereas high salinities (28 psu) induced a poorly silicified ring (Fig. 3). *Skeletonema subsalsum* grown in 3‰ salinity or less shows a reduction in the length of its connecting process, and flat-shaped valves; while at a concentration of 5‰ or more, the valves were dome-shaped and the connecting process elon-

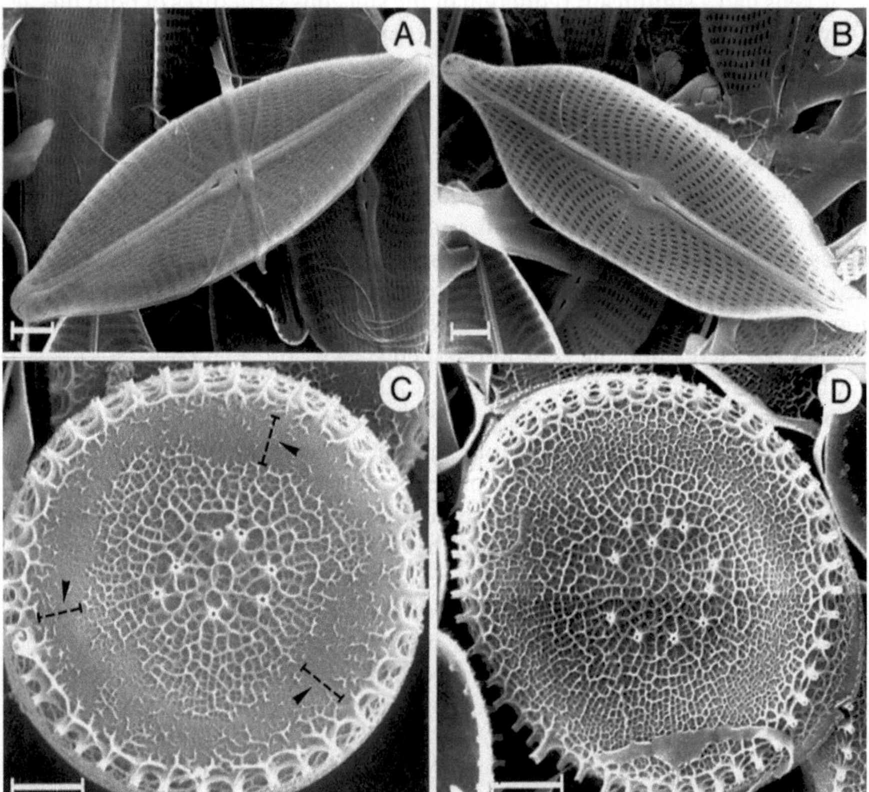


**Fig. 3A–D.** Effect of salinity on the valve morphology of *Navicula salinarum* (**A**, **B**) and *Thalassiosira weissflogii* (**C**, **D**). **A**, **C** At 28 psu and **B**, **D** at 20 psu. *Scale bars* 20 μm. (Original photographs are from Vrieling et al. 2000)

gated (Paasche et al. 1975). According to the authors, this morphology was not merely a reflection of the general physiological state because the growth rate was independent of the salinity, and the formation of the valve structure was influenced by the osmotic pressure rather than by the concentration of one or more specific ions (Paasche et al. 1975). An increase in osmolarity induces a fasciculated arrangement of areolae or valves, consisting of independent scales in *Thalassiosira eccentrica* (Schmid 1984), and aberrations in the raphe structure and the striae-interstriae pattern in *Anamoeoneis sphaerophora* (Schmid 1980). However, variability in the morphology could emerge from the combination of different factors. For example, Håkansson and Chepurnov (1999) determined that in *Cyclotella meneghiniana* the morphological variability was linked to different salinities or to different stages of the life cycle. Over a salinity range of 2–16ä, they found two size groups, which are parts of the same life cycle with a slight variation in morphological features (pattern of striation and number of fultoportulae). On the other hand, the change in the morphology of the internal tubulus of the mantle fultoportulae was seen in higher salinities and could be a response to osmotic stress. The alterations caused by salinity on frustule characters such as shape, striations density and valve dimension were also observed in some other species (Mann 1999). Salinity affects both osmotic and ionic components; one explanation is that changes in water and ion balance act on the cell turgor as well as on the regulation of the physicochemical conditions and pH inside the SDV, and thus could affect both the shape of the cell and the silicification processes (Olsen and Paasche 1986; Schmid 1979; Mann 1999; Vrieling et al. 1999a). Another explanation could be that the stabilization of tubulin by microtubule-associated proteins is suppressed by high $Na^+$ concentration or high osmotic pressure (Schmid 1980). Other salt solutions of Ca, Mg, Cl, and S at varying concentrations can also induce different forms in size, shape, pattern of interstriae and length, and marginal ribs as shown in *Surirella* spp. (Schmid 1979).

### *5.2.3*
### *Trace Elements and Pollutants*

Processes of cell division and morphogenesis are disrupted under heavy metal pollution, and teratological forms of diatoms have been observed and described. Most observations of abnormal diatoms involved samples collected from waters associated with acid mine drainage. One of the problems encountered has been to separate the effects of elevated concentration of metals from those of pH, both of which are known to cause abnormalities in diatoms (Carter 1990; Gensemer et al. 1995). However, at ambient pH values ranging from 7.77 to 8.5, and in Cd-, Cu-, Fe-, and Zn-rich waters, several valve patterns were found for *Fragilaria* (McFarland et al. 1997). They (1) were asymmetric and exhibit bent valves; (2) had notched or incised valves; (3) had valves with abnormal striae and patterned axial areas. These abnormal patterns seem to be general under conditions of metal pollution, and were also found in benthic diatoms in metal-contaminated (Cd, Cr, Cu, Ni, Pb, and Zi) sediments (Dickman 1998).

It is not very clear how metals might act on morphogenesis. Three groups of metal were distinguished on the basis of their toxic effects (Thomas et al. 1980): (1) Cu, Zn and Ge affect, at 1 mg/l, the biochemical pathway of silicon metabolism, increase the length of the cellular chain, and affect the separation of cells; (2) at the same concentration Hg, Cd and Pb provoke some disruption of cell division and cause elongation or bending of cells; and (3) Cr, Ni, Se and Sb have no effect at or below 1 mg/l. Some metals, such as copper, may impair normal membrane function, reducing silicic acid uptake. It was proposed that the initial binding site of copper might be to carboxylic and amino residues of membrane proteins, which might interfere with cell membrane permeability or interact with other essential metals (see Fischer et al. 1981; Stauber and Florence 1987). Another effect is on intracellular processes and cell division. The uncoupling of cytokinesis from other metabolic processes has been noted in several phyla under different metal treatments such as cobalt, mercury, selenium and lead (see Fischer et al. 1981 and references therein). Metals such as copper and mercury may, in some cases, inhibit the process of cell division independently of any effect on the production of new cell material. Copper and zinc induce swollen and distorted cells with large vacuoles in *Asterionella japonica*, and PCMB-treated cells (p-chloromercuribenzoate binds to SH-groups in the cell membrane) look like the copper-treated cells (Fischer et al. 1981; Stauber and Florence 1987). Mercury, copper, and silver are very reactive with SH-groups and it was speculated that copper or other toxic metals might bind to –SH groups and interfere with a number of associated metabolic pathways (Fischer et al. 1981). Copper toxicity may also result from an intracellular reaction with reduced glutathione (GSH), leading to a suppression of mitosis via inhibition of spindle formation (see Stauber and Florence 1987 and references therein). Copper and mercury might also prevent the production of methionine, which appears necessary for cell division (Davies 1976).

The teratology as described for high concentrations of heavy metals seems different to that which seems to occur under metal-deficient conditions, such as a culture medium lacking manganese. In this case, the *Achnantes longipes* cells lost their ability to form new valves (Lewin 1962). The cytoplasm appeared to detach from the wall and ultimately emerged from it altogether. In this way, nearly all the smaller cells became naked protoplasts, while the larger cells gave rise to swollen, spherical plasmodia containing hundreds of nuclei. When Mn was supplied to such cells, they produced silicified valves, although the valves were abnormal in structure. Normal silica valves typical of the species were produced only after auxospore formation (cf. Lewin 1962).

The impact of pollution on diatom morphology is quite variable and seems to depend on the genus involved. Barber and Carter (1981) observed in British diatom populations that some forms are more affected than others. For example, genera of *Nitzschia*, *Eunotia* and *Fragilaria* are very susceptible to chemical imbalance, in contrast to *Gomphonema*. Moreover, within the same genus species can react differently. This was observed in *Stephanodiscus niagarae* and *S. parvus* exposed to eutrophication and heavy metal pollution (Yang and Duthie 1993). The former species was weakly silicified, with irregularly shaped satellite pores; *S. parvus* had one valve underdeveloped with only a siliceous layer and ribs, while the other one was strongly silicified with the areolae almost completely occluded (Yang and Duthie 1993).

## 6 Regulatory Mechanisms in Silicon Metabolism and Morphogenesis

This survey on the involvement of silicon in diatoms reveals their potential to adapt to the constraints they can meet during their life and allows us to propose key features for the role of silicon in the overall metabolism and its relationships with cell division and morphogenesis.

Diatoms show plasticity for the requirement of silicon even if this element is critical in their cell cycle. Hence, this plasticity that is based on different capacities for uptake and storage allows them to grow in water very low in silicate. Despite the current poor knowledge of the impact of environmental factors on the silicon transporters, the amount of silicon taken up appears relatively independent of external factors limiting growth, as suggested by the close relationship between the total silicon-content and the growth rate. Moreover, the independence of the silicon transporter activity of cell division allows the cells to incorporate silicon even under conditions which disturb cytokinesis and morphogenesis, leading to the production of abnormal frustules which sometimes have more silica than in normal cells.

Diatoms are somehow able to acclimatize with aberrant morphological forms which are generally associated with the raphe (absent, distorted or angular), the central area (displaced), or the striae for which configurations change (radiate, parallel, or convergent). The regulation processes imposed by the environmental conditions are elusive, but some interpretations could be done on the phenotypic plasticity, based on the actual knowledge of cell growth and morphogenesis. The first effect of the environmental conditions on the morphotype (i.e., the different patterns observed in the abnormal cells under osmotic stress or cell size reduction) is similar to those occurring after treatment with microtubule inhibitors (cf. Sect. 4). This important effect could be related to stages of the silicification processes and/or linked to the cell division, and is likely to involve the cytoskeleton. Deformations of the areola pattern as well as other abnormalities occur late in the developmental stage and are probably additional effects controlled by the location and structure of the abnormal raphe. The secondary effect (i.e., alteration of the silicification pattern) is clearly linked to the disturbance of the overall process, but it is unclear if the abnormal silicification is due to change in macromorphogenesis, or to a change in the steps of silica-particle formation and deposition (micromorphogenesis), although total silicon transport is not disrupted. Silica nodules, additional or incomplete silicification means, however that some processes of the silica deposition were disturbed, perhaps linked to pH regulation inside the SDV and the ionic action on the biomineralization processes.

The overall process of morphogenesis seems to be governed in complex ways, the involvement of several steps for silicon uptake, its transport within the cell and its addition into the frustule, which are all sensitive to the modification of environmental conditions during the life cycle. This multifaceted aspect of morphogenesis might ensure accurate pattern formation and make the overall process generally robust. Genetic knowledge of the silicon metabolism and roles in diatoms is still in its infancy, but the recent development of molecular tools predicts a promising future.

*Acknowledgements.* We are grateful to Prof. John Raven, Prof. Susan Ford, Dr. Jean Houmard and Dr. Anne-Lise Etienne for critical reading of the manuscript. We also would like to thank Dr.Edouardo Morales for communicating unpublished observations; Dr. Vrieling and the *Journal of Phycology* for original photographs; Prof. Round, Dr. Crawford and Prof. Mann for original photographs from *Diatoms, Biology and Morphology of the Genera*, Cambridge University Press. We thank Pierre Huonnic, Alexandre Korber, and Monique Briand for their assistance in the preparation of the manuscript. V.M.J. is funded by the Centre National de la Recherche Scientifique (CNRS) and Université de Bretagne Occidentale. The investigation has been funded by the Ecole Normale Supérieure and the CNRS to P.J.L.

## References

Abraham RT, Acquarone M, Andersen A, Asensi A, Belle R, Berger F, Bergounioux C, Brunn G, Buquet-Fagot C, Fagot D et al. (1995) Cellular effects of olomoucine, an inhibitor of cyclin-dependent kinases. Biol Cell 83:105–120

Azam F, Chisholm SW (1976) Silicic acid uptake and incorporation by natural marine phytoplankton populations. Limnol Oceanogr 21:427–435

Azam F, Hemmingsen BB, Volcani BE (1974) Role of silicon in diatom metabolism. V. Silicic acid transport and metabolism in the heterotrophic diatom *Nitschia alba*. Arch Microbiol 97: 103–114

Barber HG, Carter JR (1981) Observations on some deformities found in British diatoms. Microscope 34:214–226

Bhattacharyya P, Volcani BE (1980) Sodium-dependent silica transport in the apochlorotic marine diatom *Nitzschia alba*. Proc Natl Acad Sci USA 77:6386–6390

Bhattacharyya P, Volcani BE (1983) Isolation of silicate ionophore(s) from the apochlorotic diatom *Nitzschia alba*. Biochem Biophys Res Commun 114:365–372

Binder BJ, Chisholm SW (1980) Changes in the soluble silicon pool size in the marine daitom *Thalassiosira weissflogii*. Mar Biol Lett 1:205–212

Blank GS, Sullivan CW (1983a) Diatom mineralization of silicic acid. VI. The effects of microtubule inhibitors on silicic acid metabolism in *Navicula saprophila*. J Phycol 19:39–44

Blank GS, Sullivan CW (1983b) Diatom mineralization of silicic acid. VIII. Influence of microtubule drugs on symmetry and pattern formation in valves of *Navicula saprophyla* during morphogenesis. J Phycol 19:294–301

Blank GS, Robinson DH, Sullivan CW (1986) Diatom mineralization of silicic acid. VIII. Metabolic requirements and the timing of protein synthesis. J Phycol 22:382–389
Borowitzka MA, Volcani BE (1978) The polymorphic diatom *Phaeodactylum tricornutum*: ultrastructure of its morphotypes. J Phycol 14:10–21
Bothwell ML, Sherbot D, Roberge AC, Daley RJ (1993) Influence of natural ultraviolet radiation on lotic periphytic diatom community growth, biomass accrual, and species composition: short-term versus long-term effects. J Phycol 29:24–35
Brzezinski MA (1985) The Si:C:N ratio of the marine diatoms: interspecific variability and the effect of some environmental variables. J Phycol 21:347–357
Brzezinski MA (1992) Cell-cycle effects on the kinetics of silicic acid uptake and resource competition among diatoms. J Plankton Res 14:1511–1539
Brzezinski MA, Conley DJ (1994) Silicon deposition during the cell cycle of *Thalassiosira weissflogii* (Bacillariophyceae) determined using dual rhodamine 123 and propidium iodide staining. J Phycol 30:45–55
Brzezinski MA, Olson RJ, Chisholm SW (1990) Silicon availability and cell-cycle progression in marine diatoms. Mar Ecol Prog Ser 67:83–96
Carter J (1990) A new *Eunotia* and its great morphological variations under stress caused by a habitat loaded with copper salts. In: Ricard M (ed) Ouvrage dédié à H. Germain. Koeltz Scientific Books, Koenigstein, pp 13–17
Chiappino ML, Volcani BE (1977) Studies on the biochemistry and fine structure of silica shell formation in diatoms. VIII. Sequential cell wall development in the pennate *Naviculosa pelliculosa*. Protoplasma 93:205–221
Chisholm SW, Azam F, Eppley RW (1978) Silicic acid incorporation in marine diatoms on light:dark cycles: use as an essay for phased cell division. Limnol Oceanogr 23:518–529
Claquin P, Martin-Jézéquel V, Kromkamp JC, Velduis MJW, Kraay GW (2002) Uncoupling of silicon compared to carbon and nitrogen metabolisms, and role of the cell cycle, in continuous cultures of *Thalassiosira pseudonana* (Bacillariophyceae) under light, nitrogen and phosphorus control. J Phycol 38:922–930
Conley DJ, Kilham SS, Theriot E (1989) Differences in silica content between marine and freshwater diatoms. Limnol Oceanogr 34:205–213
Coombs J, Volcani BE (1968) Studies on the biochemistry and fine structure of silica shell formation in diatoms. Silicon induced metabolic transients in *Navicula pelliculosa*. Planta 80:264–279
Coombs J, Spanis C, Volcani BE (1967a) Studies on the biochemistry and fine structure of silica shell formation in diatoms. Chemical composition of *Navicula pelliculosa* during silicon-starvation synchrony. Plant Physiol 42:1601–1606
Coombs J, Spanis C, Volcani BE (1967b) Studies on the biochemistry and fine structure of silica shell formation in diatoms. Photosynthesis and respiration in silicon starvation synchrony in *Navicula pelliculosa*. Plant Physiol 42:1607–1611
Cox EJ (1999) Variation in patterns of valve morphogenesis between representatives of six raphid diatom genera (Bacillariophyceae). J Phycol 35:1297–1312
Crawford SA, Higgins MJ, Mulvaney P, Wetherbee R (2001) Nanostructure of the diatom frustule as revealed by atomic force and scanning electron microscopy. J Phycol 37:543–554
Darley WM, Volcani BE (1969) Role of silicon in diatom metabolism. A silicon requirement for deoxyribonucleic acid synthesis in the diatom *Cylindrotheca fusiformis* Reimann and Lewin. Exp Cell Res 58:334–342
Davies AG (1976) An assessment of the basis of mercury tolerance in *Dunaliella tertiolecta*. J Mar Biol Assoc UK 56:39–57
Davis CO (1976) Continuous culture of marine diatoms under silicate limitation. II. Effect of light intensity on growth and nutrient uptake of *Skeletonema costatum*. J Phycol 12:291–300
De La Rocha CL, Hutchins DA, Brzezinski MA (2000) Effects of iron and zinc deficiency on elemental composition and silica production by diatoms. Mar Ecol Prog Ser 195:71–79

Del Amo Y, Brzezinski MA (1999) The chemical form of dissolved Si taken up by marine diatoms. J Phycol 35:1162–1170
Dickman MD (1998) Benthic marine diatom deformities associated with contaminated sediments in Hong Kong. Environ Int 24:749–759
Estes A, Dute RR (1994) Valve abnormalities in diatom clones maintained in long-term culture. Diatom Res 9:249–258
Falconi M, Higgins NP, Spurio R, Pon CL, Gualerzi CO (1993) Expression of the gene encoding the major bacterial nucleotide protein H-NS is subject to transcriptional auto-repression. Mol Microbiol 10:273–282
Fischer NS, Jones GJ, Nelson DM (1981) Effects of copper and zinc on growth, morphology, and metabolism of *Asterionella japonica* (Cleve). J Exp Mar Biol Ecol 51:37–56
Furnas M (1978) Influence of temperature and cell size on the division rate and chemical content of the diatom *Chaetoceros curvisetum* Cleve. J Exp Mar Biol Ecol 34:97–109
Gensemer RW (1990) Role of aluminum on silica and growth rate on changes in cell size and silica content of silica-limited populations of *Astrerionella valsii* var. *americana* (Bacillariophyceae). J Phycol 26:250–258
Gensemer RW, Smith REH, Duthie HC (1995) Interactions of pH and aluminium on cell length reduction in *Asterionnella ralfsii* var. *americana* Körner. In: Marino D, Montresor M (eds) Proceeding of 13th international diatom symposium, 1–7 Sept 1994. Koelts Scientific Books, Koenigstein, pp 39–46
Gordon R, Drum RW (1994) The chemical basis of diatom morphogenesis. Int J Cytol 150:243–372
Gutenbrunner SA, Thalhamer J, Schmid A-MM (1994) Proteinaceaous and immunochemical distinctions between the oval and fusiform morphotypes of *Phaeodactylum tricornutum* (Bacillariophyceae). J Phycol 30:129–136
Håkansson H, Chepurnov V (1999) A study of variation in the valve morphology of the diatom *Cyclotella meneghiniana* in monoclonal cultures: effects of auxospore formation and different salinity conditions. Diatom Res 14:251–272
Harrison PJ, Conway HL, Holmes RW (1977) Marine diatoms grown in chemostats under silicate or ammonium limitation. III. Cellular composition and morphology of *Chaetoceros debilis*, *Skeletonema costatum*, and *Thalassiosira gravida*. Mar Biol 43:19–31
Hildebrand M, Volcani BE, Gassmann W, Schroeder JI (1997) A gene family of silicon transporters. Nature 385:688–689
Hildebrand M, Dahlin K, Volcani BE (1998) Characterization of a silicon transporter gene family in *Cylindrotheca fusiformis*: sequences, expression analysis, and identification of homologs in other diatoms. Mol Gen Genet 260:480–486
Hostetter HP, Hoshaw RW (1972) Asexual development patterns of the diatom *Stauroneis anceps* in culture. J Phycol 8:289–296
Hostetter HP, Rutherford KD (1976) Polymorphism of the diatom *Pinnularia brebissonii* in culture and field collection. J Phycol 12:140–146
Hwang SP, Wang SK, Wei SF, Cheng LC, Chang J (1999) Identification and expression pattern of DNA polymerase alpha gene in a marine diatom, *Skeletonema costatum*. Mar Biotechnol 1:200–206
Iler RK (1979) The chemistry of silica: solubility, polymerisation, colloid and surface properties, and biochemistry. Wiley-Interscience, New York
Kociolek JP, Williams DM (1987) Unicellular ontogeny and phylogeny: examples from the diatoms. Cladistics 3:274–284
Lewin JC (1962) Silicification. In: Lewin RA (ed) Physiology and biochemistry of algae. Academic Press, New York, pp 445–453
Lewin JC, Reimann BEF (1969) Silicon and plant growth. Annu Rev Plant Physiol 20:289–304
Lewin JC, Lewin RA, Philpott DE (1958) Observations on *Phaeodactylum tricornutum*. J Gen Microbiol 18:418–426
Mann DG (1999) The species concept in diatoms. Phycologia 38:437–495
Martin-Jézéquel V, Hildebrand M, Brzezinski M (2000) Silicon metabolism in diatoms: implications for growth. J Phycol 36:791–720

McLaughlin RB (1988) Diatoms. Microscope 36:261–271
McFarland BH, Hill BH, Willingham WT (1997) Abnormal *Fragilaria* spp. (Bacillariophyceae) in streams impacted by mine drainage. J Freshwater Ecol 12:141–149
Morel NML, Rueter JG, Morel FMM (1978) Copper toxicity to *Skeletonema costatum*. J Phycol 14:43–48
Okita TW, Volcani BE (1978) Role of silicon on diatom metabolism. IX. Differential synthesis of DNA polymerases and DNA-binding proteins during silicate starvation and recovery in *Cylindrotheca fusiformis*. Biochim Biophys Acta 519:76–86
Olsen S, Paasche E (1986) Variable kinetics of silicon-limited growth in *Thalassiosira pseudonana* (Bacillariophycea) in response to charged chemical composition of the growth medium. Br Phycol J 21:183–190
Paasche E (1973) Silicon and the ecology of marine plankton diatoms. II. Silicate-uptake kinetics in five diatom species. Mar Biol 19:262–269
Paasche E (1980) Silicon. In: Morris I (ed) The physiological ecology of plankton. Studies in ecology. Univ California Press, Berkeley, pp 259–284
Paasche E, Johansson S, Evensen DL (1975) An effect of osmotic pressure on the valve morphology of the diatom *Skeletonema subsalsum* (A. Cleve) Bethge. Phycologia 14:205–211
Pickett-Heaps J, Schmid A-MM, Edgar LA (1990) The cell biology of the diatom valve formation. In: Round FE, Chapman DJ (eds) Progress in phycology research, vol 7. Biopress, Bristol, pp 2–168
Raven JA (1983) The transport and function of silicon in plants. Biol Rev 58:179–207
Round FE, Crawford RM, Mann DG (1990) The diatoms. Cambridge Univ Press, Cambridge
Rueter JJG (1983) Effect of copper on the growth, silicic acid uptake and soluble pools of silicic acid in the marine diatom, *Thalassiosira weisflogii* (Bacillariophyceae). J Phycol 19:101–104
Rueter JJG, Morel FMM (1981) The interaction between zinc deficiency and copper toxicity as it affects the silicic acid uptake mechanisms in *Thalassiosira pseudonana*. Limnol Oceanogr 26:67–73
Sandgren CD, Hall SA, Barlow SB (1996) Siliceous scale production in Chrysophyte and Synurophyte algae. I. Effects of silica-limited growth on cell silica content, scale morphology, and the construction of the scale layer of *Synura petersenii*. J Phycol 32:675–692
Schmid A-MM (1979) Influence of environmental factors on the development of the valve in diatoms. Protoplasma 99:99–115
Schmid A-MM (1980) Valve morphogenesis in diatoms: a pattern-related filamentous system in pennates and the effect of APM, colchicine and osmotic pressure. Nova Hedw 33:811–847
Schmid A-MM (1984) Wall morphogenesis in *Thalassiosira eccentrica*: comparison of auxospore formation and the effect of MT-inhibitors. In: Mann DG (ed) Proceedings of the 7th international diatom symposium. Koeltz, Koenigstein, pp 47–70
Schmid A-MM (1994) Aspects of morphogenesis and function of diatom cell walls with implications for taxonomy. Protoplasma 181:43–60
Schmid A-MM (1997) Intraclonal variation of the tripolar pennate diatom "*Centronella reicheltii*" in culture: strategies of reversion to the bipolar *Fragilarscia*-form. Nova Hedw 65:27–45
Schmid A-MM, Schulz D (1979) Wall morphogenesis in diatoms: deposition of silica by cytoplasmic vesicles. Protoplasma 100:267–288
Stauber JL, Florence TM (1987) Mechanism of toxicity of ionic copper and copper complexes to algae. Mar Biol 94:511–519
Stoermer EF (1967) Polymorphism in *Mastogloia*. J Phycol 3:73–77
Subba Rao DV, Wohlgeschaffen G (1990) Morphological variants of *Nitzschia pungens* Grunow f. *multiseries* Hasle. Bot Mar 33:545–550
Sullivan CW (1976) Diatom mineralization of silicic acid. I. $Si(OH)_4$ transport rates during the cell cycle of *Navicula pelliculosa*. J Phycol 13:86–91
Sullivan CW (1977) Diatom mineralization of silicic acid. II. Regulation of Si(OH) transport rates during the cell cycle of *Navicula Pelliculosa*. J Phycol 13:86–91

Sullivan CW (1980) Diatom mineralization of silicic acid. V. Energetic and macromolecular requirements for Si(OH) mineralization events during the cell cycle of *Navicula Pelliculosa.* J Phycol 16:321–328

Sullivan CW (1986) Silicification by diatoms. Ciba Found Symp 121:58–89

Sullivan CW, Volcani BE (1981) Silicon in the cellular metabolism of diatoms. In: Simpson TL, Volcani BE (eds) Silicon and siliceous structures in biological systems. Springer, Berlin Heidelberg New York, pp 15–42

Taylor NJ (1985) Silica incorporation in the diatom *Coscinodiscus granii* as affected by light intensity. Br Phycol J 20:365–374

Thomas WH, Hollibraugh JT, Seibert DLR (1980) Effects of heavy metals on the morphology of some marine phytoplankton. Phycologia 19:202–209

Vaulot D, Olson RJ, Merkel S, Chisholm SW (1987) Cell-cycle response to nutrient starvation in two phytoplankton species, *Thalassiosira weissflogii* and *Hymenomonas carterae.* Mar Biol 95:625–630

Volcani BE (1981) Cell wall formation in diatoms: morphogenesis and biochemistry. In: Simpson TL, Volcani BE (eds) Silicon and siliceous structures in biological systems. Springer, Berlin Heidelberg New York, pp 157–200

Vrieling EG, Gieskes WWC, Beelen TPM (1999a) Silicon deposition in diatoms: control by the pH inside the silicon deposition vesicle. J Phycol 35:548–559

Vrieling EG, Poort L, Beelen TPM, Gieskes WWC (1999b) Growth and silica content of the diatoms *Thalassiosira weissflogii* and *Navicula salinarum* at different salinities and enrichments with aluminium. Eur J Phycol 34:307–316

Vrieling EG, Beelen TPM, van Santen RA, Gieskes WWC (2000) Nanoscale uniformity of pore architecture in diatomaceous silica: a combined small and wide angle X-ray scattering study. J Phycol 36:146–159

Werner D (1977) Silicate metabolism. In: Werner D (ed) The biology of diatoms, vol 13. Univ California Press, Berkeley, pp 110–149

Yang J-R, Duthie H (1993) Morphology and ultrastructure of the teratological forms of the diatoms *Stephanodiscus niagarae* and *S. parvus* (Bacillariophyceae) from Hamilton Harbourg (Lake Ontario, Canada). Hydrobiologia 269/270:57–66

# Organisms: Higher Plants

# Functions of Silicon in Higher Plants

Jian Feng Ma

## 1 Introduction

Silicon (Si) is the second most abundant element, either on the basis of weight or number of atoms, in the earth's crust. Because of its strong affinity with oxygen, in nature Si always exists as silica (Si dioxide) or silicates which are combined with various metals. Si dioxide comprises about 60% of the earth's crust. In soil, Si dioxide accounts for more than 50% and Si concentration in soil solution is between 3.5 and 40 mg Si $L^{-1}$ in the form of silicic acid (Marscner 1995). Therefore, all plants rooting in soil contain Si in their tissues. However, because of its universal existence, earlier researchers paid little attention to the impact of Si on plant growth. Sacks (1862) first asked the following question in his article on Si nutrition "... whether silicic acid is an indispensable substance for those plants that contain silica, whether it takes part in nutritional processes, and what is the relationship that exists between silicic acid and the life of the plant." However, Sachs concluded that Si was insignificant for the nutritional process of maize when he found that water-cultured maize which contained 0.3% Si did not show any abnormality in growth compared to maize containing 9% Si. Since then, a number of studies have been carried out on the effect of Si on plant grwoth. However, until now Si has not been put in the list of essential elements for higher plants. According to the criteria proposed by Arnon and Stout (1939) for essential elements, a given plant must be unable to complete its life cycle in the absence of the element. However, no evidence has yet been shown that plant is unable to complete its life cycle in the absence of Si. One argument about this is that Si may function as a micronutrient and that it is not possible to completely remove Si from the growth medium by currently available techniques because of various contaminations. However, the fact that a larger effect is gained when more Si accumulates in the shoots, suggests that a quite large amount of Si is required for functions of Si. Another criterion for essentiality is that the element must be directly involved in plant metabolism. However, evidence is lacking on the involvement of Si in plant metabolism. Nevertheless, the beneficial effects of Si have been observed in a wide variety of plant species. In this chapter, the functions of Si in higher plants are described.

Faculty of Agriculture, Kagawa University, Ikenobe 2393, Miki-cho, Kita-gun, Kagawa 761-0795, Japan

Progress in Molecular and Subcellular Biology, Vol. 33
W. E. G. Müller (Ed.)

# 2 Beneficial Effects of Silicon in Different Plant Species

Although all plants contain silicon (Si), the Si concentration of plant shoots varies greatly among plant species; ranging from about 0.1 to 10% (w/w) Si on a dry weight basis (Takahashi et al. 1990; Ma et al. 2001a). One characteristic of Si functions is that the beneficial effects differ with the plant species. Beneficial effects are usually obvious in plants that accumulate high Si in their shoots (Ma et al. 2001a).

## 2.1 Si-Accumulating Plants Versus Si-Nonaccumulating Plants

Takahashi et al. investigated the Si content of a number of plant species grown on the same soil and found that plants can be distinguished into Si-accumulating plants and Si-nonaccumulating plants based on the Si content of the top (mostly leaves) and mol. ratio of Si/Ca (for a review, see Takahashi et al. 1990). Plant species with a Si content in the top higher than 1.0%, and Si/Ca mol. ratio higher than 1.0, are termed Si-accumulating plants, while plant species having less than 0.5% Si and less than 0.5 of Si/Ca mol. ratio are termed Si-nonaccumulating plants. Intermediate plants are those having 0.5–1% Si or higher than 1.0% Si, but a Si/Ca mol. ratio less than 1. According to these criteria, a characteristic distribution of Si-accumulating plants is observed in the phylogenetic tree (Fig. 1). Plant species in *Bryophyta*, *Lycopsida*, *Sphenopsida*, and part of *Pteropsida* classes among *Pteridophyta* belong to Si-accumulating plants. In higher plants, only species in *Gramineae* and *Cyperaceae* families are Si-accumulating plants. Plant species in *Cucurbitales* and *Urticales* show intermediate accumulation of Si and other species are Si-nonaccumulating plants.

The variation in the degree of Si accumulation is largely due to different capacities for Si uptake by plant roots. Three uptake modes have been suggested; active, passive, and rejective uptake, for Si-accumulating plants, intermediate type, and Si-nonaccumulating plants, respectively (Takahashi et al. 1990; Ma et al. 2001a). However, the mechanisms responsible for these different uptake modes are not understood. A specific transport system for the active uptake of silicic acid seems to exist in the roots of Si-accumulating plants.

## 2.2 Accumulation Process of Si in Si-Accumulating Plants

In contrast to Si-nonaccumulating plants, Si taken up by the roots of Si-accumulating plants is translocated to the top and deposited in different tissues, which are associated with functions of Si. The accumulation process of Si in rice, a typical Si-accumulator, has been intensively studied. The form of Si taken up by the roots of rice is silicic acid, an uncharged molecule (Takahashi and Hino 1978). After uptake, silicic acid is rapidly translocated to the top with the transpiration stream. With loss of water due to transpiration, silicic acid is

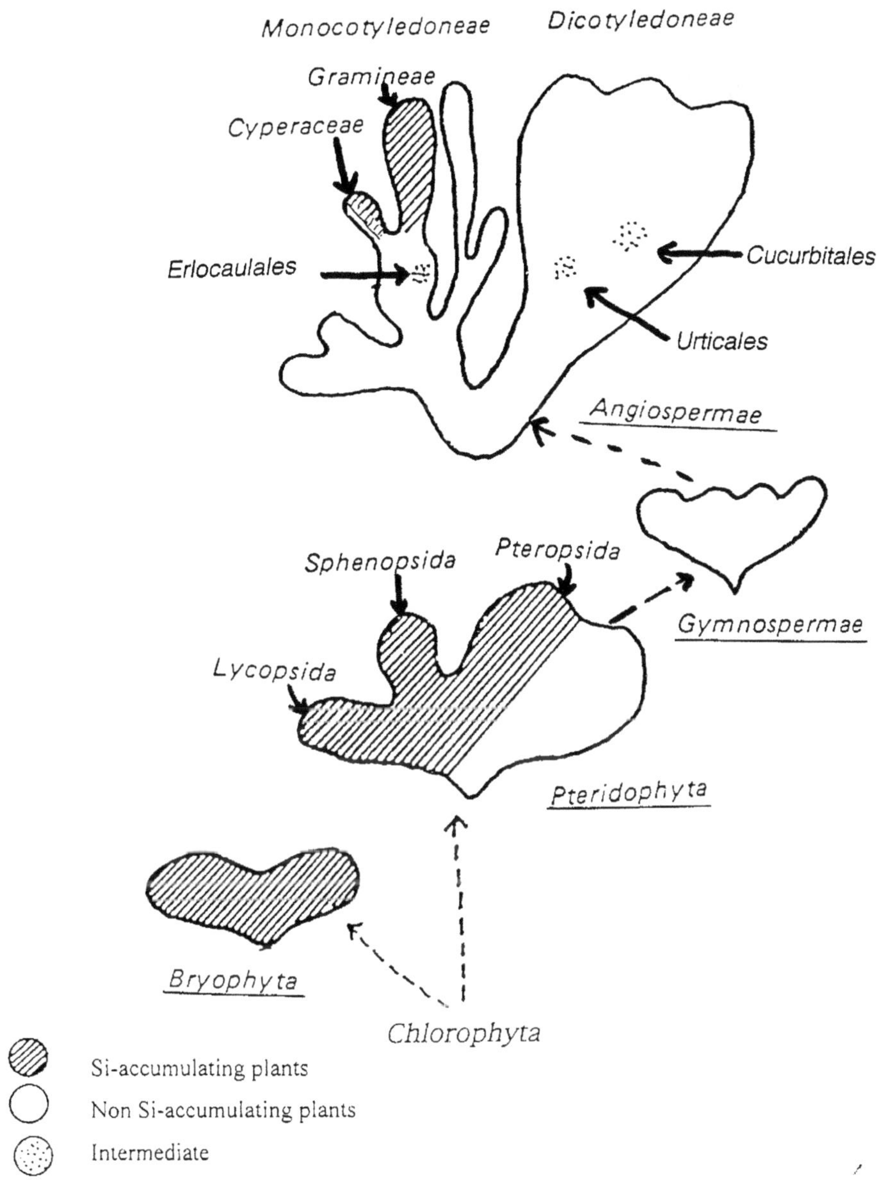


**Fig. 1.** Distribution of Si-accumulating plants in the plant phylogenetic tree

gradually concentrated and polymerized to a colloidal, and finally to a gel form (silica). In leaf blades, Si is deposited as a 2.5-μm-thick layer in the space immediately beneath the thin (0.1 μm) cuticle layer, forming a cuticle-Si double layer (Fig. 2; Yoshida et al. 1962). In addition, Si is also deposited on some specific cells. There are two types of silicified cells; silica cell and silica body or silica bulliform cell. Silica cells are located on vascular bundles, showing a dumb-

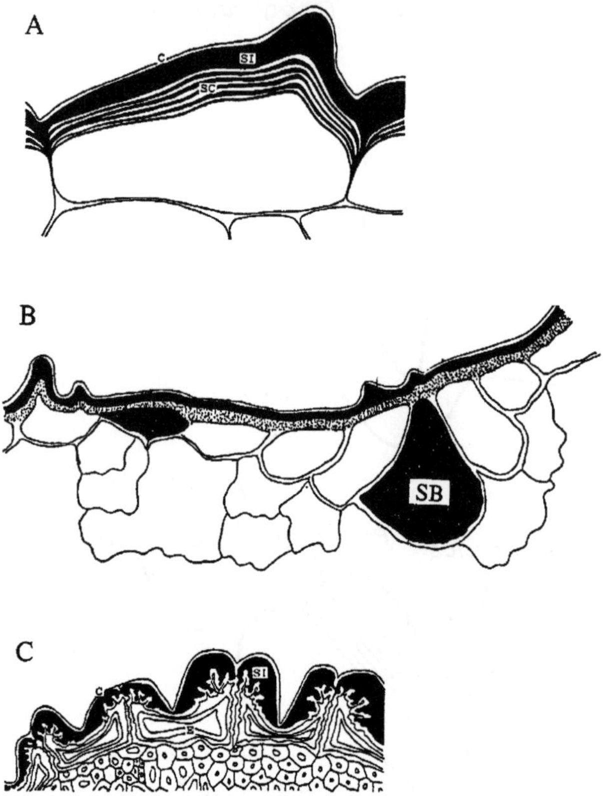


**Fig. 2A–C.** Diagram of Si deposition in different tissues of rice. **A** Cuticle-silica double layer in the leaf blade, **B** silicified cells, **C** cuticle-silica double layer in hull of rice grain. *Blackened area* represents location of Si deposition. *C* Cuticle, *SI* silica layer, *E* epidermis, *SC* silica cellulose membranes, *SB* silica body

bell-shape, while silica bodies are in bulliform cells of rice leaves (Ma et al. 2001a). When the Si content is low, only silica cells are formed. When the Si content of rice shoots increases to more than 2.3%, silica bodies start to form. The silica bodies could be detected by soft X-ray and the number of the silica bodies increases with increasing Si content in the shoot (Fig. 3; Ma 1990).

In addition, Si is also deposited on the epidermis and vascular tissues of stem, leaf sheath, and hull (Fig. 2), which contributes to enhance tissue strength and reduce transpiration as described below.

## 2.3 Effect of Si on the Growth of Si-Accumulating Plants

Important crops such as rice, barley, wheat, and sugarcane are Si accumulators. The effect of Si on the plant growth has been investigated widely, especially in rice. Rice accumulates up to 10% Si in the shoot, which is several times that of

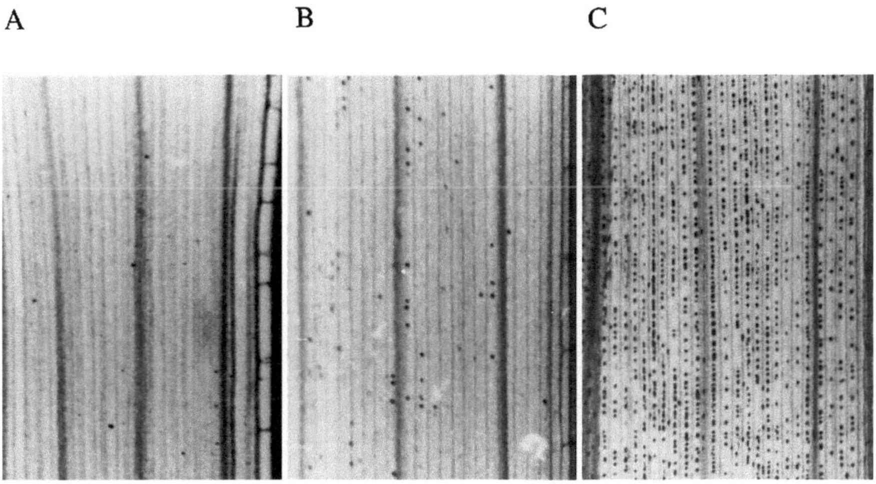


**Fig. 3A–C.** Silica bodies detected by soft X-ray in the rice leaf. The Si content is 0.06 (**A**), 1.5 (**B**), 5.3% Si (**C**), respectively

macronutrients such as N, P, and K (Epstein 1994; Savant et al. 1997). When rice is grown in a nutrient solution free of Si, the leaf blades droop (Fig. 4). Necrotic spots on the leaf blades will also appear if the Fe or Mn concentration in the nutrient solution is high.

Table 1 shows the effect of Si on dry matter production and yield of rice (Ma 1990). The rice was cultured very carefully in a nutrient solution with or without 1.67 mM silicic acid under conditions free of lodging, disease and pest damage, mutual shading and mineral disorders. At harvest after a growth period of 5 months, the dry weight of the shoot (except the grain) was increased by 30% in Si-supplied rice (Table 1). The yield (dry weight of the grain) of Si-supplied plants was more than double that of Si-free plants. Even after deducting the weight of Si from the Si-supplied plants, the increase in the dry weight of both shoot and yield is significant (Table 1). Among the yield components, especially the percentage of ripened spikelets is significantly increased (Fig. 5). A split application study of Si revealed that the most effective period of application for increased rice yield is during the reproductive stage in which dry weight production and Si uptake are most vigorous (Ma et al. 1989).

The effect of Si on the growth of barley was also investigated in a nutrient solution (Okuda and Takahashi 1961). Compared to rice, barley accumulates less Si. The dry weight of shoot (stems + leaves) is increased by 15% in the presence of 0.83 mM silicic acid. In particular, the yield of Si-supplied barley is almost three times that of Si-free barley. Similarly to rice, the percentage of ripening is significantly increased by Si addition.

Beneficial effects of Si were also observed in other Si-accumulating plants such as sugarcane, wheat, and sorghum. For example, in the case of sugarcane,

**Fig. 4.** Symptoms of Si deficiency in rice. Rice is grown in a nutrient solution containing 0 (*left*) or 1.67 mM (*right*) Si as silicic acid

**Table 1.** Dry weight and Si content of shoot and grain of rice at harvest. The plants were cultured in a nutrient solution with or without 1.67 mM Si as silicic acid for 5 months. (Ma 1990)

| | Root | Shoot | Grain | Hull |
|---|---|---|---|---|
| Dry weight (g) | | | | |
| +Si | 12.1 | 129.5 | 27.7 | |
| –Si | 13.9 | 103.6 | 11.2 | |
| Dry weight after deducing Si (g) | | | | |
| +Si | 12.0 | 122.2 | 25.7 | |
| –Si | 13.9 | 103.5 | 11.2 | |
| Si content (%) | | | | |
| +Si | 0.97 | 5.63 | | 7.10 |
| –Si | 0.02 | 0.02 | | 0.18 |

at least 1.0% Si is required for the optimal cane yield in the leaf dry matter and at 0.25% Si the yield drops to about one half (Anderson et al. 1991).

Not only water culture experiments in the laboratories, but also a number of field trials demonstrate that application of Si results in increased yield. Therefore, Si as a fertilizer is widely applied in the world including Japan,

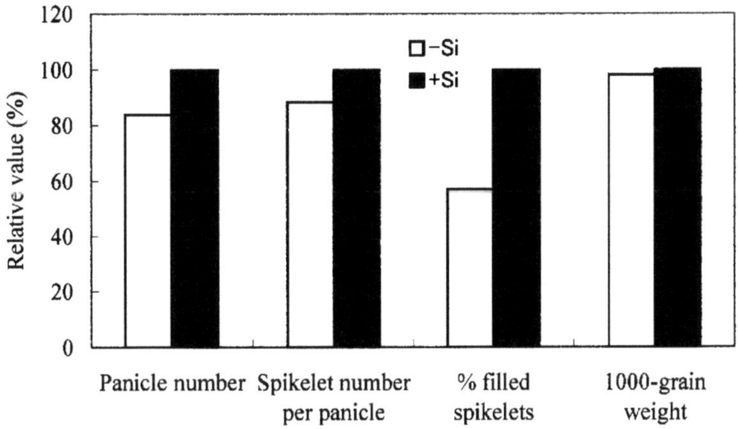


**Fig. 5.** Effect of Si on yield components in rice. Rice is cultured in a nutrient solution with or without 1.67 mM Si as silicic acid. Values shown are relative to Si-supplied plants (as 100)

China, USA, and South Korea. Most silicate fertilizer are applied to paddy soil for the sustainable production of rice (Savant et al. 1997).

## 2.4 Effect of Si on the Growth of Si-NonAccumulating Plants

Compared to Si-accumulating plants, nonaccumulating plants respond to a weekly Si supply. However, Si-deficiency symptoms were reported in Si-nonaccumulating plants including tomato (Miyake and Takahashi 1978), soybean (Miyake and Takahashi 1985), and strawberry (Miyake and Takahashi 1986). Specific conditions such as light intensity, day length and temperature, seem to be required to induce Si deficiency symptoms (Miyake and Takahashi 1978). In tomato, Si deficiency causes a malformation of new leaves and the failure of pollination and fruit formation. Such symptoms appear after the formation of the first flower bud, prior to which the growth of Si-free tomato is normal. However, the mechanisms responsible for these effects of Si remain to be examined.

Cucumber shows an intermediate degree of Si accumulation. The growth (both dry matter production and fruit yield) of this plant increases with increasing Si supply in a nutrient solution (Table 2; Miyake and Takahashi 1982a). Similarly to tomato, the effect of Si is obvious after flowering and the pollen fertility is significantly reduced by Si deficiency (Miyake and Takahashi 1982a).

Although the expression of Si effects in Si-nonaccumulating plants is not as strong as that in Si-accumulating plants under normal growth conditions, the effect of Si becomes larger under stressed conditions as described below. It is interesting to notice that Si supply during the reproductive stage

**Table 2.** Effect of various Si concentrations on the growth of the cucumber plant. (Miyake and Takahashi 1982a)

| Si conc. ($SiO_2$, ppm) | Top-length (cm) | Top dry wt. (g) | Root dry wt. (g) | Fruit fresh wt. (g) | Si (%) |
|---|---|---|---|---|---|
| 0 | 180 | 64.1 | 5.6 | 8 | 0.03 |
| 5 | 205 | 70.9 | 4.5 | 67 | 0.21 |
| 20 | 231 | 78.8 | 5.0 | 142 | 0.64 |
| 100 | 238 | 94.6 | 5.4 | 261 | 2.60 |

is more effective in both Si-accumulating plants and Si-nonaccumulating plants. Silicon may play an important role in some processes involved in reproduction.

## 3 Functions of Si in Higher Plants

Although evidence is still lacking on the involvement of Si in plant metabolism, Si has a number of functions such as stimulation of photosynthesis, enhancement of tissue strength, and reduction of plant transpiration. All these functions contribute to increased dry matter production and resistance of plants to physical, chemical, and biological stresses.

### 3.1 Stimulation of Photosynthesis

Silicon may stimulate photosynthesis by improving light interception, light transmission and reducing excess transpiration. Silica deposited on the leaf blade keep the leaf erect (Fig. 4), thereby improving light interception by the leaf blades and stimulating canopy photosynthesis. This is particularly important in dense plant stands and when nitrogen fertilizers are heavily applied, to minimize mutual shading (Ma and Takahashi 2002).

Silicon was once hypothesized to increase photosynthesis by facilitating the transmission of light (Kaufman et al. 1979). Silica bodies in the leaf epidermal system might act as a "window" to facilitate the transmission of light to photosynthetic mesophyll tissue. However, evidence supporting this hypothesis has not been obtained. The optical properties of leaf transmittance, reflectance and absorbance spectra have been reported to be almost equal in Si-supplied and nonsupplied rice leaves (Agarie et al. 1996). Furthermore, the efficiency of light energy use and quantum yield of leaves with Si is less than those of leaves without Si, suggesting that silica bodies do not function as windows in rice leaves.

**Table 3.** Effect of Si application on transpiration in rice. (Ma 1990)

| Shoot Si content (%) | Transpiration (g $H_2O$ $g^{-1}$ dry wt. $72 h^{-1}$) |
|---|---|
| 0.01 | 200.3 |
| 0.74 | 181.7 |
| 4.81 | 168.0 |
| 6.18 | 154.4 |

Excess water loss (transpiration) causes stomata closure and, therefore, decreases the photosynthetic rate. Transpiration from the leaves is mediated through the stomata and cuticle. As Si is deposited beneath the cuticle of the leaves, forming a Si-cuticle double layer (Fig. 2), the transpiration from the cuticle would be decreased by Si deposition. Rice has a thin cuticle and Si deposition in the leaf blades significantly decreases the transpiration by 30% during the whole growth period (Ma 1990). Moreover, the transpiration decreases with increasing Si content in the shoot (Table 3).

## 3.2 Alleviation of Physical Stress

Physical stress includes stresses of temperature, light, wind, water, drought, freeze, radiation, ultraviolet ray and so on. A number of studies show that Si functions in alleviating these stresses. A few examples are described here.

### *3.2.1 Radiation Damage*

Radiation causes injury to the plant. Silicon seems to protect the plant from radiation injury. When rice seedlings (30 days old) were irradiated with different doses of γ-rays, the decrease in dry weight was less in Si-supplied plants than in Si-nonsupplied plants (Fig. 6; Takahashi 1966), suggesting that Si increases resistance of rice to radiation stress. Furthermore, when the plant was supplied with Si after radiation treatment, the growth recovery was faster compared to plants without Si supply (Fig. 6).

### *3.2.2 Water Stress*

Water-deficiency (drought stress) causes reduction of plant growth. Silicon can alleviate water stress by decreasing transpiration as mentioned above (Table

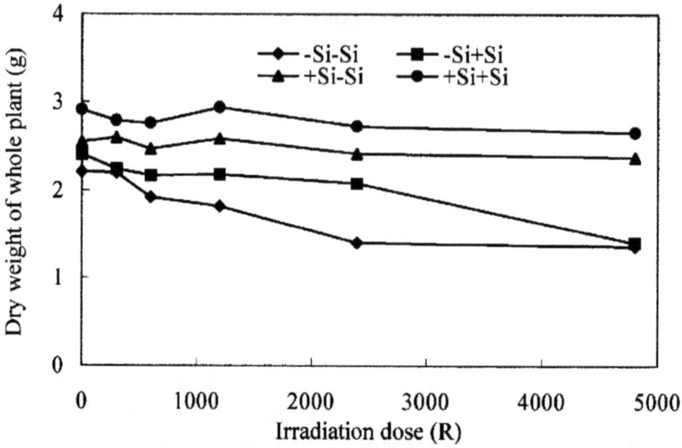


**Fig. 6.** Effect of Si on resistance of rice to radiation stress. Silicon as silicic acid is supplied before and after radiation

**Table 4.** Effect of Si application on the growth and transpiration rate of rice grown under two different humidities. Data are means ($n = 4$) ± SD. (Ma 1990)

| Treatment | Shoot dry wt. ($g\,pot^{-1}$) | Transpiration rate ($g\ H_2O\ g^{-1}$ dry wt.) |
|---|---|---|
| | Relative humidity at 40%[a] | |
| +Si | 0.91 ± 0.06 | 471.1 ± 28.7 |
| −Si | 0.73 ± 0.04 | 635.9 ± 21.4 |
| +Si/−Si | 1.25 | 0.74 |
| | Relative humidity at 90%[b] | |
| +Si | 4.40 ± 0.18 | 297.6 ± 3.5 |
| −Si | 4.05 ± 0.21 | 323.3 ± 8.5 |
| +Si/−Si | 1.09 | 0.92 |

[a] Grown for 10 days.
[b] Grown for 30 days.

3). Table 4 shows the effect of Si on rice growth under different humidity (Ma 1990). Under a water-stressed condition (low humidity), the effect of Si on rice growth is greater than when cultivated under a nonstressed condition (high humidity). The transpiration of Si-supplied rice is 30% less than that of Si non-supplied rice at a low humidity (Table 4).

As mentioned above, the percentage of ripening among yield components is most affected by Si in both rice and barley (Fig. 5). This function of Si may be attributed to Si deposited on the hull (Fig. 2). One important factor for normal development of spikelets is to keep a high moisture condition within the hull (Seo and Ota 1982). The Si content of the hull of the rice grain reaches as much as 7% Si (Table 1) and that of the barley grain is 1.5% (Ma, unpubl. data). Silicon in the hull also deposits between the epidermis cell wall and the

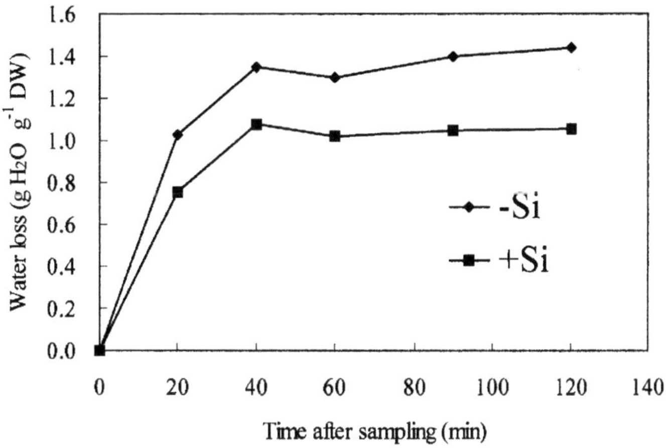


**Fig. 7.** Effect of Si on water loss from the rice spikelets at milky stage. Spikelets are sampled on the tenth day after heading and water loss is measured in an incubator with 30 °C, 30% relative humidity

cuticle, forming a cuticle-Si double layer as in the leaf blades (Fig. 2). However, in contrast to the leaves, the transpiration occurs only from the cuticle because the hull does not have stoma. Silicon is effective in decreasing transpiration from the hull. Figure 7 shows water loss from the excised spikelets at the milky stage, which were sampled at 10 days after heading. The rate of water loss from Si-free spikelets is about 20% more than that from Si-containing (7% Si) spikelets. Therefore, Si plays an important role in keeping a high moisture condition within the hull by decreasing the transpiration from the hull. This is especially important under water deficiency stress and climate stress as discussed below.

### 3.2.3
### *Climatic Stress*

In the field, it can be observed that Si application is effective in alleviating damage caused by climate stress such as typhoons, low temperature and insufficient sunshine during the summer season (Ma et al. 2001a). A typhoon attack usually causes lodging and sterility of rice, resulting in a fatal reduction of rice yield. Deposition of Si in rice enhances the strength of the stem by increasing the thickness of the culm wall and the size of the vascular bundle (Shimoyama 1958), thereby preventing lodging. Strong winds also cause excess water loss from the spikelets, resulting in sterility. Silicon deposited on the hull is effective in preventing excess water loss (Fig. 7). In addition, the effect of Si on the rice yield is also obvious under low temperatures and insufficient sunshine stress (for a review, see Ma et al. 2001a).

## 3.3 Improvement of Resistance to Chemical Stress

Chemical stresses include nutrient deficiency and excess, low and high soil pH, metal toxicity, pesticide, herbicide, and so on. Silicon has been reported to have important functions in improving the resistance of plants to these chemical stress factors.

### *3.3.1 Nutrient Imbalance Stress*

#### 3.3.1.1 Phosphorus Deficiency and Excess

A deficiency of phosphorus (P) in soil is a worldwide problem. The beneficial effects of Si under P-deficiency stress have been observed in many plants including rice and barley. Early observations from a long-term field experiment, conducted at Rothamsted Experimental Station, showed that the barley yield is higher in a field applied with Si than in a field without Si application when P fertilizers were not applied. In an experiment with nutrient solution, Si supply increases the dry weight of rice shoots more at a low P level (14 μM P) than at a mediate level (210 μM; Ma and Takahashi 1990a).

Such beneficial effects of Si were previously explained as a partial substitute of Si for P or an improvement of P availability in soil. However, later experiments showed that Si is unable to affect the P availability in soil. In a P-deficient soil, the previous addition of silicic acid at various concentrations does not affect the P fixation capacity of soil (Fig. 8A). Fixed phosphorus is also not desorbed by various concentrations of silicic acid (Fig. 8B; Ma and Takahashi 1990b, 1991). Silicon is present in the form of silicic acid in soil solution, which does not dissociate at a pH of less than 9. Therefore, it is unlikely that an interaction between silicic acid and phosphate (anionic form) occurs in the soil.

The uptake of P is also unaffected by Si supply at a low P level in both soil and solution culture (Table 5; Ma and Takahashi 1990a, b, 1991). However, the uptake of Fe and Mn is significantly decreased in Si-supplied plants. Phosphorus is translocated and redistributed in plants as an inorganic form. Since P has a high affinity with metals such as Fe and Mn, the internal availability of P could be controlled by the level of Mn, Fe and other metals when P concentration is low. Therefore, the larger beneficial effect of Si on plant growth under P-deficiency stress may be attributed to enhanced availability of internal P by decreasing excess Fe and Mn uptake. This is supported by the fact that Si supply increases the rate of P translocation to the panicles in rice (Nagaoka 1998).

Excess P stress hardly occurs in natural soils, but in some house soils where P fertilizers are heavily applied or in nutrient solution culture where high P

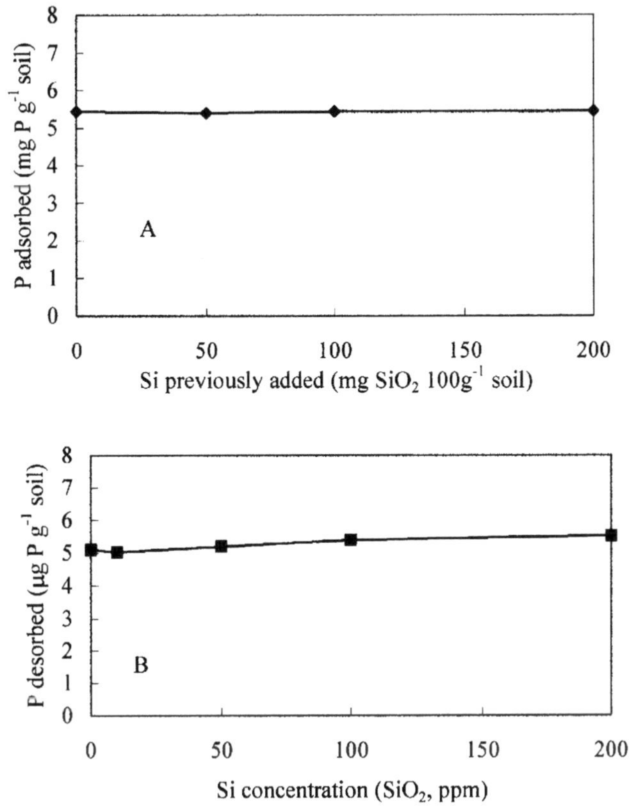


**Fig. 8A,B.** Effect of Si on P availability in a P-deficient soil. **A** Effect of Si on P fixation capacity, **B** effect of Si on P displacement

concentration is supplied. Excess P causes chlorosis or necrosis in leaves, probably due to decreased availability of essential metals such as Fe and Zn. Silicon can alleviate P excess-induced damage (Ma and Takahashi 1990a). The mechanism is that Si decreases excessive uptake of P, resulting in a decrease in internal inorganic P concentration (Table 5). Silicon deposited on the roots and/or Si-induced decrease of transpiration may be responsible for the decreased uptake of P when P in the medium is high. The Si-induced decrease of P uptake is also observed in some Si-nonaccumulating plants including tomato, soybean, strawberry and cucumber (for a review, see Ma et al. 2001a).

#### 3.3.1.2
#### N Excess

The application of nitrogen fertilizers is an important practice for increasing yield. However, excess N causes lodging, mutual shading, susceptibility

**Table 5.** Effect of Si on uptake of P, Mn, and Fe in rice at different P levels. (Ma and Takahashi 1990a)

| | P level (mM) | | | | | |
|---|---|---|---|---|---|---|
| | –Si | | | +Si | | |
| | 0.014 ppm | 0.21 | 0.70 | 0.014 | 0.21 | 0.70 |
| Inorganic P | 356 | 6930 | 14242 | 248 | 3401 | 5472 |
| Organic P | 1648 | 2083 | 2139 | 1493 | 1669 | 1586 |
| Mn | 615 | 708 | 705 | 194 | 257 | 246 |
| Fe | 166 | 151 | 143 | 98 | 94 | 88 |
| P/Mn | 3 | 13 | 23 | 9 | 19 | 29 |
| P/Fe | 12 | 59 | 114 | 17 | 54 | 80 |

to diseases and so on. Silicon deposited on the stems and leaf blades prevents lodging and mutual shading as stated above (Fig. 2). The occurrence of blast disease is significantly inhibited by Si application in the field, especially when N application is heavy (Ohyama 1985). These functions of Si are especially important in the culture system with dense planting and high N applications.

## 3.3.2 *Metal Toxicity Stress*

### 3.3.2.1 Mn and Fe Toxicity

An alleviating function of Si in Mn toxicity has been observed in water-cultured rice (Okuda and Takahashi 1962), barley (Williams and Vlamis 1957; Horiguchi and Morita 1987), bean (Horst and Marschner 1978), and pumpkin (Iwasaki and Matsumura 1999). Three different mechanisms seem to be involved depending on the plant species. In rice, Si reduces Mn uptake by promoting Mn oxidizing power of the roots (Okuda and Takahashi 1962). In bean (Horst and Marschner 1978) and barley (Williams and Vlamis 1957), Si does not reduce Mn uptake, but causes homogeneous distribution of Mn in the leaf blade. Although the mechanism for this homogeneous distribution is still unknown, Horst et al. (1999) found that Si led to a lower apoplasmic Mn concentration in cowpea and suggest that Si modifies the cation-binding properties of the cell wall. By contrast, Si caused localized accumulation of Mn around the base of trichomes in pumpkin (Iwasaki and Matsushima 1999). The uptake of Mn is also unaffected by Si in this plant.

Silicon is also effective in alleviating Fe excess toxicity in rice (Okuda and Takahashi 1962). Silicon enhances the oxidative power of rice roots, resulting in enhanced oxidation of Fe from ferrous iron to insoluble ferric iron. Therefore, excess Fe uptake is indirectly prevented by Si. For upland plants, excess Fe stress is not the problem.

#### 3.3.2.2
#### Na Excess

A beneficial effect of Si under salt stress has been observed in rice (Matoh et al. 1986; Yeo et al. 1999), wheat (Ahmad et al. 1992) and barley (Liang et al. 1996). In rice, shoot and root growth was inhibited by 60% in the presence of 100 mM NaCl for 3 weeks, but Si addition significantly alleviates salt-induced injury (Matoh et al. 1986). The Na concentration in the shoot is decreased to about half by Si addition. This function of Si may be ascribed to a Si-induced decrease of transpiration (Matoh et al. 1986) and to partial blockage of the transpirational bypass flow, the pathway by which a large proportion of the uptake of Na in rice occurs (Yeo et al. 1999).

#### 3.3.2.3
#### Al Toxicity

Al toxicity is a major factor limiting crop production in acid soils. Ionic Al inhibits root growth and nutrient uptake (Ma et al. 2001b). An alleviating effect of Si on Al toxicity has been reported in many crops including maize, cotton, rice, teosinte, sorghum and wheat (for a review, see Cocker et al. 1998). In an experiment with maize, Si addition as silicic acid significantly alleviated Al-induced inhibition of root elongation (Fig. 9; Ma et al. 1997). The alleviating effect is more apparent with increasing Si concentration. The concentration of toxic $Al^{3+}$ is found to decrease by the addition of silicic acid. These results suggest that the interaction between Si and Al occurs in solution, probably by formation of Al–Si complexes, a nontoxic form. However, interaction between Al and Si within the plant has also been suggested (Cocker et al. 1998).

## 3.4
## Increase in Resistance to Abiotic Stress

### *3.4.1*
### *Disease*

It is well known that Si application reduces the severity of some diseases. A list of diseases where occurrence is suppressed by Si is shown in Table 6. Rice blast caused by the fungus *Magnaporthe grisea* (Hebert) Barr is one of the most serious threats to rice production in the world. A number of studies have

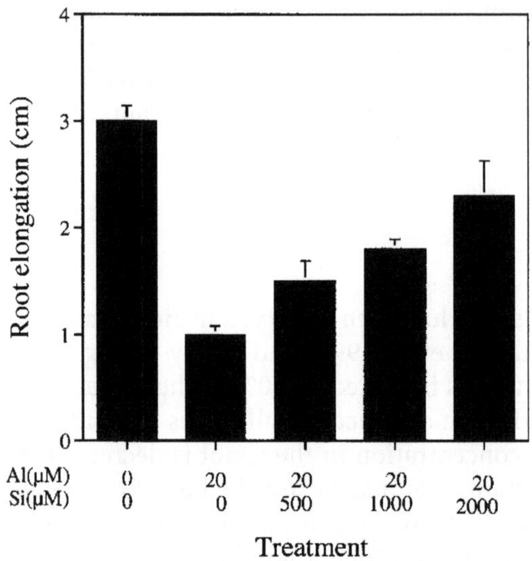


**Fig. 9.** Effect of Si on Al-induced inhibition of root elongation in maize. Roots were exposed to 20 μM $AlCl_3$ for 20 h in the presence of various concentrations of silicic acid

**Table 6.** Diseases whose occurrence is suppressed by Si

| Plant species | Disease |
|---|---|
| Rice | Blast |
| | Sheath bright |
| | Brown spot |
| | Stem rot |
| | Leaf scald |
| | Grain discoloration |
| Barley | Powdery mildew |
| Wheat | Powdery mildew |
| Cucumber | Powdery mildew |
| | Stem rot |
| Muskmelon | Powdery mildew |

shown that Si confers resistance to rice blast. In Florida, where soil is deficient in Si, application of silicate fertilizer is as effective as fungicide in controlling rice blast (Fig. 10; Datnoff et al. 1997). In addition to Si-accumulating plants, Si has also been reported to prevent powdery mildew and stem rot diseases of cucumber (Miyake and Takahashi 1982b). With increasing Si concentration in the culture solution, the Si content in the cucumber shoot is increased, resulting in resistance to powdery mildew disease (Table 7). Silicon is also effective in increasing the resistance to fungal pathogen in cucumber roots (Cherif et al. 1994).

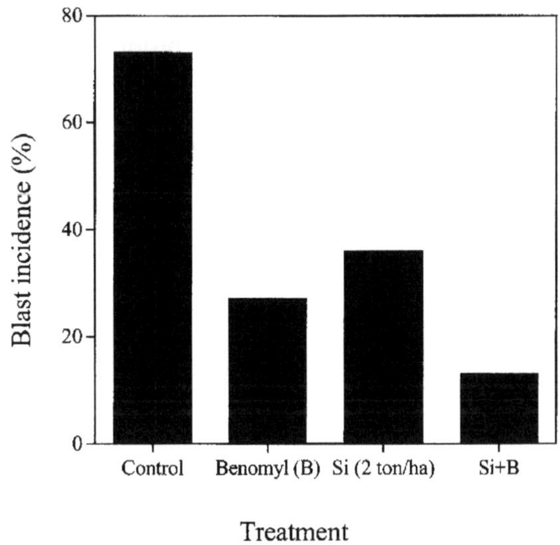


**Fig. 10.** Effect of Si on suppression of rice blast. Silicate fertilizer is applied at 2 mg ha$^{-1}$, and fungicides (Benomyl) at 1.68 kg ha$^{-1}$

**Table 7.** Effect of Si on the resistance of cucumber plants to powdery mildew. (Miyake and Takahashi 1982b)

| | Si supply concentration ($SiO_2$ ppm) | | | | |
|---|---|---|---|---|---|
| | 0 | 5 | 20 | 50 | 100 |
| Si content in dry leaves (%) | 0.04 | 0.06 | 0.33 | 1.02 | 1.51 |
| Leaf position | Severity of powdery mildew[a] | | | | |
| 1 (top) | +++ | ++ | + | + | – |
| 2 | ++++ | ++ | ++ | + | – |
| 3 | +++ | ++++ | ++ | + | – |
| 4 | + | +++ | ++ | – | – |
| 5 | ++ | ++ | + | + | – |

[a] Number of colonies per cm$^2$ of leaf area: – 0, + <0.1, ++ 0.2–0.5, +++ 1.0, ++++ >1.0

Foliar application of Si has been reported to be effective in inhibiting powdery mildew development on cucumber, muskmelon, and grape leaves (Menzies et al. 1992; Bowen et al. 1992). Si applied to leaves may deposit on the surface of leaves and play a similar role as Si taken up from the roots. This approach may be useful for the crops with a passive or rejective uptake mode.

Two hypotheses for the Si-induced resistance to diseases have been proposed. One is that Si deposited on the tissue surface functions as a physical barrier. It prevents physical penetration and/or makes the plant cell less susceptible to enzymatic degradation by fungal pathogens. The other one is that Si functions as a signal to induce production of phytoalexin (Cherif et al. 1994). Si application to cucumber results in stimulation of chitinase activity and rapid activation of peroxidases and polyphenoloxidases after infection with *Pythium* spp. Glycosidically bound phenolics extracted from Si-treated plants when subjected to acid or beta-glucosidase hydrolysis display a strong fungistatic activity.

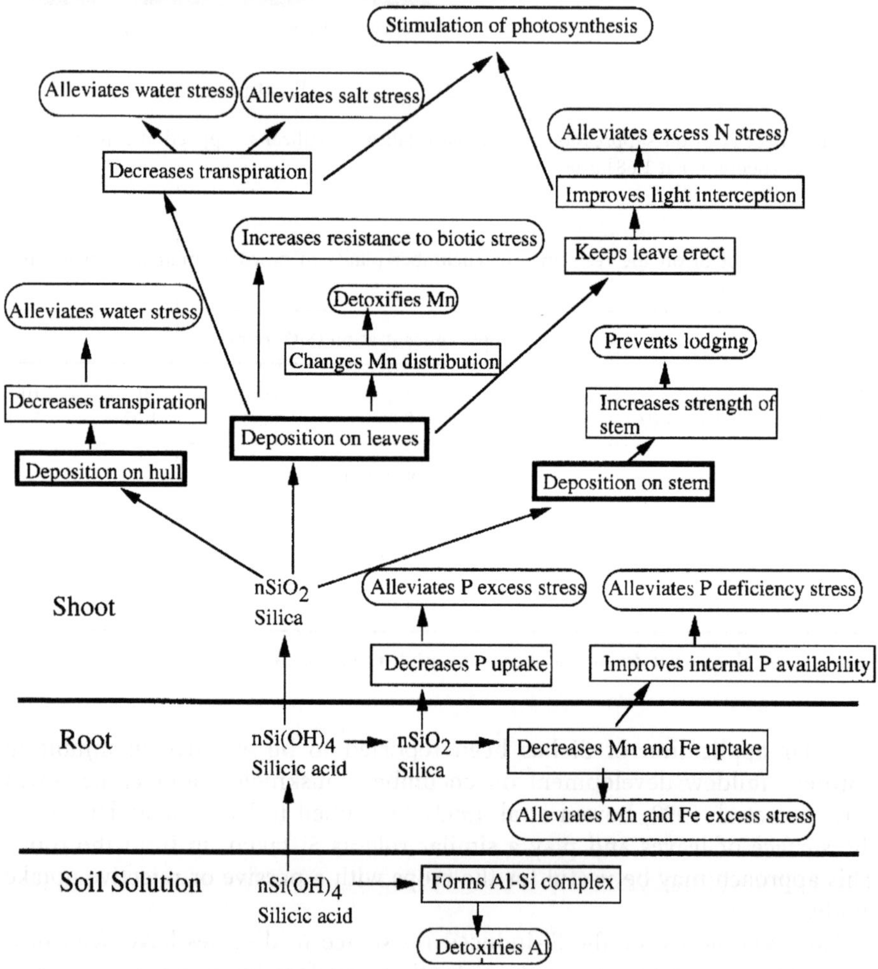


**Fig. 11.** Functions of Si in higher plants

### 3.4.2
### *Pests*

Silicon suppresses insect pests such as the stem borer, brown plant-hopper, rice green leafhopper, and white-backed plant-hopper, and noninsect pests such as leaf spider and mites (Savant et al. 1997). Stems attacked by the rice stem borer were found to contain less Si (Sasamoto 1961). In a field study, a positive relationship between Si content of rice and resistance to brown plant-hopper has been observed (Sujatha et al. 1987). These protective functions of Si may be attributed to Si deposited on the tissue surface.

## 4
## Conclusion

The functions of Si in higher plants are summarized in Fig. 11. It is clear that most functions of Si are expressed through Si deposited on the leaves, stems, and hulls. Therefore, in contrast to essential elements, the functions of Si in plants are mechanical ones, but not physiological ones.

As stated above, Si effects are more obvious under stress conditions. Plants are exposed to various biotic and abiotic stresses in the field during the growth period. Therefore, Si in plants certainly plays an important role in maintaining healthy growth. It should be kept in mind that an absence of an apparent response to Si fertilizers does not mean that Si does not function in plant growth. This is because Si is abundant in soil and the requirement of Si for healthy growth of most plant species is supplied by the soil.

*Acknowledgements.* The author would like to thank Prof. Eiichi Takahashi for his critical reading of this manuscript.

## References

Agarie S, Agata W, Uchida H, Kubota F, Kaufman PB (1996) Function of silica bodies in the epidermal system of rice (*Oryza sativa* L.): testing the window hypothesis. J Exp Bot 47:655–660

Ahmad R, Zaheer SH, Ismail S (1992) Role of silicon in salt tolerance of wheat (*Triticum aestivum* L.). Plant Sci 85:43–50

Anderson DL, Snyder GH, Martin FG (1991) Multi-year response of sugarcane to calcium silicate slag on Everglades Histosols. Agron J 83:870–874

Arnon DI, Stout PR (1939) The essentiality of certain elements in minute quantity for plants with special reference to copper. Plant Physiol 14:371–375

Bowen P, Menzies J, Ehret D (1992) Soluble silicon sprays inhibit powdery mildew development on grape leaves. J Am Soc Hortic Sci 117:906–912

Cherif M, Asselin A, Belanger RR (1994) Defense responses induced by soluble silicon in cucumber roots infected by *Pythium* spp. Phytopathology 84:236–242

Cocker KM, Evans DE, Hodson MJ (1998) The amelioration of aluminium toxicity by silicon in higher plants: solution chemistry or an in planta mechanism? Physiol Plant 104:608–614

Datnoff LE, Deren CW, Snyder GH (1997) Silicon fertilization for disease management of rice in Florida. Crop Protect 16:525–531
Epstein E (1994) The anomaly of silicon in plant biology. Proc Natl Acad Sci USA 91:11–17
Hall AD, Morison CG (1906) On the function of silica in the nutrition of cereals, part 1. Proc R Soc Lond B 77:455–477
Horiguchi T, Morita S (1987) Mechanism of manganese toxicity and tolerance of plants. IV. Effect of silicon on alleviation of manganese toxicity of barley. J Plant Nutr 10:2299–2310
Horst WJ, Marschner H (1978) Effect of silicon on manganese tolerance of bean plants (*Phaseolus vulgaris* L.). Plant Soil 50:287–303
Horst WJ, Fecht M, Naumann A, Wissemeier AH, Maier P (1999) Physiology of manganese toxicity and tolerance in *Vigna unguiculata* (L.) Walp. J Plant Nutr Soil Sci (Z Pflanzenernahr) 162:263–274
Iwasaki K, Matsumura A (1999) Effect of silicon on alleviation of manganese toxicity in pumpkin (*Cucurbita moschata* Duch cv. Shintosa). Soil Sci Plant Nutr 45:909–920
Kaufman PB, Takeoka Y, Carlson TJ, Bigelow WC, Jones JD, Moore PH, Ghosheh NS (1979) Studies on silica deposition in sugarcane using scanning electron microscopy, energy-dispersive X-ray analysis, neutron activation analysis and light microscopy. Phytomorphology 29:185–193
Liang Y, Shen Q, Shen Z, Ma T (1996) Effects of silicon on salinity tolerance of two barley cultivars. J Plant Nutr 19:173–183
Ma JF (1990) Studies on beneficial effects of silicon on rice plants. PhD Thesis, Kyoto Univ
Ma JF, Takahashi E (1990a) Effect of silicon on the growth and phosphorus uptake of rice. Plant Soil 126:115–119
Ma JF, Takahashi E (1990b) The effect of silicic acid on rice in a P-deficient soil. Plant Soil 126: 121–125
Ma JF, Takahashi E (1991) Effect of silicate on phosphate availability of rice in a P-deficient soil. Plant Soil 133:151–155
Ma JF, Takahashi E (2002) Soil, fertilizer, and plant silicon research in Japan. Elsevier, Amsterdam
Ma JF, Nishimura K, Takahashi E (1989) Effect of silicon on the growth of rice plant at different growth stages. Soil Sci Plant Nutr 35:347–356
Ma JF, Sasaki M, Matsumoto (1997) Al-induced inhibition of root elongation in corn, *Zea mays* L. is overcome by Si addition. Plant Soil 188:171–176
Ma JF, Miyake Y, Takahashi E (2001a) Silicon as a beneficial element for crop plants. In: Datnoff LE, Snyder GH, Korndorfer GH (eds) Silicon in agriculture. Elsevier, Amsterdam, pp 17–39
Ma JF, Ryan PR, Delhaize E (2001b) Aluminum tolerance in plants and the complexing role of organic acids. Trends Plant Sci 6:273–278
Marschner H (1995) Beneficial mineral elements. In: Marschner H (ed) Mineral nutrition of higher plants. Academic Press, San Diego, pp 405–435
Matoh T, Kairusmee P, Takahashi E (1986) Salt-induced damage to rice plants and alleviation effect of silicate. Soil Sci Plant Nutr 32:295–304
Menzies J, Bowen P, Ehret D (1992) Foliar application of potassium silicate reduces severity of powdery mildew on cucumber, muskmelon, and zucchini squash. J Am Soc Hortic Sci 117: 902–905
Miyake Y, Takahashi E (1978) Silicon deficiency of tomato plants. Soil Sci Plant Nutr 24:175–189
Miyake Y, Takahashi E (1982a) Effect of silicon on the growth of cucumber plants in a solution culture. Jpn J Soil Sci Plant Nutr 53:15–22
Miyake Y, Takahashi E (1982b) Effect of silicon on the resistance of cucumber plants to the microbial disease. Jpn J Soil Sci Plant Nutr 53:106–110
Miyake Y, Takahashi E (1985) Effect of silicon on the growth of soybean plants in a solution culture. Soil Sci Plant Nutr 31:625–634
Miyake Y, Takahashi E (1986) Effect of silicon on the growth and fruit production of strawberry plants in a solution culture. Soil Sci Plant Nutr 32:321–326
Nagaoka K (1998) Study on interaction between P and Si in rice plants. Graduation Thesis, Kinki Univ

Ohyama N (1985) Amelioration of cold weather damage of rice by silicate fertilizer application. Agric Hortic 60:1385–1389

Okuda A, Takahashi E (1961) Effect of silicon deficiency on plant growth and nutrient uptake in barley, tomato, radish, Welsh onion, and Chinese cabbage. Jpn J Soil Sci Plant Nutr 32:623–626

Okuda A, Takahashi E (1962) Effect of silicon supply on the injuries due to excessive amounts of Fe, Mn, Cu, As, Al, Co of barley and rice plant. Jpn J Soil Sci Plant Nutr 33:1–8

Savant NK, Snyder GH, Datnoff LE (1997) Silicon management and sustainable rice production. Adv Agron 58:151–199

Saches J (1862) Ergebnisse einiger neuerer Untersuchungen über die in Pflanzen enthaltene Kieselsaure. Flora 33:65–71

Sasamoto K (1961) Resistance of the rice plant applied with silicate and nitrogen fertilizers to the rice stem borer *Chito suppressalis* WALKER. Proc Faculty Liberal Arts Educ, Yamanashi Univ, vol 3, pp 1–73

Seo SW, Ota Y (1982) Role of the hull in the ripening of rice plants. V. Water loss in hull and development of rice kernel. Jpn J Crop Sci 51:529–534

Shimoyama S (1958) Effect of silicon on lodging and wind damage in rice. Report for the Research Funds Granted by Ministry of Agriculture, Japan, p 82

Sujatha G, Reddy GPV, Murthy MMK (1987) Effect of certain biochemical factors on expression of resistance of rice varieties to brown planthopper (*Nilaparvata lugens* Stal). J Res Andhra Pradesh Agric Univ 15:124–128

Takahashi E (1966) Effect of silicon on resistance of rice to radiation. Jpn J Soil Sci Plant Nutr 37:183–188

Takahashi E, Hino K (1978) Effect of Si forms on Si uptake by rice plants. Jpn J Soil Sci Plant Nutr 49:357–360

Takahashi E, Ma JF, Miyake Y (1990) The possibility of silicon as an essential element for higher plants. Comments Agric Food Chem 2:99–122

Williams DE, Vlamis J (1957) The effect of silicon on yield and manganese-54 uptake and distribution in the leaves of barley grown in culture solutions. Plant Physiol 32:404–409

Yeo AR, Flowers SA, Rao G, Welfare K, Senanayake N, Flowers TJ (1999) Silicon reduces sodium uptake in rice (*Oryza sativa* L.) in saline conditions and this is accounted for by a reduction in the transpirational bypass flow. Plant Cell Environ 22:559–565

Yoshida S, Ohnishi Y, Kitagishi K (1962) Histochemistry of Si in rice tissues. III. The presence of cuticle-silica double layer in the epidermal tissue. Soil Sci Plant Nutr 8:1–5

# Silicon in Plants

Dieter Neumann

## 1 Introduction

With only a few exceptions, soils contain high amounts of Si, however, only a small part is available for uptake into plants. The concentration of soluble Si in the soil depends on the pH in the rhizosphere and is in the range of 0.1–0.6 mM (Epstein 1994, 1999). Higher plants differ in their ability to take up and accumulate Si on a dry wt. basis (ranging from 0.1–15%). Depending on their Si content, they can roughly be divided into:

- Si accumulators – members of the Cyperaceae and wetland Gramineae (10–15% Si on a dry wt. basis), dryland grasses, most of the other cereals and some members of the dicots (1–3% Si on a dry wt. basis).
- Si nonaccumulators – many of the dicots (<0.5% Si on a dry wt. basis).

In addition, the Si content varies vastly in different ecotypes (Deren et al. 1992) and different parts of the plants. In the sedge *Schoenus nigricans*, the highest Si values were found in the pericarp (34%), whereas the Si content of the vegetative parts of the shoot was lower by more than one order of magnitude (Ernst et al. 1995). Such differences are also observed between roots and shoots (Hodson and Sangster 1984, 1989; Parry et al. 1984; Hodson and Evans 1995).

## 2 Silicon in Monocots

It has been known for a long time that Si is accumulated in Gramineae (Grob 1896). Later SEM studies showed "silica bodies" in different species, but hot chromic acid, nitric acid and hydrogen peroxide used in these experiments destroyed the cell structures and, hence, prevented the acquisition of information on compartmentation and chemical nature of the silica deposits

Institute of Plant Biochemistry, Weinberg 3, 06120 Halle, Germany

Progress in Molecular and Subcellular Biology, Vol. 33
W. E. G. Müller (Ed.)

(Sangster 1968, 1970; Parry and Smithson 1958, 1964). In shoot internodes of *Avena*, epidermal idioblasts were found to accumulate Si (Kaufman et al. 1970, 1971). First EM studies combined with microanalysis showed Si deposits in the endodermis of *Sorghum* (Sangster and Parry 1976). They seem to consist of 100-nm particles in close connection with cellulose microfibrils. In the hollow stems of bamboo, large amounts of "opal phytoliths" (tabashir) occur, which consist of 10-nm particles containing more water and smaller amounts of alkalis and alkaline earth than natural opals (Jones et al. 1965). With more sophisticated analytical electron microscopice methods, Si was found in trichomes and epidermal cells of *Deschampsia*, *Festuca* and *Schoenus* (Bode et al. 1994; Neumann et al. 1999).

## 2.1 $SiO_2$ Deposits in Monocots

Electron translucent structures were found in different compartments in *Festuca lemanii* and *Deschampsia caespitosa*. Beside thick layers on epidermal cell walls (Fig. 1A) and in vacuoles on the tonoplast membrane (Fig. 1B), crystal-like deposits were frequently observed in the vacuoles (Fig. 1C) and the cytoplasm, sometimes obviously penetrating the plasmalemma and organellar envelopes (Fig. 1E). Nevertheless, in these cells, the ultrastructure looks quite normal (Fig. 1E), an indication that the membrane function is not disturbed. EDX (EDX-Energy dispersive analysis of X-rays) spectra of these structures contain only Si (Fig. 1D). Information on the chemical coordination of Si comes from EEL spectra (electron energy loss spectroscopy) of the Si-$L_{2,3}$ and the O–K edge of such structures. The comparison with corresponding standard spectra (Fig. 2A, B) and quantum-chemical calculations for interpretation of the spectra (Lichtenberger and Neumann 1995, 1997) confirm that in all cases investigated so far the electron-translucent Si deposits consist of $SiO_2$. Although $SiO_2$ deposits very often look like crystals with sharp borders, they are not surrounded by a membrane and do not damage the membranes (Fig. 1E). Obviously, these precipitates are $SiO_2$ gels with a high water content, masking the membranes and structures only and preventing their staining in electron micrographs, without affecting the membrane function.

## 2.2 Silicic Acid in Monocots

Some monocots are able to accumulate Si, but not in the form of $SiO_2$. In epidermal cells of *Deschampsia caespitosa*, only Si-containing deposits were found (Fig. 3A). In intercellular spaces of the same plant, needle-like structures were observed, also only containing Si (Fig. 3B). The EEL spectra of the Si-$L_{2,3}$ edge are identical to the spectra of $SiO_2$, however, the spectra of the O–K edge are

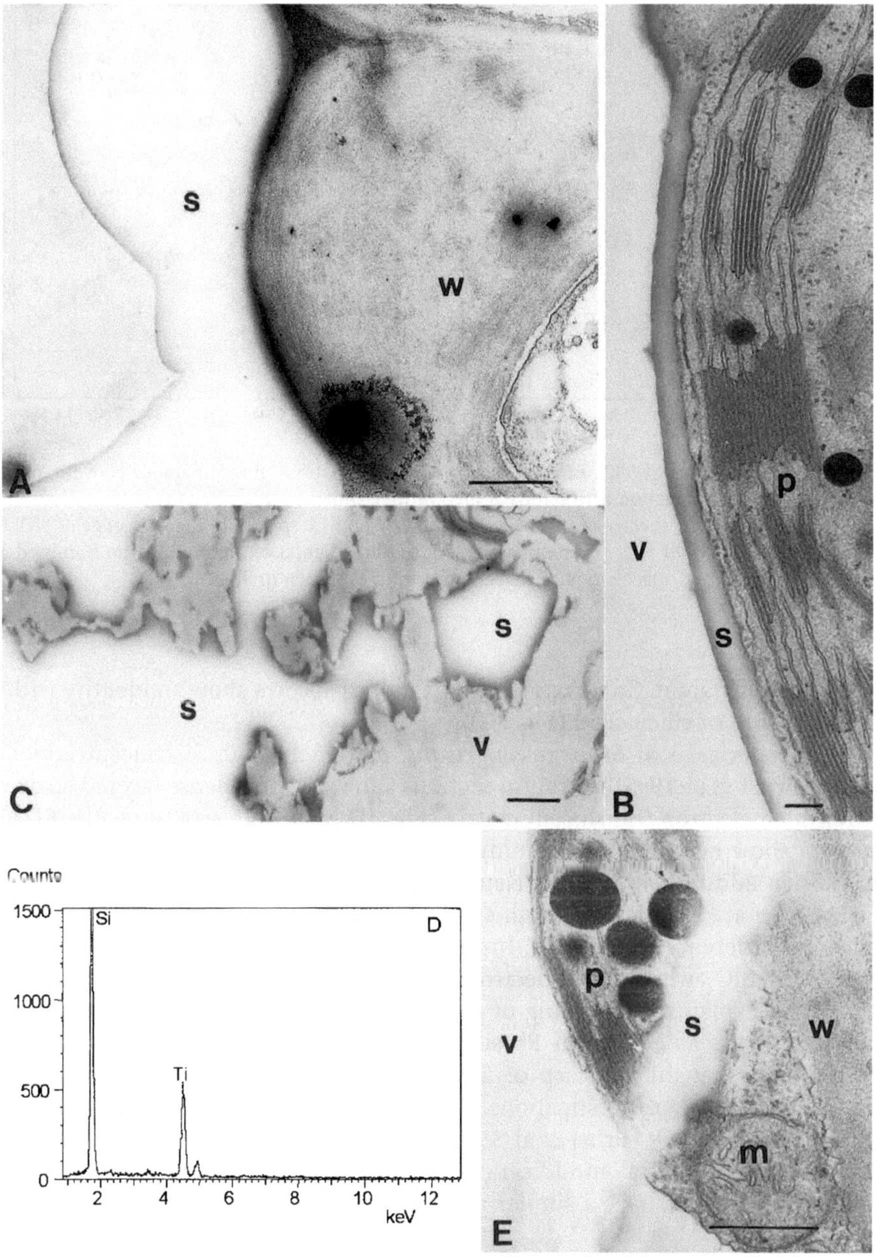


**Fig. 1A–E.** $SiO_2$ deposits in monocots. **A** Thick layer of $SiO_2$ (*s*) on the outer epidermal cell wall (*w*) of *Festuca*, **B** $SiO_2$ deposit (*s*) inside the vacuole on the tonoplast membrane of *Deschampsia*, **C** crystal-like $SiO_2$ precipitates (*s*) in the vacuole (*v*) of *Festuca*, **D** The EDX spectra of all of these Si deposits show Si only, the Ti peak is generated by the grid, **E** $SiO_2$ precipitate in *Festuca* obviously penetrating the cytoplasm and the plastid (*p*). *m* Mitochondrium. *Bar* = 0.1 µm in **B**, 0.5 µm in **A**, **C**, **E**

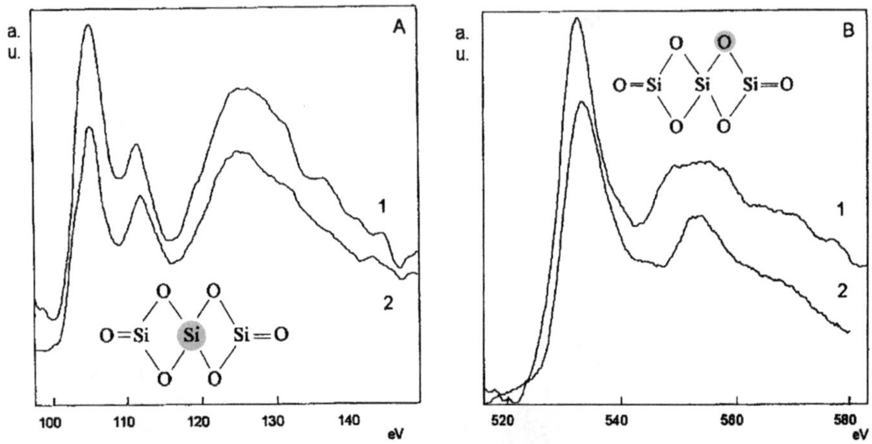


**Fig. 2A,B.** EEL spectra of $SiO_2$. EEL spectra contain information on the chemical coordination of an element. In comparison with standard spectra, they allow conclusion on the chemical nature of a compound. **A** EEL spectra of the Si-$L_{2,3}$ edge of $SiO_2$, **B** EEL spectra of the O–K edge of $SiO_2$, *1* standard spectrum of $SiO_2$ [crystalline $SiO_2$ with (100) orientation thinned by ion bombardment], *2* experimental spectra typical for $SiO_2$ precipitates shown in Fig. 1

completely different. Comparison with standard spectra show an identity with the spectrum of silicic acid (Fig. 3C).

In the pericarp of *Schoenus nigricans*, extremely high Si concentrations occur (Ernst et al. 1995). Ultrathin sections show electron-dense layered structures in an electron translucent matrix (Fig. 3D). For both structures the EDX spectra show only Si. After staining with Pb-citrate, only the dark structures exhibit an additional Pb peak (Neumann et al. 1999). The EEL spectra of the Si-$L_{2,3}$ edge are identical for both structures, whereas the spectra of the O–K edge are different (see Fig. 3C). In the electron-translucent parts, the spectra agree with $SiO_2$, while in the electron-dense part they are identical with silicic acid. This explains the binding of Pb in these structures, free OH groups of silicic acid are able to form Pb-silicate. These data are consistent with the assumption that the pericarp of *Schoenus* consists of silicic acid in a $SiO_2$ matrix. From earlier investigations, it was assumed that the pericarp does not only consists of $SiO_2$ (Ernst et al. 1995). Such layered structures of silicic acid are unusual. Chemical modifications of silicic acid (lepidoic silicic acids), synthesized earlier, have a similar structure (Kautsky and Irnich 1958). They

**Fig. 3A–D.** Silicic acids in monocots. **A** Epidermal cell of *Deschampsia* filled with silicic acid deposits (*si*). **B** Silicic acid needle-like precipitates (*si*) in the intercellular space of *Deschampsia* (*i*). **C** EEL spectra of the O–K edge of silicic acid (*1*) and of the electron-dense structures (*si*) shown in **A, B, D** (*2*). **D** Lepidoic silicic acid (*si*) in a $SiO_2$ (*s*) matrix of the pericarp of *Schoenus*. *Bar* = 1 μm in **A**, 0.1 μm in **B, D**

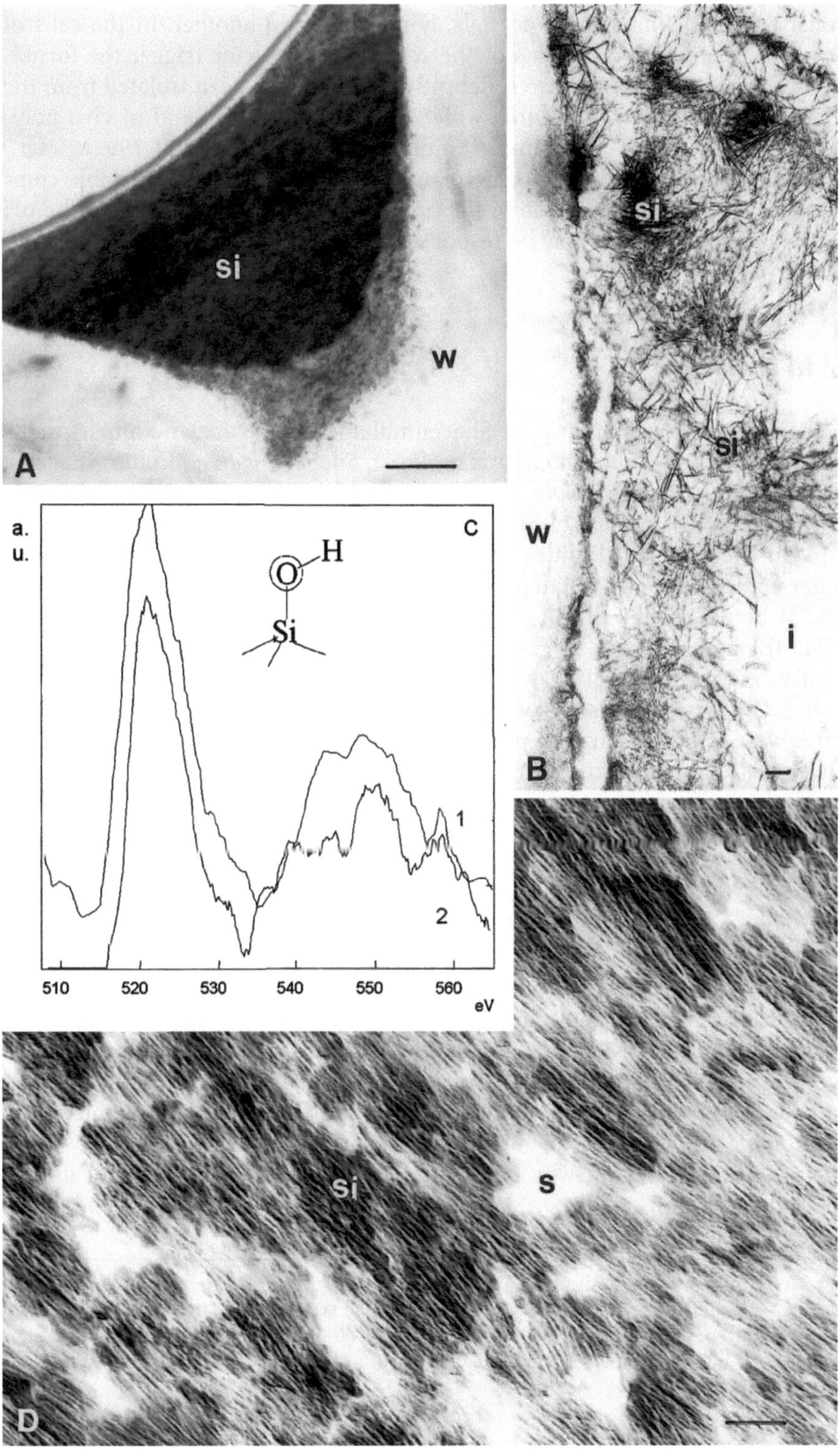
A
si
w
B
si
si
w
i
C
a.
u.
H
O
Si
1
2
510
520
530
540
550
560
eV
D
si
s

consist of two-dimensional crystals, lying one upon another. In the cells of *Schoenus*, specific components of the cell wall or proteins trigger the formation of such unusual structures. Recently, a protein has been isolated from the marine sponge, *Tethya aurantia*, which directs the in vitro and in vivo polymerization of silicon compounds (Shimizu et al. 1998; Cha et al. 1999).

It is not clear so far, why some plants are able to store the unstable compound silicic acid. In addition to the proteins or cell wall compounds, the pH or the ion environment may be responsible for the stabilization of silicic acid.

## 3 Si in Dicots

Although dicots are rather pure Si accumulators, in particular compartments of *Nicotiana tabacum*, *Arabidopsis thaliana*, *Silene vulgaris*, *Minuartia verna*, *Cardaminopsis halleri*, *Viola calaminaria*, *Thlaspi caerulescens*,and *Armeria maritima*, high amounts of Si were found (Neumann et al. 1997a; Neumann and zur Nieden 2001). Besides the accumulation in cell walls – especially in the outer epidermal walls – Si deposits were found in the intercellular space (Fig. 4A, D), between cell wall and plasma membrane (Fig. 4B), inside the cytoplasm (Fig. 4E) and vacuoles (Fig. 4C). In all cases, the Si deposits are electron-translucent and their Si-$L_{2,3}$ and O–K edges of the EEL spectra are identical with $SiO_2$ (for comparison, see Fig. 2A, B). As in monocots, crystal-like structures were found, apparently penetrating the plasma membrane or organellar envelopes. These electron-translucent structures in dicots are also $SiO_2$ gels with a high water content.

The vacuolar Si accumulation seems to be restricted to particular cells of the leaf parenchyma. This is possibly due to an enhanced uptake of Si into these cells by a quite unusual mechanism. In Si-accumulating cells, invaginations were found which are formed by two membranes, the plasmalemma and the tonoplast, including a small border of cytoplasm (Fig. 5A). In the same cells, vacuolar vesicles can be observed, obviously derived from these invaginations (Fig. 5B). Both structures contain Si originating from the extracellular space (Neumann and zur Nieden 2001). In this way, Si is taken up into the vacuole without a membrane or cytoplasm passage. This novel mechanism allows the uptake of condensed Si compounds with a higher molecular weight, which cannot be translocated via membrane-integrated transporters. Up to now, there is no explanation why only some cells show such an unusual uptake mechanism for Si.

**Fig. 4A–E.** $SiO_2$ deposits in dicots. **A** Crystal-like $SiO_2$ deposit (*s*) in the intercellular space (*i*) of *Minuartia*. **B** $SiO_2$ precipitate (*s*) between the cell wall (*w*) and the cytoplasm (*c*) of *Minuartia*. **C** $SiO_2$ layer (*s*) in the vacuole (*v*) of *Thlaspi*. **D** Crystal-like $SiO_2$ deposit (*s*) in the intercellular space (*i*) of *Silene*. **E** $SiO_2$ (*s*) in the cytoplasm (*c*) of *Cardaminopsis*. *Bar* = 1 μm in **A, C**, 0.5 μm in **B, D, E**

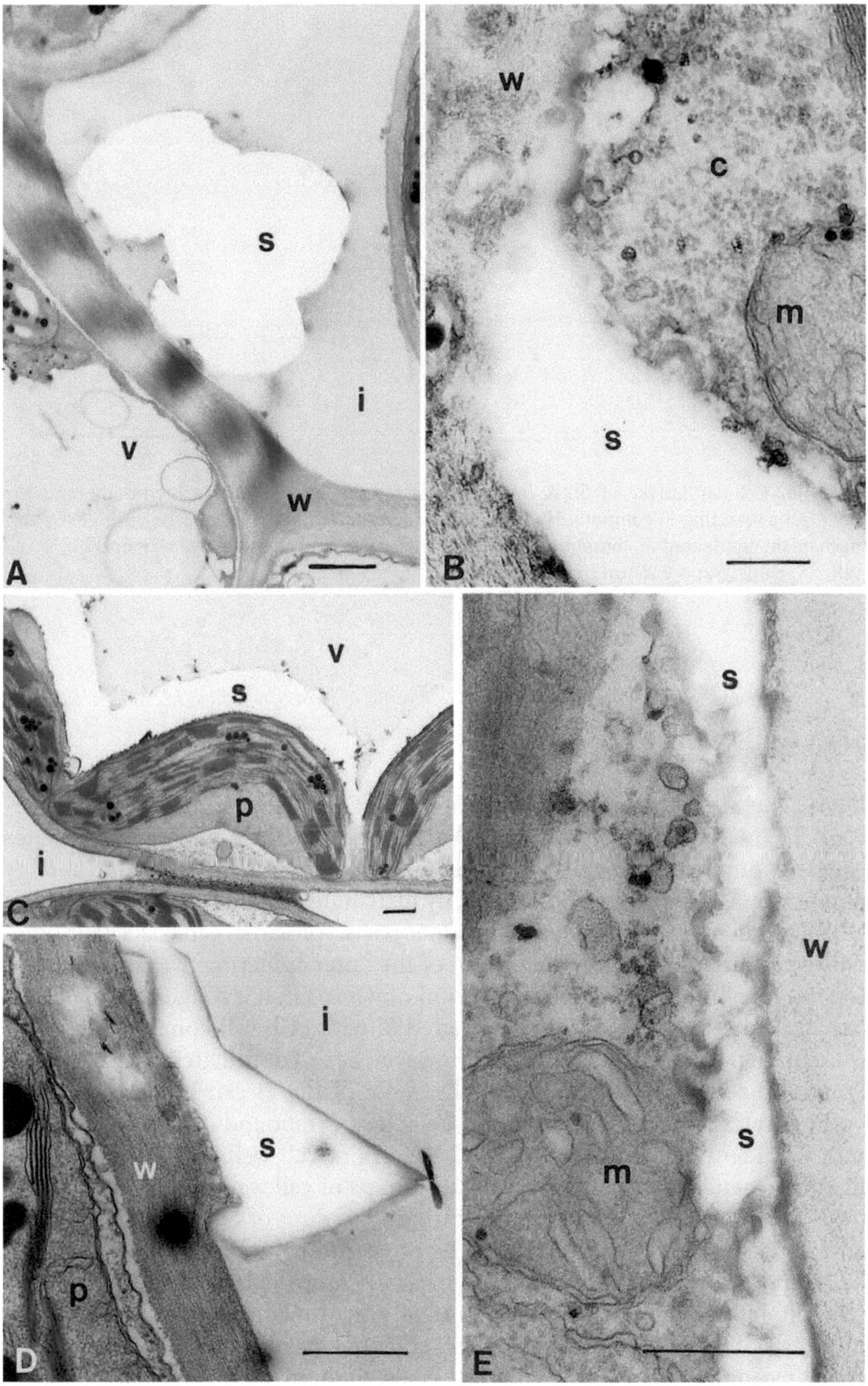
w
s
i
v
w
A
w
c
m
s
B
v
s
p
i
C
s
s
w
s
m
E
i
s
w
p
D

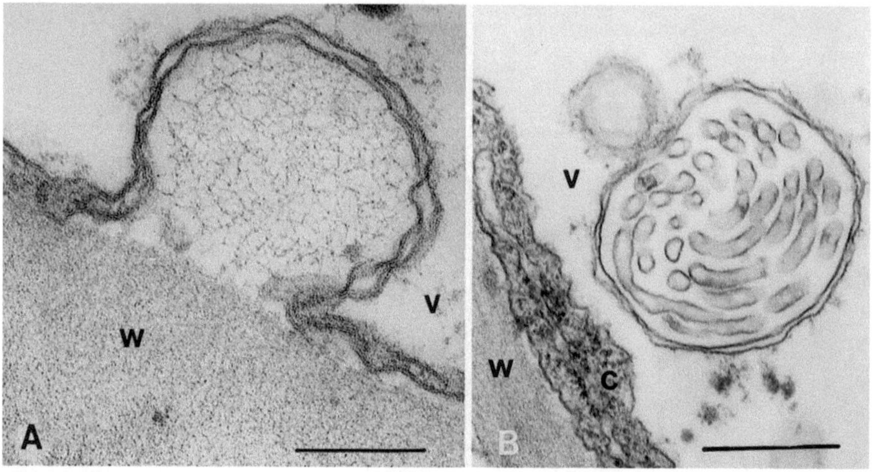


**Fig. 5A,B.** Cellular uptake of Si. **A** Invaginations formed by the plasma membrane and the tonoplast containing Si compounds of the extracellular space. **B** Vesicles in the vacuole. By this mechanism condensed Si compounds can be translocated into the vacuole. *c* Cytoplasm, *w* cell wall, *v* vacuole. *Bar* = 0.25 μm in **A**, 0.5 μm in **B**

## 4 Si in Cell Walls

It is a general observation that most of the Si is accumulated in cell walls, especially in the outer epidermal walls. A deposition in form of $SiO_2$ precipitates is often found in cell walls of the epidermis and the palisade and spongy parenchyma. Most of the Si is uniformly distributed in the epidermal cell walls with a gradient from the outer to the inner parts. Anticlinal cell walls have often only one tenth of the Si concentration of the outer epidermal walls. The chemical nature of the Si in the cell walls is unknown so far. It is suggested that ester-like derivatives of hexaedric-coordinated Si with cell wall components occur (for summary, see Marschner 1995), however, hexa-coordinated Si compounds are very unstable. From rye cell walls, most of the Si can be removed with boiling water and in the extract a Si-galactose compound of unknown structure was identified (Engel 1953). EEL spectra of cell walls show that most of the peaks generated by OH and HOOC groups of cell wall components disappear after incrustation with Si, indicating a binding of Si to such groups. In animal extracellular polysaccharides and peptidoglucans, Si was found to be bound to uronic acid residues (Schwarz 1973), which is an indication of a possible interaction with plant pectins or glucuronoarabinoxylans. However, up to now it is not clear which cell wall components are involved and in which coordination Si is present in the cell walls. Only in the epidermal cell walls of the heavy metal-tolerant *Minuartia verna*, is Si found to be accumulated as

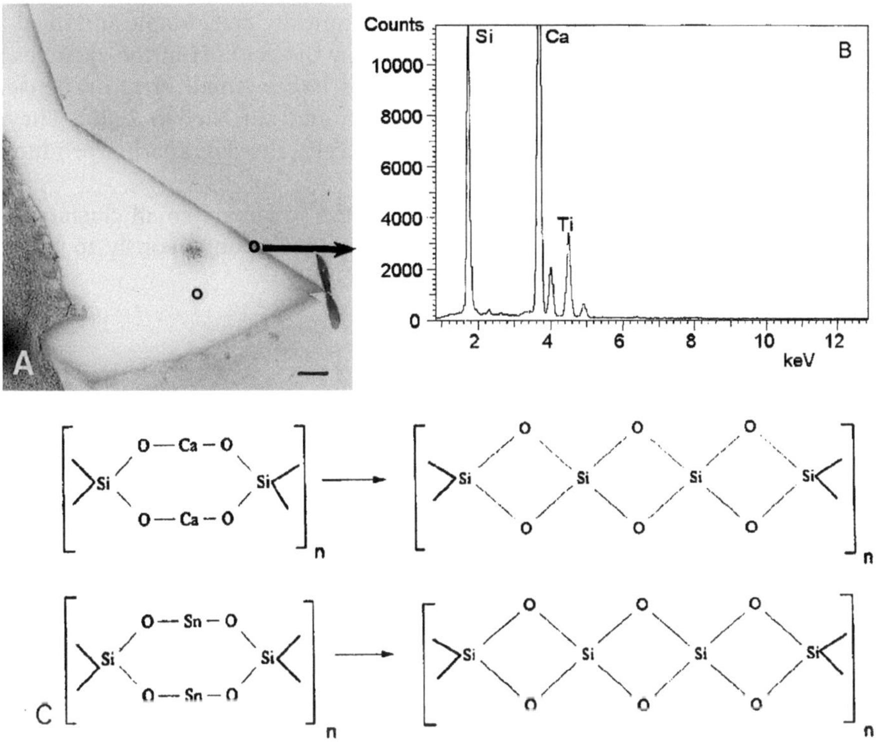


**Fig. 6A–C.** Formation of $SiO_2$ in plants. **A** Crystal-like $SiO_2$ deposit in the intercellular space of *Silene*; **B** The EDX spectrum of the electron dense border region shows Si and Ca, while the electron translucent central part (*small open dot*) contains only Si (for the spectrum, see Fig. 1D). **C** Formation of the insoluble $SiO_2$ deposits from small clusters of soluble Ca- or Sn-silicate. The figure shows the principle only, because the water content of the $SiO_2$ and the Ca and Sn contents of the silicates are unknown. *Bar* = 0.1 μm

Zn-silicate (Neumann et al. 1997b) and in the pericarp of *Schoenus nigricans* a special form of silicic acid in a $SiO_2$ matrix occurs (Neumann et al. 1999).

## 5
## Formation of $SiO_2$ Deposits in Plants

In many cases, the insoluble $SiO_2$ precipitates contain information on their formation. The border regions of the crystal-like structures have an enhanced contrast in the electron microscope (Fig. 6A) and the corresponding EDX spectra of this region show, in addition to Si, Ca (Fig. 6B) in *Minuartia*, *Nicotiana* and *Pisum* or Sn in *Viola*, *Thlaspi*, *Silene*, *Deschampsia* and *Festuca*. The EEL spectra of the border regions correspond to the spectra of Ca- and Sn-silicate. These silicates are unstable compounds (Schwieger et al. 1987) which are spontaneously converted to the thermodynamically more stable and insoluble $SiO_2$ (Fig. 6C). This process is accelerated by the acid pH in the vacuoles.

In the Zn-hyperaccumulator *Cardaminopsis halleri* small electron-dense precipitates occur in the cytoplasm (Neumann and zur Nieden 2001). They were identified as Zn-silicate and, in the same cells, the degradation product $SiO_2$ was found in the cytoplasm.

The transport form of Si in plants investigated so far, are small clusters of soluble Ca-, Sn- or Zn-silicates, which are degraded spontaneously to $SiO_2$, forming insoluble water containing gels (Fig. 6C).

## 6
## Uptake and Long-Distance Transport

The concentration of soluble Si compounds in the soil available for uptake by the root system depends on the pH in the rhizosphere. Acidification increases the concentration of soluble Si compounds which may be one reason for the high Si content in hyperaccumulators. For the radial transport of Si in the cortex, an apoplastic route is assumed. As for other compounds and ions, it is assumed that this free space pathway is blocked by the suberin incrustation in the endodermal cell walls. Indications are a strong endodermal silicification in *Sorghum* and no correlation between Si uptake and transpiration rate in rice and other hyperaccumulators (Sangster and Perry 1976; Hodson and Sangster 1998). On the other hand, in wheat, a close correlation exists between transpiration rate and Si uptake, speaking for an apoplastic transport into the xylem. The presence of passage cells in the endodermis and the disruption of the endodermal layer during branch root formation could be responsible for this apoplastic transport. It seems that both transport routes, the apoplastic and the symplastic, contribute to the Si distribution in plants.

The distribution of Si in the shoot takes place apoplastically via the xylem vessels.

On xylem cell walls Si deposits can frequently be found in hyperaccumulators. The driving force for Si transport in the vessels is the transpiration stream, it accumulates often at its terminal sites, e.g., the outer epidermal cell walls and trichomes.

In all plant species investigated so far, Si was also found in the cytoplasm and in vacuoles, presuming a transport through the plasma membrane and the tonoplast. However, specific Si transporters were found in diatoms only (Hildebrand et al. 1993, 1997, 1998). In higher plants comparable proteins were not detected up till now.

## References

Bode E, Kozik S, Lehmann H (1994) Vergleichende Untersuchungen zur Lokalisation von Silizium in Blättern zweier verschiedener Gräserarten. Dtsch Tierarztl Wochenschr 101:367–372

Cha J, Shimizu K, Zhou Y, Christiansen SC, Chmelka B, Stucky GD, Morse DE (1999) Silicatein filaments and subunits from the marine sponge direct the polymerization of silica and silicones in vitro. Proc Natl Acad Sci USA 96:316–365

Deren CW, Datnoff LE, Snyder GH (1992) Variable silicon content of rice cultivars grown on Everglades Histosols. J Plant Nutr 15:2363–2368

Engel W (1953) Untersuchungen über Kieselsäureverbindungen im Roggenhalm. Planta 41: 358–390

Epstein E (1994) The anomaly of silicon in plant biology. Proc Natl Acad Sci USA 91:11–17

Epstein E (1999) Silicon. Annu Rev Plant Mol Biol 50:641–664

Ernst WHO, Vis RD, Piccoli F (1995) Silicon in developing nuts of the sedge *Schoenus nigricans*. J Plant Physiol 146:481–488

Grob A (1896) Beiträge zur Anatomie der Epidermis der Gramineenblätter. Bibliotheka Bot 7: 1–122

Hildebrand M, Higgins DR, Busser K, Volcani BE (1993) Silicon responsive cDNA clones isolated from the marine diatom *Cylindrotheca fusiformis*. Gene 132:213–218

Hildebrand M, Volcani BE, Gassmann W, Schroeder JI (1997) A gene family of silicon transporters. Nature 385:688–689

Hildebrand M, Dahlin K, Volcani BE (1998) Characterization of a silicon transporter gene family in *Cylindrotheca fusiformis*: sequences, expression analysis, and identification of homologs in other diatoms. Mol Gen Genet 260:480–486

Hodson MJ, Evans DE (1995) Aluminium/silicon interactions in higher plants. J Exp Bot 46: 161–171

Hodson MJ, Sangster AG (1984) Observations on the distribution of mineral elements in the leaf of wheat (*Triticum aestivum* L.) with particular reference to silicon. Ann Bot 62:463–471

Hodson MJ, Sangster AG (1989) X-ray microanalysis of the seminal root of *Sorghum bicolor* with particular reference to silicon. Ann Bot 64:659–667

Hodson MJ, Sangster AG (1998) Mineral deposition in the needles of white spruce [*Picea glauca* (Moench.) Voss]. Ann Bot 82:375–385

Jones LHP, Milne AA, Sanders JV (1965) Tabashir: an opal of plant origin. Science 151:464–466

Kaufman PB, Petering LB, Smith JG (1970) Ultrastructural development of cork-silica cell pairs in *Avena sativa* internodal epidermis. Bot Gaz 131:173–185

Kaufman PB, Petering LB, Soni SL (1971) Ultrastructural studies on cellular differentiation in the internodal epidermis of *Avena sativa*. Phytomorphology 20:281–309

Kautsky H, Irnich R (1958) Das Netz der Kieselsäure $H_2Si_2O_5$ als Basis lepidoider Silikate. Z Anorg Allg Chem 295:193–205

Lichtenberger O, Neumann D (1995) A study of Si-L and O-K ELNES in plant material: $SiO_2$, Ca- and Zn-silicate in *Minuartia*. J Microsc 183:45–52

Lichtenberger O, Neumann D (1997) Analytical electron microscopy as a powerful tool in plant cell biology: examples using electron energy loss spectroscopy and X-ray microanalysis. Eur J Cell Biol 73:378–386

Marschner H (1995) Mineral nutrition of higher plants. Academic Press, London

Neumann D, zur Nieden U (2001) Silicon and heavy metal tolerance of higher plants. Phytochemistry 56:685–692

Neumann D, Lichtenberger O, Schwieger W, zur Nieden U (1997a) Silicon storage in selected dicotyledons. Bot Acta 110:282–290

Neumann D, zur Nieden U, Schwieger W, Leopold I, Lichtenberger O (1997b) Heavy metal tolerance of *Minuartia verna*. J Plant Physiol 151:101–108

Neumann D, Schwieger W, Lichtenberger O (1999) Accumulation of silicon in the monocotyledons *Deschampsia caespitosa, Festuca lemanii and Schoenus nigricans*. Plant Biol 1:290–298

Parry DW, Smithson F (1958) Techniques for studying silica in grass leaves. Ann Bot 22:543–549

Parry DW, Smithson F (1964) Types of opaline silica depositions in the leaves of British grasses. Ann Bot 28:169–185

Parry DW, Hodson MJ, Sangster AG (1984) Some recent advances in studies of silicon in higher plants. Philos Trans R Soc Lond Ser B 304:537–549

Sangster AG (1968) Studies of opaline silica deposits in the leaf of *Sieglingia decumbens* L. "Bernh.", using the scanning electron microscope. Ann Bot 32:237–240
Sangster AG (1970) Intracellular silica deposition in immature leaves in three species of the *Gramineae.* Ann Bot 34:245–270
Sangster AG, Parry DW (1976) The ultrastructure and electron-probe microassay of silicon deposits in the endodermis of the seminal roots of *Sorghum bicolor* (L.) Moench. Ann Bot 40:447–459
Schwarz K (1973) A bound form of silicon in glycosaminoglycans and polyuronides. Proc Natl Acad Sci USA 70:1608–1612
Schwieger W, Heyer W, Wolf F, Bergk K-H (1987) Zur Synthese von kristallinen Metallsilikathydraten mit Schichtstruktur. Z Anorg Chem 548:204–216
Shimizu K, Cha J, Stucky GD, Morse DE (1998) Silicatein α: Cathepsin L-like protein in sponge biosilica. Proc Natl Acad Sci USA 95:6234–6238

# Organisms: Sponges

# Silica Deposition in Demosponges

Maria J. Uriz,[1] Xavier Turon,[2] and Mikel A. Becerro[1]

## 1 Introduction

Demosponges are the most widespread class of the phylum Porifera. They secrete siliceous (spicules) and proteinaceous (fibers) elements, which constitute the skeleton that shapes the sponge growth and allow cell organization and establishment of the aquiferous system through which the surrounding water accesses the internal cells.

The siliceous spicules of demosponges display very different shapes and sizes, ranging from micrometers to millimeters. They are widespread in the sponge mesohyl or arranged in bundles, tracts or networks tending to form three-dimensional structures as the sponge grows massive, erect or branching (Fig. 1). Spongin, a collagen-like protein, accompanies the siliceous skeletons to a variable extent and may even embody the spicules totally (Fig. 2).

The high diversity of spicule shapes was early shown by sponge paleontologists (Hinde and Holmes 1892) and taxonomists and, more recently, has been illustrated by scanning electron microscope images (e.g., Hartman 1981; Simpson 1984) and drawings (Wiedenmayer 1977; Boury-Esnault and Rützler 1997). In short, spicules are conventionally divided into microscleres and megascleres (Fig. 3) according to their shape, size or skeletal function (but see Uriz et al. 2003). Demosponge megascleres show one (monaxons) or four (tetraxons) symmetry axes (Fig. 4). Yet instances of incipient polyaxonid symmetry have been reported in some apparently monaxonid spicules (Rützler and Smith 1993; Uriz and Maldonado 1995), which supports a relationship between the subclasses Tetractinomorpha (with tetraxon spicules) and Ceratinomorpha (with monaxon spicules; Lévi 1973) within demosponges. Microscleres can also range from monaxons to polyaxons (star-like forms; Fig. 5). Hypersilified representatives of both spicule categories, which are named desmas, are present in several genera belonging to different orders of demosponges. Moreover, some genera and families of desma-bearing sponges are conventionally

[1] Centre d'Estudis Avançats de Blanes (CEAB, CSIC), Accés a la Cala Sant Francesc 14, 17300 Blanes (Girona), Spain
[2] Department of Animal Biology (Invertebrates), Faculty of Biology, University of Barcelona, 645 Diagonal Ave, 08028 Barcelona, Spain

Progress in Molecular and Subcellular Biology, Vol. 33
W. E. G. Müller (Ed.)

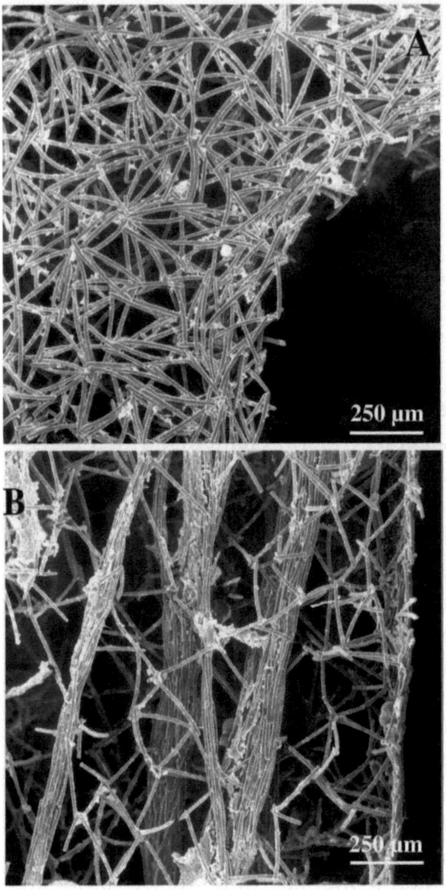


**Fig. 1A,B.** *Haliclona submonilifera* (Uriz 1988). **A** Spicule network at the sponge surface. **B** Combination of spicule tracks and uni-spiculated networks at the sponge choanosome. (Modified from Uriz 1988)

included in the "lithistids", a group with no phylogenetic meaning (Pisera and Lévi 2002).

Megascleres and microscleres appear to play several functions in sponges. Megascleres have a clear supportive function (Uriz et al. 2003), whereas the role of microscleres is difficult to ascertain (Fig. 6). Only the anisocheloid microscleres of the carnivorous sponge *Abestopluma hypogea* (Vacelet and Boury-Esnault 1996), have been shown experimentally to play a role in capturing micropreys (Vacelet and Boury-Esnault 1995).

The polymerization of the silicic acid around the axial filament to build a sponge spicule was reviewed ca. 20 years ago (Garrone et al. 1981; Simpson 1984). However, our knowledge of this process has currently improved. Demosponge spicules usually have a central core, the axial canal (but see Rützler and Macyntyre 1978), which in intact forms is occupied by an organic filament (the

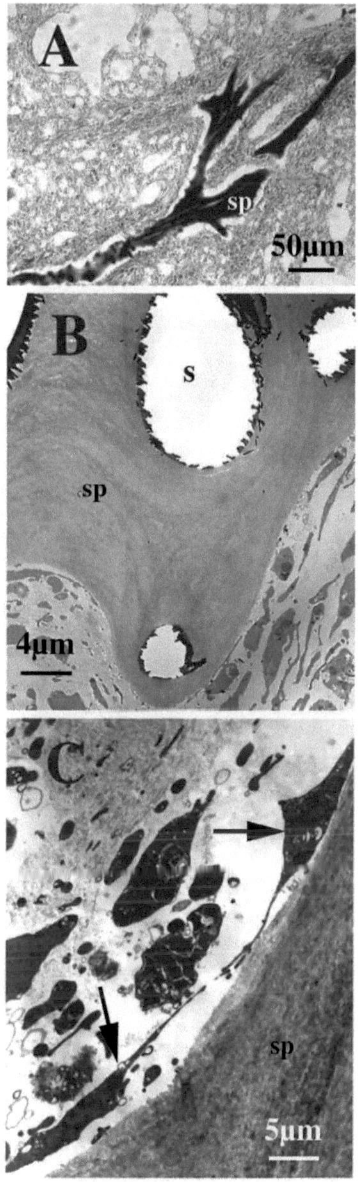


**Fig. 2A–C.** Instances of spongin surrounding spicules in *Crambe crambe*. **A** Light microscope image of a plumose spicule tract completely surrounded by spongin (sp). **B** TEM image of a transversal section of a spicule bundle. See the empty rounded spaces, which correspond to the spicules (*s*) and the fibrillar aspect of the surrounding spongin (*sp*). **C** TEM image of the spongocytes (*arrows*) that secrete the perispicular spongine (*s*). (**A**, **B** modified after Uriz et al. 2000)

axial filament; Fig. 7). A clear relationship between branching axial filaments and polyaxonid spicules has been shown in TEM studies of three-dimensional carbon replicas (Drum 1968), in SEM observation of corrosion casts of mature spicules (Rützler and Muzik 1993; Rützler and Smith 1993) and in SEM images

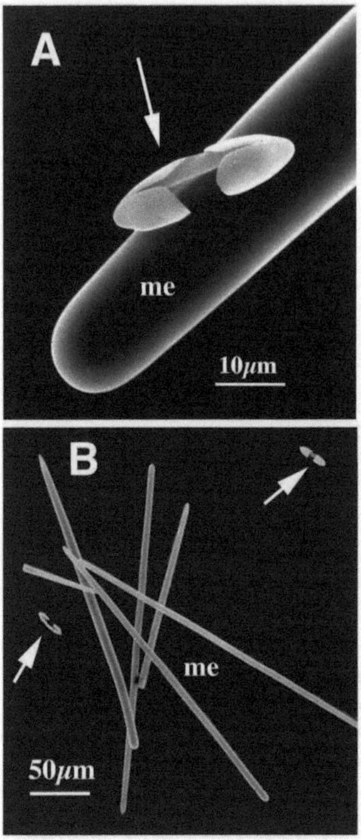


**Fig. 3A,B.** Scanning electron micrographs showing the comparative size of typical megascleres and microscleres from the same species (*Esperiopsis lesliei*, Uriz 1988). *me* Megasclere (style), *arrows* indicate microsclera (*palmate isochelae*). (Modified after Uriz 1988)

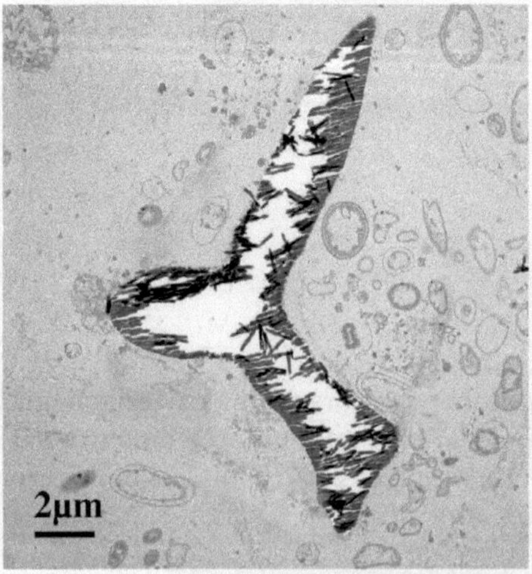


**Fig. 4.** Transmission electron micrograph of three actines of a tetraxon spicule of *Corticium candelabrum*

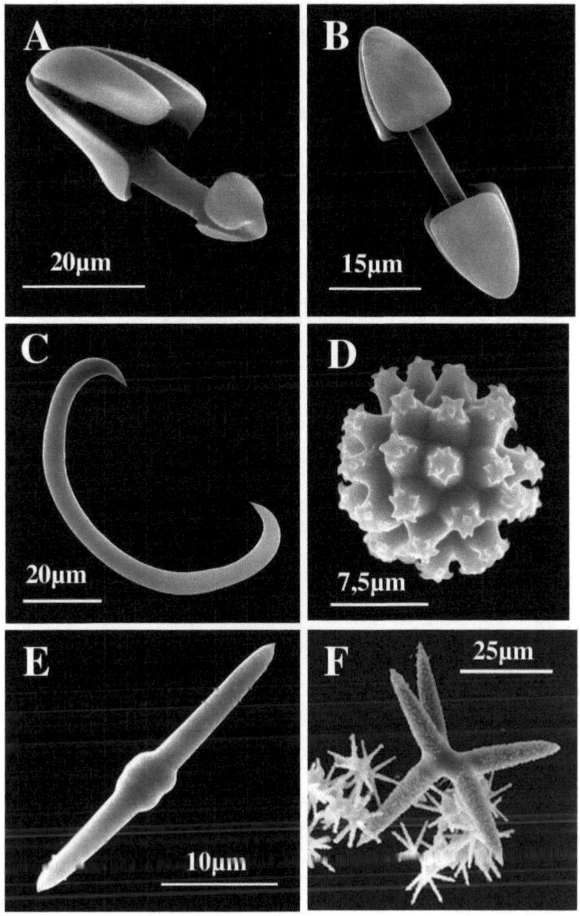


**Fig. 5A–F.** Various shapes of microsclere. **A–C, E** Monaxons: **A** anisochela, **B** palmate isochela, **C** sigma, **E** microxea. **D, F** Polyaxons: **D** spheraster, **F** metaster and spirasters. (**A–C** Modified after Uriz 1988; **D, F** modified after Boury-Esnault et al. 1994)

of young spicules (Uriz and Maldonado 1995). Although a relevant role of the axial filament in determining the spicule symmetry was early proposed (e.g., Reiswig 1971), and its organic (proteinaceous) nature was reported long ago (e.g., Schwab and Shore 1971a; Shore 1972), only the molecular studies performed at the end of the 1990s (Shimizu et al. 1998; Cha et al. 1999; Kraso et al. 2000), revealed the true nature and function of this filament in governing silica deposition. Notwithstanding, other related topics such as the role of the membranes, which surround the growing spicules, in determining the final spicule shape or how organic molecules, which do not constitute a discrete filament induce silica polymerization (Pisera 2003), remain poorly understood.

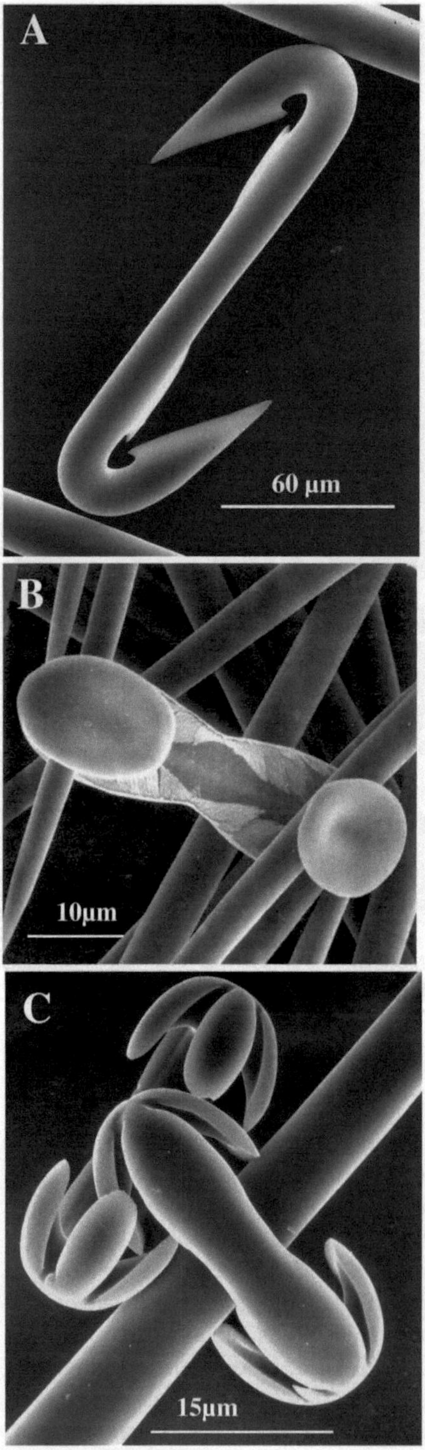


**Fig. 6A–C.** Instances of microscleres, which might play a role in framing the skeleton by joining monactine megascleres. **A** Diancister of *Hamacantha esperioides* (Ridley and Dendy) between two monactine megascleres (modified after Uriz 1988). **B** Plachochela of *Guitarra* sp. **C** Isochelae of *Crambe acuata* (Lévi). (Modified after Uriz and Rützler 1993)

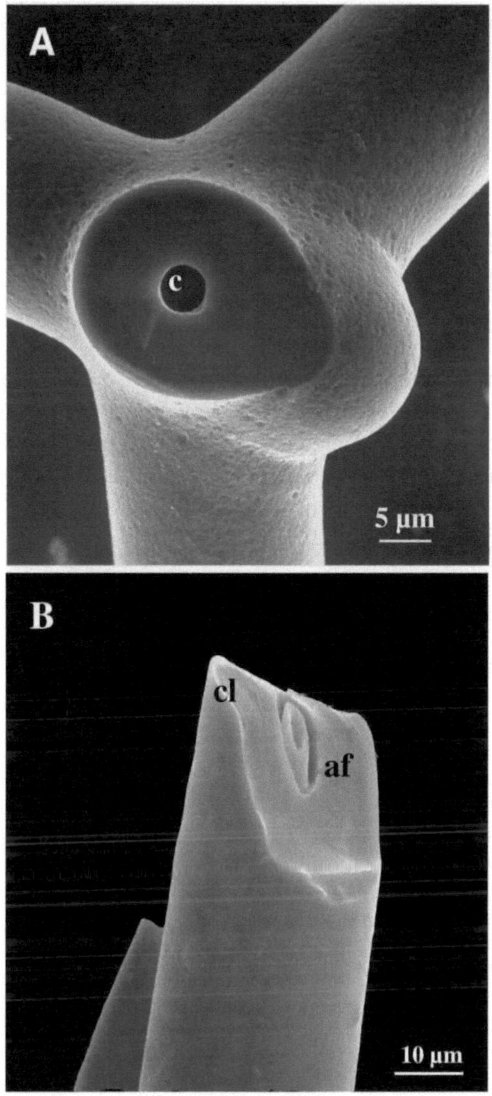


**Fig. 7.** **A** Scanning electron micrograph of a tetraxon spicule showing the axial canal (*c*). **B** SEM image of an axial filament in a monaxon spicule

Spicule formation is a complex process controlled by numerous variables. Chemical, physiological, biological, and environmental issues may all modulate at one or several stages spicule formation. We review the current knowledge on silification and spicule formation in demosponges, and we specifically examine the cells involved and the role of the axial filament, membranes, and environmental factors in modulating silica deposition. Although recent advances in molecular and genetic mechanisms have significantly contributed to better understand the process of spicule formation, many other questions outlined several decades ago are still unresolved. Spiculogenesis in demosponges faces both new and traditional challenges that reveal a promising future.

## 2 The Cells Involved

Sclerocytes, the cells that secrete spicules, are conventionally divided into megasclerocytes and microsclerocytes, depending on whether they secrete megascleres or microscleres. Megasclerocytes are large ameboid cells with a nucleolate nucleus, easily distinguishable from archeocytes in both adults and larvae (Fig. 8) thanks to their abundant small clear vesicles, microtubules,

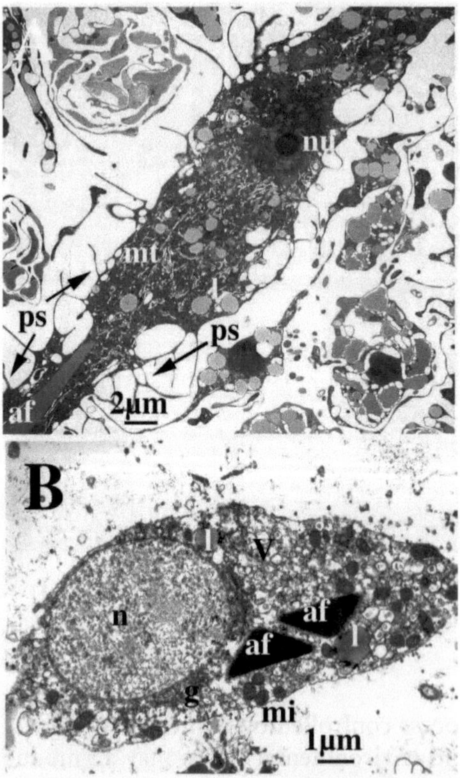


**Fig. 8A,B.** Larval sclerocytes containing axial filaments. A Megasclerocyte from *Scopalina lophyropoda* larva. B Megasclerocyte of *Crambe crambe* larva (modified after Uriz et al. 2000). *af* Axial filament, *g* Golgi apparatus, *l* lipidic inclusions, *mi* mitochondria, *n* nucleus, *nu* nucleolus, *ps* pseudopodia, *v* vesicles mitochondria, and conspicuous Golgi complex. Ultrastructural differences between megasclerocytes and microsclerocytes, as outlined in several papers, rely mainly on cell and nuclei sizes, which are notably smaller in microsclerocytes, and on the absence of a clear nucleolus in the latter (Wilkinson and Garrone 1980).

Yourassowsky and Rasmont (1983) demonstrated that at Si concentrations as low as 15 μg/l, sclerocytes of the freshwater sponge *Ephydatia fluviatilis* differentiate and axial filaments are secreted, although spicules are not produced. However, these authors were not able to assess whether traces of silica are necessary for sclerocyte differentiation since total elimination of Si from the experimental cultures was not possible due to methodological constraints. Later, Kraso et al. (2000) found that the presence of silicate in the water contributes to regulate the expression of the silicatein genes and thus, appears to be crucial for the formation of the axial filament and the spicule.

Maldonado et al. (1999) showed that different spicule types are produced, depending on the experimental silicon concentration in the water, and suggested that each spicule type may be secreted by a specific sclerocyte type, which would be activated by a particular $Si(OH)_4$ threshold. According to the experimental results of Maldonado et al., sclerocytes involved in desma production would require higher Si concentrations to secrete spicules than isochelae-producing microsclerocytes, and these in turn would require higher concentrations than the megasclerocytes that secrete the styles. There are several field studies reporting variation in the spicule complement (i.e., types of spicules present) in individuals of the same species from different habitats (e.g., Carballo and Uriz 1998; Cristobo 1998), but the instances are notably scarcer for megascleres than for microscleres. However, although some degree of environmentally determined spicule variation is now widely recognized, new species continue to be described on the basis of the presence or absence of a particular microsclere (e.g., *Hymedesmia* spp.).

Microsclerocytes responded differently than megasclerocytes to changes in silicic acid concentration in experimental cultures of *Spongilla lacustris* (Jorgensen 1944). Microscleres of *Spongilla lacustris* also showed different features than megascleres when germanium was added to the culture medium (Simpson 1990). Simpson's results showed that germanium induced the formation of two terminal bulbs in the birotules of *E. fluviatilis* while it produced a central swelling in oxeas. Simpson interpreted these bulbs as centers for $Si(OH)_4$ processing, which would be placed at the extremities of the microscleres and at the center of the megascleres of *E. fluviatilis*, respectively. All the above mentioned investigations indicate that higher concentrations of Si are necessary for microsclerocytes than for megasclerocytes to secrete their respective spicules, which in the species *C. crambe* has been related to the extracellular (megascleres) or intracellular (microscleres) spicule secretion (Uriz et al. 2000).

The question remains as to whether clusters of sclerocytes develop simultaneously in a particular zone of the sponge and thus spicules are formed in their definite place or spicules are secreted anywhere in the sponge mesohyl and then transported to the appropriate site to form the skeletal framework (see Uriz et al. 2003). The second option appears to be more reliable for megasclerocytes according to the video images that recorded sclerocytes actively transporting megascleres (Bond and Harris 1988; Bond 1992; TokyoCinema 1996). However, the process may be different in the case of microsclerocytes. Instances of microscleres forming particular structures (e.g., rosette formations of *Mycale anisochelae*) suggest that several microsclerocytes should be in close contact to secrete these microscleres. Microsclerocytes secreting sterrasters have been found in their definite position within the sponge cortex of *Stelletta* (Simpson et al. 1985). Moreover, numerous side-by-side placed sclerocytes elaborating a bundle of raphids have been reported by Wilkinson and Garrone (1980) in an ultrastructural study of the poecilosclerid *Neofibularia irata.*

The formation of the desmoid (hypersilified) spicules appears as a special case (Fig. 9). These spicules are reported to have an axial filament (Fig. 9A) which does not reach the final part of the spicule arms (Pisera and Lévi 2002). The terminal arms of these spicules interlock each other without fusing (Fig. 10) in such a way that a simultaneous formation of several spicules with secreting sclerocytes in close contact must be envisaged. Silification of the terminal arms of desmas, and spines or other ornamentation of nondesmoid spicules is likely modulated by the sclerocyte membrane (see below).

## 3 The Axial Filament

As stated above, the axial filament has been assigned an important role in determining spicule morphology in sponges (e.g., Reiswig 1971; Simpson et al. 1985). Although several amino acids were identified from the axial filaments of the species *Acarnus erithacus* (Shore 1972), the molecular structure and functioning of the axial filament remained elusive until recently (Shimizu et al. 1998; Cha et al. 1999; Kraso et al. 2000). Cha et al. (1999) demonstrated that the axial filament of *Tethya aurantium* guides Si polymerization along its length. This finding represents the first experimental demonstration of a role in determining the spicule shape for the axial filament.

Filaments triangular in cross section have been reported in tetraxon megascleres of astrophorids and spirophorids (Bütschli 1901; Reiswig 1971; Simpson et al. 1985) and hadromerids (Bütschli 1901), all them belonging to the subclass Tetractinomorpha. More recently, Uriz et al. (2000) showed that the axial filaments of the styles of the poecilosclerid *Crambe crambe* are always triangular before silica deposition starts and in young spicules (Fig. 11A), but that the filament is retracted in mature spicules and the angles of the triangle may become cut off in parallel with a decrease in Si content, given rise to hexago-

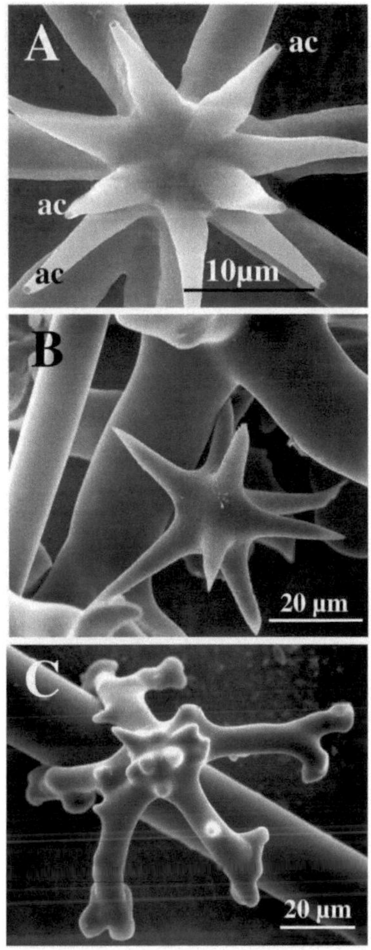


**Fig. 9A–C.** Scanning electron micrographs of different steps in the formation of a desma. **A** A very young desma showing the actines, which have a terminal orifice corresponding to the end of an axial canal (modified after Uriz and Maldonado 1995). **B** Desma of *C. acuata* in an intermediate stage of silification. **C** A completely formed desma of *C. tuberosa*. (Modified after Uriz and Maldonado 1995)

nal shapes. Filament retraction may account for the "false" hexagonal shapes occasionally reported (e.g., Simpson et al. 1985). Yet, true hexagonal shapes in cross section are reported for filaments of haplosclerid megascleres (Garrone 1969), which are thinner than those triangular shapes and were not resolvable through light microscope (Reiswig 1971; Wilkinson and Garrone 1980). Amazingly, only in hexagonal filaments a paracrystalline structure has been conspicuously observed (Garrone 1969; Wilkinson and Garrone 1980; Donadey et al. 1990).

A particular case has been observed in a representative of the subclass Homosclerophora with tetractinal spicules of a size intermediate between megascleres and microscleres. Cross sections of spicules of *Corticium candelabrum* show an irregular, narrow axial filament (Fig. 11B–D) which appear to

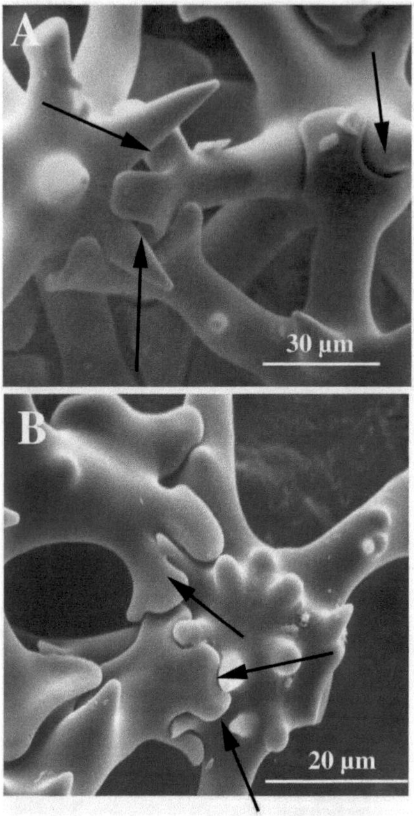


**Fig. 10A,B.** Scanning electron micrographs of interlocked desmas: see the complementary shape of their terminal arms called zygomes (*arrows*). **A** *C. acuata* (modified after Uriz 1988). **B** *C. tuberosa*. (Modified after Uriz and Maldonado 1995)

be embedded within the silica of the spicule walls instead of being discrete (Fig. 9D).

Although the tetraxon spicule has traditionally been considered ancestral to those diactinal and monactinal (e.g., Dendy 1921; Reiswig 1971), the fossil record shows evidence that the monaxonid type appeared first and may represent the origin for other demosponge spicule shapes (Finks 1970). However, a polyaxonid origin has been demonstrated for some apparently monaxonid spicules (Fig. 12) such as the tylostyles of *Terpios* spp. (Rützler and Muzic 1993; Rützler and Smith 1993) and the tylostyles-subtylostyles of the poecilosclerids *Discorhabdella* and *Crambe* (Uriz and Maldonado 1995). In the latter genera, mature tylostyles are tuberose (Fig. 13A), but young forms have a clear appearance of acanthostyles with conical spines at the proximal end. However, at higher magnifications the spines showed an axial canal, which proves the existence of an axial filament and that the "spines" were in fact short spicule rays, and thus the spicules were polyaxones instead of true monaxones (Fig. 13B). Consequently, polyaxonid spicules are present in representatives of the subclasses Tetractinomorpha and Ceractinomorpha (Lévi 1973) and the

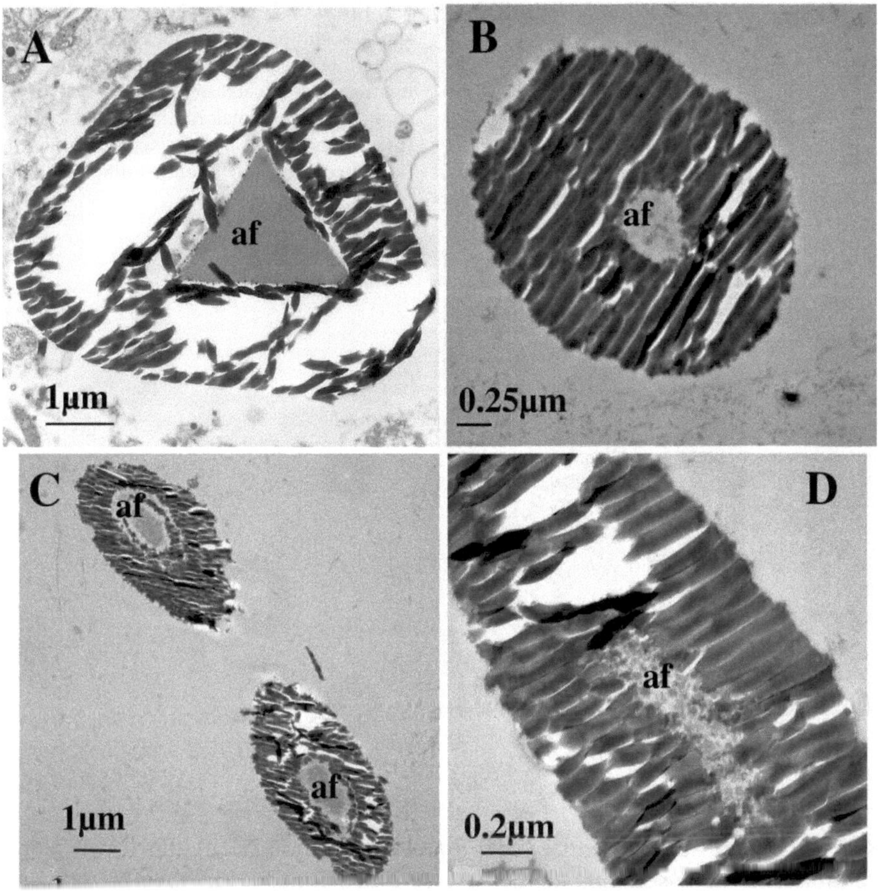


**Fig. 11A–D.** Transmission electron micrographs of axial filaments in demosponges. **A** Cross section of the axial filaments (*af*) of a *C. crambe* spicule, clearly triangular (modified after Uriz et al. 2000). **B, C** Transversal sections of spicules of *Corticium candelabrum* showing an irregular axial filament (*af*). **D** Longitudinal section of a spicule of *C. candelabrum* showing a narrow, very irregular filament

polyaxial symmetry might be the plesiomorphic condition within the Demospongiae

In contrast to monactinal forms (styles to tylostyles), all the diactinal spicules (oxeas and strongyles) so far examined have axial filaments unequivocally hexagonal in cross sections (Garrone 1969; Simpson and Vaccaro 1974; Wilkinson and Garrone 1980; Weissenfels and Langenbruch 1985). This fact has been reported for both marine (*Haliclona* spp.) and freshwater (*Ephydatia* spp.) haplosclerids, the poecilosclerid *Neofibularia irata* (Wilkinson and Garrone 1980) and the rhaphids of the axinellid *Axinella polypoides* (Donadey et al. 1990). Simpson et al. (1985) assumed that both oxeas and triaenes of

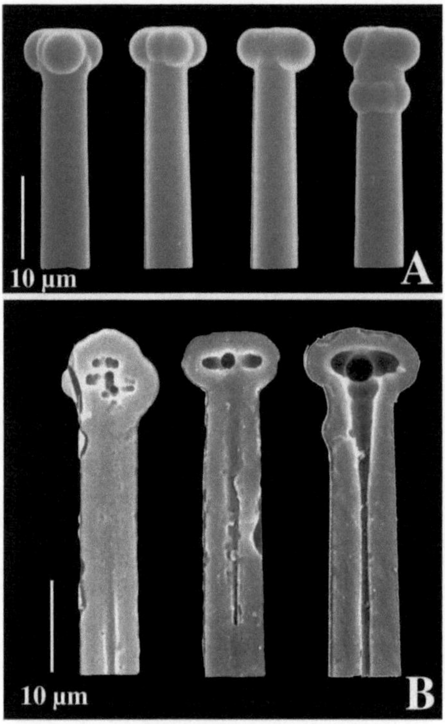


**Fig. 12A,B.** Polyaxial tylostyles of *Terpios* spp. **A** Scanning electron micrographs of the polytylote basal zone. **B** Scanning electron micrographs of corrosion casts of tylostyles showing the internal ramified axial canals (previously filled by the axial filament). (Modified after Rützler and Smith 1993)

*Stelletta grubii* had triangular axial filaments because triangular forms were abundant in TEM sections of the sponge internal zone, but they recognized that their observations do not allow them to ensure whether these filaments belonged to one or both spicule types. The shape of axial filaments in halichondrids (sponges also bearing diactines) has not yet been examined through TEM, but their shape, although unresolved through light microscopy, was suspected not to be triangular in cross section (Reiswig 1971). On the other hand, homosclerophorid spicules appear to have an irregular, poorly defined axial filament.

The two cross-sectional shapes of the axial filaments, triangular and hexagonal, appear to correlate with the two different models of spicule secretion postulated by Simpson (1990) for monactines and diactines. Oxeas of *E. fluviatilis* cultured in a synthetic medium containing high concentrations of silicon (1.0 mM) produced central bulbous structures (Simpson 1981) similar to those produced in oxeas of *Spongilla lacustris*, cultured in the presence of germanium (Simpson 1990) and in natural populations of *Dendroxea* spp. (Fig. 14; e.g., Corriero et al. 1996). In contrast, monactines start silification at the proximal end (Simpson 1990), which in tylostyles and subtylostyles present a terminal or subterminal swelling, respectively. Pulsed exposures to germanium

**Fig. 13A,B.** Tylostyles of *Crambe tuberosa* (SEM). **A** Mature tylostyle with a tuberose tyle. **B** Young polyaxial tylostyle showing the origin of the tubercles, which consists in short actines containing an axial filament (*arrows* point to the orifices corresponding to the axial canals). (Modified after Uriz and Maldonado 1995)

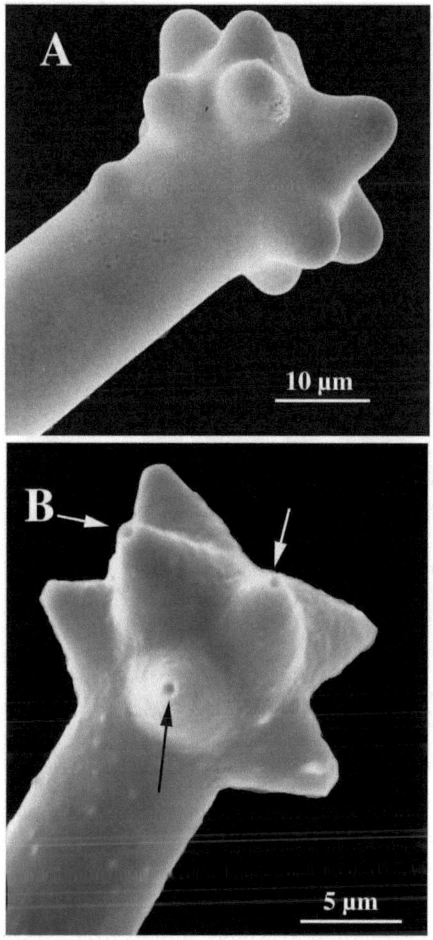


allowed Simpson (1990) to demonstrate that diactines of *E. muelleri*, *E. fluviatilis* and *Spongilla lacustris* grow bidirectionally from a central $Si(OH)_4$ processing region, whereas the growth of the monactines of *Suberites domuncula* (tylostyles) is unidirectional from the spicule head. This fact is supported by the many instances of very young tylostyles with a marked tyle and a thin shaft observed in this species (Uriz 1983). However, the existence of true silification centered along the axial filament has not been unambiguously demonstrated. Likewise, the formation of silica bulbs seems to be regulated by the unit-type membrane that surrounds the growing spicule (Simpson 1981).

A particularly appealing research topic is the formation of the articulate framework of desmas, and more concretely the final part of the desmoid arms, which are devoid of axial filament (Pisera and Lévi 2002). It has been demonstrated that silica polymerization in sponges requires the presence of enzymes

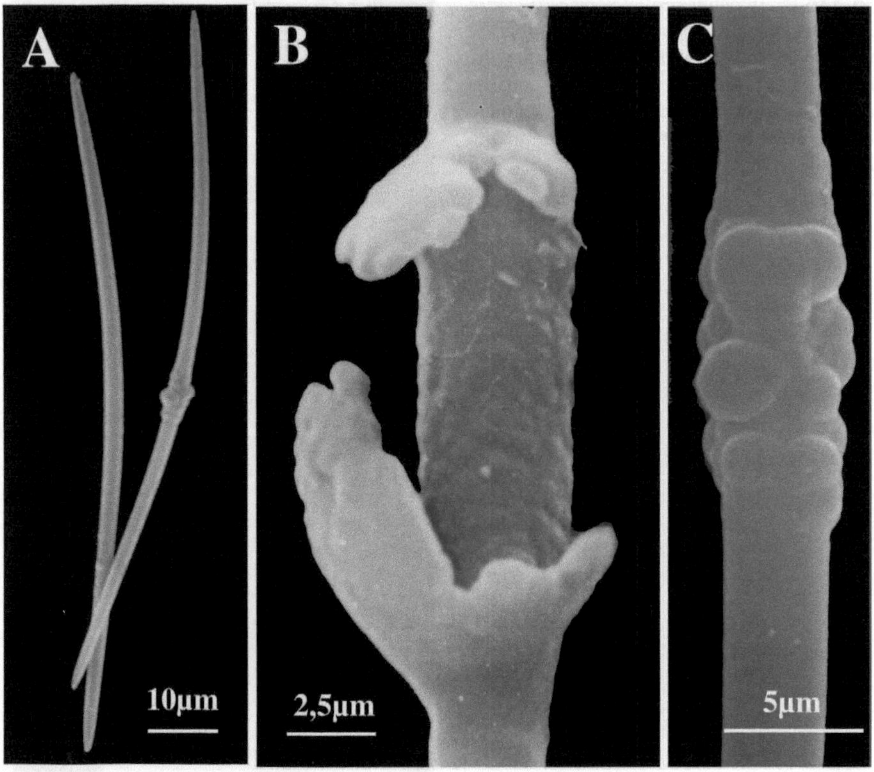


**Fig. 14A–C.** Central swelling in oxeas of *Dendroxea* spp. (SEM). **A, C** *D. adumbrata* (Corriero et al. 1996). **B** *D. pseudodiscoides.* (Modified after Corriero et al. 1996)

(Shimizu et al. 1998), which first catalyze the hydrolysis of Si compounds and then their polymerization. Organic molecules (proteins), which do not form a concrete macrofilament, may be responsible for silification in the terminal arms of desmas, according to the outstanding observations by Pisera (2003).

## 4 Extracellular Versus Intracellular Silica Deposition: the Role of Membranes

A never-ending discussion on whether the process of silica polymerization around the axial filament is intracellular or extracellular can be found in most papers dealing with spicule secretion of demosponges (see Simpson 1984 for a review). This discussion was mainly due to the attempt of extrapolating particular observations to the whole demosponge spicules and species. Most microscleres observed are secreted intracellularly and only in one case (toxas of *Microciona*) the secretion appeared to be extracellular at a light microscope level (Simpson 1978), although this instance would need verification through electron microscopy. Similarly, most previous studies showed an intracellular secretion of megascleres (Garrone 1969; Simpson and Vaccaro 1974; Garrone et al. 1981; Hartman 1981). An extracellular secretion of megascleres has been documented only once (Uriz et al. 2000), although it was previously envisaged due to the physical impossibility of a sclerocyte to completely surround a spicule several milimeters long, such as those of many astrophorid sponges. Extracellular filaments in early stages of silification surrounded by the sclerocyte pseudopodia (Fig. 15) have been reported in the demosponge *Crambe crambe* (Uriz et al. 2000), which indicates that filaments are secreted before silification starts and shows that megasclere silification proceeds extracellularly in this species. The extracellular silica secretion questions the hypothesis that filaments continue to grow even after the onset of silification (Simpson 1990) and supports Reiswig's (1971) statement that the axial filament is formed in full length and breadth before silica deposition started.

With the information at hand, it can be concluded that most microscleres are secreted intracellularly in a space surrounded by a unit-type membrane (called silicalemma) and that inter- and intra-ellular secretion processes occur in different types of megascleres and/or species. Consequently, both membrane types (the internal silicalemma and the external plasmalemma) appear to play similar roles in silica deposition and may even have the same origin since fusion between them appears likely (Uriz et al. 2000). However, freeze-fracture images (Garrone and Lethias 1990) of the internal surfaces of those membranes look very different, the silicalemma being covered by a higher density of "intra-membrane particles", although the different magnification at which the pictures of the two membranes were taken makes comparisons imprecise.

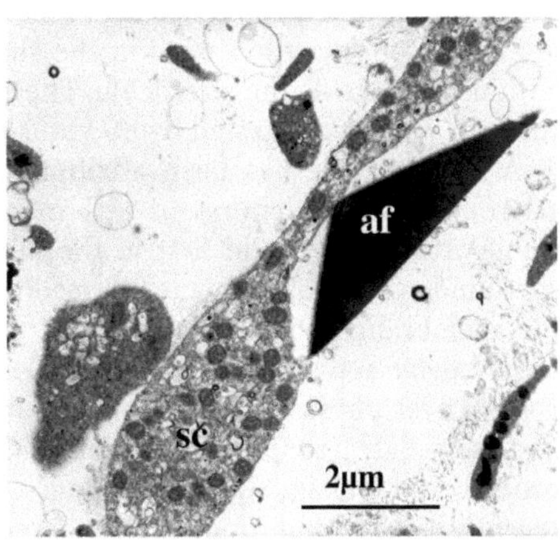


**Fig. 15.** Sclerocyte (*sc*) of *Crambe crambe*, which surrounds an axial filament (*af*) that has been released to the mesohyl (TEM)

It can be highlighted that all the instances of an intracellular secretion of megascleres reported up to now correspond to small diactines from species of the order Haplosclerida. It has been assumed that the diactines of the poecilosclerid *Neofibularia* are intracellular, but this assumption remains to be demonstrated (Wilkinson and Garrone 1980). Moreover, the secretion of long megascleres of astrophorids or spirophorids has never been examined through TEM.

## 5 The Process of Silica Polymerization

Silica polymerization occurs and is crucial in many forms of living beings from diatoms to mammals (Volcani 1981). This process is accomplished in organisms at low (ambient) temperatures, whereas silica polymerization under non-living conditions requires extreme temperature, pressure and pH (Kröger et al. 1999).

There is a great deal of speculation as to the possible pathways followed by the silicic acid after it enters the sponge until it polymerizes around the axial filament to form the spicules. There is no current evidence of cellular transport of Si from the pseudoepithelial cells (pinacocytes and choanocytes) to mesohyl cells such as the sclerocytes (Uriz et al. 2000). As a consequence, the hypothesis that suggests transport of silicic acid by simple diffusion through the ground substance to the sclerocyte (Simpson 1981) appears to be more likely.

Recent research on silica content in the different compartments of a sclerocyte secreting a spicule and its surroundings (Uriz et al. 2000) showed that mitochondria, vesicles, dense inclusions, and the cytoplasm of the sclerocyte far from the growing spicule contained less than 10% Si (relative to that contained in the spicule wall). In contrast, cytoplasm close to the growing spicule contained up to 50%, and the extracellular space between the sclerocyte and the growing spicule contained 50–65% (Fig. 16). These results proved that there is a clear concentration of Si in the surroundings of the axial filament both within and outside the sclerocyte. These data do not support the Si transport through vesicles suggested by previous hypothesis (Schröeder 1936).

A relevant role has been attributed to the unit-type membranes of the sclerocyte (plasmalemma or cell membrane and silicalemma) in actively transporting silicic acid first to the sclerocyte, and then to the space which surrounds the axial filament (Simpson 1989). Silicic acid super-saturation in the membrane-bound compartment containing the axial filament may be achieved by active molecular transporting. Hence, the local regulation and control over physico-chemical factors may be done through selectivity in ion and molecular transport. However, little is known about how cell transporters proceed and manage silicon in sponges. However, Hildebrand et al. (1997) have recently isolated and functionally characterized cDNAs encoding proteins that

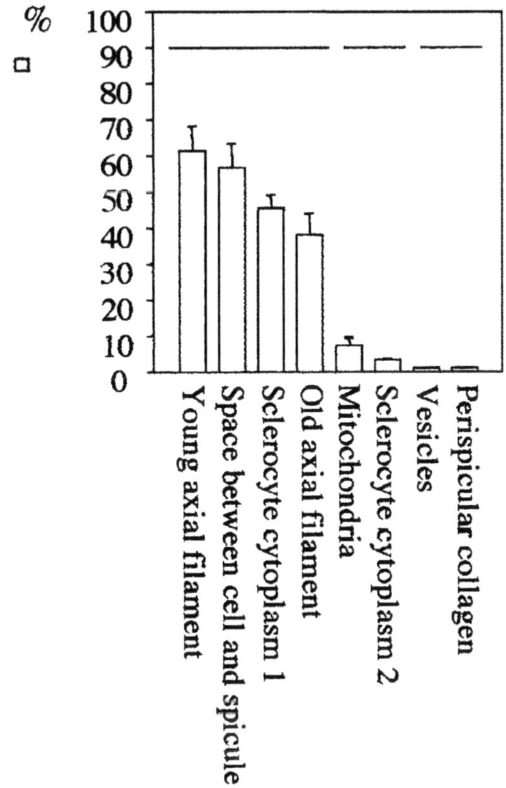


**Fig. 16.** Percentage of Si content in the different compartments of a sclerocyte and its surroundings. (Modified after Uriz et al. 2000)

interact specifically with silicon and transport it across biological membranes of the diatom *Cylindrotheca fusiformis*. The mechanism may be comparable in sponges, but silicon transporters still remain elusive in these animals.

Experiments with germanium addition to freshwater sponges have demonstrated that the membrane that surrounds the growing spicule plays a role in modulating the final spicule shape, which is initially determined by the axial filament. This would explain the cases of spines without any protein core (De Pomar 1973) or those of complicated microscleres (Simpson 1984) and, particularly the zygosis (interlocked arms) of the desmoid spicules (Fig. 10).

It has been shown that the membrane that embodies the axial filament in a sclerocyte of *Spongilla lacustris* cultured in a Ge-rich medium (Ge/Si concentration ratio of 14) forms a bulbous vacuole (Fig. 17) at the place where there will be a swelling in the mature spicule (Simpson 1981). As Simpson suggested, uncoupling of the axial filament and the silicalemma due to an excess of the silicalemma synthesis appears to account for the abnormal silica deposition and may also be responsible for the formation of spines and other spicule orna-

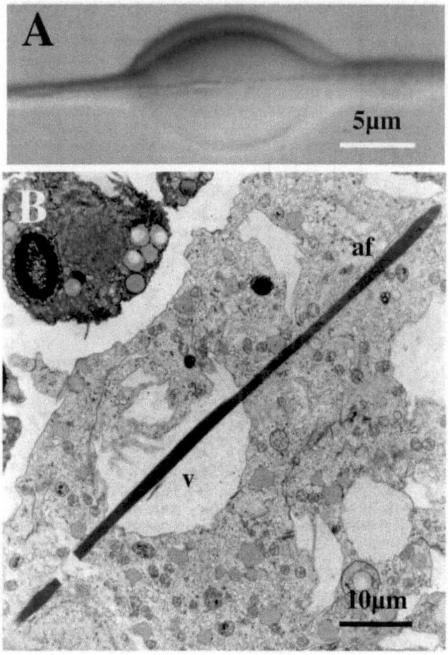


**Fig. 17A,B.** *Ephydatia fluviatilis.* **A** Central swelling of an oxea from specimens grown in a Ge-enriched medium. **B** Transmission electron micrograph of a sclerocyte showing a central vacuole (*v*) around the axial filament (*af*), which resulted in the central spicule swelling. (Modified after Simpson 1981)

mentation. An increase of the membrane surface may end in an accumulation of silica in the space delimited by the membrane.

Shimizu et al. (1998) demonstrated that the main proteins that constitute the axial filament of sponge spicules (called silicateins α, β, and γ) are enzymes highly similar to members of the papain family of proteases, and they can catalyze the hydrolysis and polycondensation of silicon alkoxides. Cha et al. (1999) showed that silicatein induced very fast silica polymerization in vitro as compared to silica polymerization without this protein. A dependence on the three-dimensional conformation of the silicatein units isolated from the axial filaments of *Tethya aurantium* has been assumed since the activity of the protein was abolished by thermal denaturation (Cha et al. 1999).

Si polymerization both in vitro (Cha et al. 1999, 2000; Kröger et al. 1999) and in vivo (e.g., Volcani 1981) produces rounded small particles, which tend to fuse to each other giving rise to a network of micro- or nanospheres. The size of spheres resulting from silicic acid polycondensation has been reported to range from 1 nm in laboratory experiments in presence of cellulose (Perry and Lu 1992) to 1200 nm in naturally occurring tylostyles of the sponge *Cliona schmidti*.

In diatoms, Kröger et al. (1999) have isolated several proteins (called silaffins), which generate networks of silica nanospheres within seconds at ambient temperature when added to a solution of silicic acid. The size of these nanospheres varied depending on the protein used: by using silaffin-1, the nanospheres reached 500–700 nm in diameter while they were <50 nm when a mixture of silaffins was utilized. Cha et al. (2000) have recently synthesized cysteine-lysine block polypeptides that mimic the properties of silicatein, by controlling at the same time hydrolysis and condensation as well as structural templating of tetraethoxysilane (TEOS). Polymerization by these polypeptides results in silica spheres ca. 100 μm in size.

Spherical silica units less than 20 nm in diameter that further coalesce can also be observed in TEM sections of higher plants (Sangster and Parry 1981) and silica particles of 40–50 nm have been shown in SEM images of diatoms (Volcani 1981).

Nanosphere formation has also been reported in early silification stages of sponge spicules (Fig. 18A; Schönberg 2001; Rosell and Uriz 2002) and the nanospheres increase their size and tend to fuse (Rosell and Uriz 2002; Pisera 2003) as late silification proceeds (Fig. 18B). In sponges inhabiting silicon-poor environments, instances of growing spicules showing an uneven surface (Fig. 19A) made of rounded protuberances as a result of incomplete silification have often been reported and are described as "malformations" (e.g., Uriz and Maldonado 1993; Fig. 19B). SEM pictures of spicule surfaces from sponges reveal nanospheres ranging from ca. 17 nm in *Spongilla lacustris* (Simpson 1981) to 100–1200 nm in growing tylostyles of several Clionidae species (Schönberg 2001; Rosell and Uriz 2002). Nanospheres from 100–120 nm in maximum diameter can be shown in TEM images of styles of *Crambe crambe* at early stages of silification (Fig. 20; Uriz et al. 2000).

Although completely different proteins appear to be involved in silification processes in diatoms (Kröger et al. 1999) and sponges (Shimizu et al. 1998; Cha et al. 1999), the resulting silica nanospheres appear to be quite similar. Consequently, nanosphere formation is the first observable feature of silica polymerization both in in vitro and in organisms; they increase in size and fuse with each other forming a network as the silification proceeds. Fusion of nanospheres by addition of silica results in an even spicule surface.

The spicules of sponges are formed by successive concentric silica layers as has been repeatedly shown in transverse sections of megascleres (e.g., Bütschli 1901; Garrone 1969; Uriz et al. 2000). Several mechanisms responsible for silica stratification have been proposed. They have been interpreted by Schwab and Shore (1971b) as the result of pauses in silification, which would allow the siliceous surface to become highly hydroxylated. It has been suggested that these pauses might be coincident with a replacement of the sclerocyte that deposits the silica (Uriz et al. 2000), although the involvement of several sclerocytes in secreting a single spicule, although conceivable in the secretion of large spicules, has not been verified.

The concentric silica layers are particularly conspicuous in spicules of hexactinellid sponges where, conversely to demosponge spicules, layers appear not to be completely adhered, according to their irregular fracture in cross section (e.g., Lévi 1989).

**Fig. 18A–C.** Different stages of silification of *Cliona schmidti* spicules. **A** Young tylostyle showing the surface of the tyle formed by silica nanospheres. **B** Tylostyle tyle in a subsequent stage of silification. **C** Irregularly silified amphiasters. (Modified after Rosell and Uriz 2002)

## 6 Environmental Factors Modulating Silica Deposition

There are several environmental factors which influence the final spicule size or even the expression or absence of one or several spicule types (particularly microscleres). Species growing in upwelling zones (e.g., Uriz 1988), in the Antarctic or at great depths are reported to have larger spicules than their conspecifics living in oligotrophic seas (e.g., Bibiloni 1990; Bavestrello et al. 1993). Spicules from eutrophic environments show a well-defined shape with conspicuous tyles, spines, points or any ornamentation typical of the spicule (Uriz

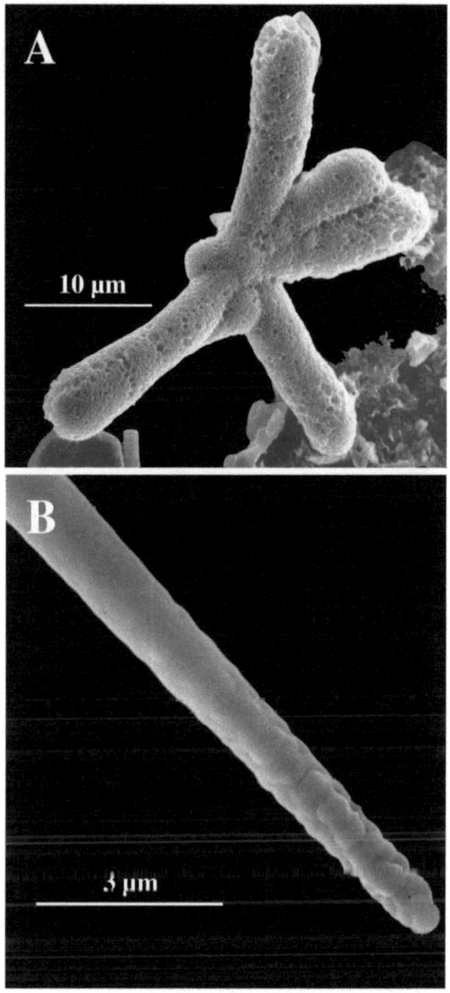


**Fig. 19.** A An incompletely silified desma of *Crambe crambe* showing the surface irregular as a result of incomplete nanosphere fusion (modified after Maldonado and Uriz 1996). B Distal zone of a monactine megasclere of *Clathria oxistyla* Ferrer-Hernàndez from a silica-poor environment. See the tuberose aspect of the spicule surface. (Modified after Uriz and Maldonado 1993)

1988). Among the environmental factors that are responsible for differences in spicule conformation, silicic acid concentration appears to be particularly important. However, low temperatures and high concentrations of particulate organic matter, which are usually associated with high amounts of silicates in upwelling areas, are also to be considered.

Manipulative experimentation in the laboratory (e.g., Pé 1973; Yourassowsky and Rasmont 1983; Maldonado et al. 1999; Le Pennec et al. 2002) has been decisive in understanding the role of silicates and other substances in the silica deposition process. Yourassowsky and Rasmont (1983) first demonstrated that silicic acid limitation drastically affects spicule secretion.

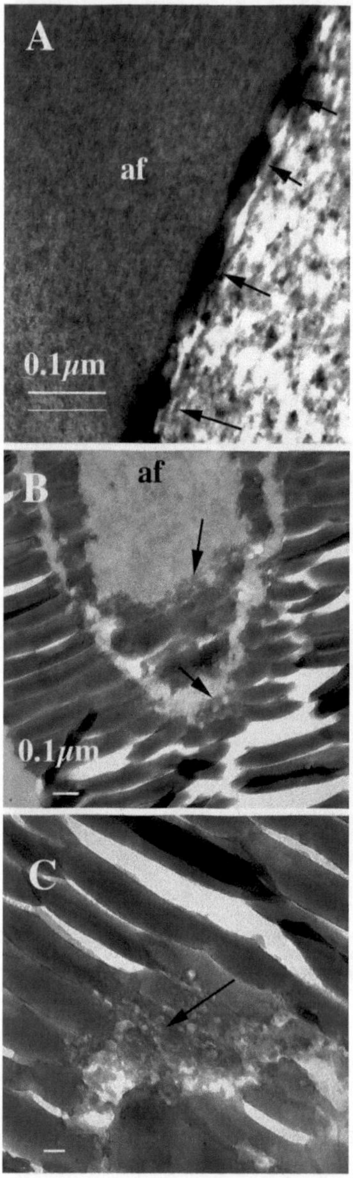


**Fig. 20A–C.** Nanospheres formation (*arrows*) in close contact to the axial filament (*af*) (TEM). **A** Early silification pattern in a spicule of *Crambe crambe* . **B, C** Nanospheres formation at the central zone of *Corticium candelabrum* spicules

When these authors cultivated freshwater sponges (*Ephydatia fluviatilis*) from gemmules in a mineral medium containing as little silicic acid as possible (i.e., less than 15 µg/l), no spicules at all were secreted, although sclerocytes differentiated and expressed the axial filament. By recording the growing sponges through time-lapse video under a light microscope, they reported that when silica was added to the medium, the sclerocytes that contained an axial filament produced normal spicules. In contrast, when the silica-poor medium was maintained, the sclerocytes degenerated and shed their filaments.

More recently, Maldonado et al. (1999) have performed the opposite experiment by growing larvae of a marine sponge in silica-enriched medium. Individuals of the sponge *Crambe crambe* inhabiting some regions of the western and central Mediterranean lack one or more of the spicule types (e.g., Boury-Esnault 1971; Bibiloni 1990; Uriz et al. 2000). In contrast, specimens from the Adriatic coasts produced the whole spicule complement of the species, including desmas and isochelae (Thiele 1899). Silica concentration in the water column was thought to be responsible for these differences in the sponge skeleton and this was experimentally tested. These authors examined the production of spicules by newly settled sponges maintained during 4.5 months under three concentrations of silicic acid (<4.5 μM, ca. 30 μM and ca. 100 μM). All the sponges cultured in the silica-poor medium (Si concentration similar to that of most Mediterranean waters) only produced one spicule type (small styles), while sponges cultured under ca. 30 μM produced small styles, large styles and isochelae. A further spicule category, the asterose desmas, appeared when the sponges were cultured under ca. 100 μM $Si(OH)_4$. These results show that *C. crambe* is genetically capable of producing several spicule types that are normally absent from natural populations inhabiting Si-poor environments.

Increasing silicic acid concentration resulted in longer and wider megascleres in cultured freshwater sponges (Pé 1973). Furthermore, high silicic acid concentrations produce bulbous expansions in both megascleres (Pé 1973) and microscleres (Jorgersen 1944) of freshwater sponges and other abnormalities such as tuberose ornamentation in megascleres (styles) and supplementary spines or teeth in isochelae of the marine sponge *Crambe crambe* (Maldonado et al. 1999). Other contrasting effects such as a decrease in the growth rate of the spicules length have also been described in *Ephydatia fluviatilis* cultured under 25 μM $Si(OH)_4$ (Elvin 1971). This may be explained by possible saturation of the sites of Si polymerization, which is in concordance with the decrease in Si uptake observed in the sponge *C. crambe* when $Si(OH)_4$ increased from 30 μM to 100 μM (Maldonado et al. 1999).

Other mineral elements besides silicon have also been reported to influence spicule secretion in demosponges. The addition of a $Fe^{2+}$ chelator (2,2′-Bipyridin) to the culture medium at concentrations lower than 4 μg/ml affected skeletogenesis of freshwater sponges, which slowed down silification, produced bulbous shorter spicules and reduced the total production of spicules by 50% compared to the controls (Holvoet and Van der Vyver 1990). These results indicate that $Fe^{2+}$ is necessary for silica polymerization and allowed these authors to suggest that enzymes with $Fe^{2+}$ cofactors appeared to be directly involved in the biosilification of the axial filament. More recently, it has been suggested (Le Pennec et al. 2002; Müller et al. 2003) that $Fe^{2+}$ in the water would be taken up by sponge cells and transferred after oxidation to $Fe^{3+}$ to the axial filament where they would possibly activate the enzymes that catalyze silica polymerization to form the spicules.

Studies dealing with the effects of physical parameters on silica deposition are scarce, particularly for marine sponges. However, they have shown that temperature also affects silica deposition even when it is maintained within the range found in natural sponge habitats. In *Haliclona oculata*, spicules are larger in specimens inhabiting colder regions (Hartman 1958). Simpson (1978) reported that *Microciona prolifera* produced wider megascleres at lower temperatures. However, experimental cultures of *Ephydatia muelleri* (Lieberkühn) from asexual gemmules under temperatures ranging from 10 and 25 °C and different silicic acid concentrations (Elvin 1971) showed that the rate of growth in spicule length, and the rate of silica deposition increased with increasing temperatures. Low temperatures may induce a decrease in the silification rate as a consequence of a slow-down of the sponge metabolism, but a slower spicule growth appears to result in the formation of larger spicules.

## 7 The Future

Although studies on demosponge spiculogenesis have been scarce in the last decade, the few available instances have greatly contributed to the progress of our knowledge on the whole process. Several unresolved questions were outlined in ca. 20-year-old revisions (Garrone et al. 1981; Simpson 1984). Some of those questions have been resolved, but others still remain unanswered.

The identification and characterization of protein subunits in the axial filament of *T. aurantium* (Shimizu et al. 1998; Cha et al. 1999) and the identification of the genes that encode these proteins (Kraso et al. 2000) have represented an extraordinary advance in our knowledge of the silification process. Knowledge of the role of chemical elements and physical factors in the silica deposition of demosponges has experienced important progress in the last decade. The effects of silicate concentration and other inorganic ions (i.e., Ge and $Fe^{2+}$) on gene and spicule expression has allowed the identification of a different physiological behavior in microsclerocytes and megasclerocytes. The extracellular secretion of a potentially polyaxonid megasclere has reinforced the hypothesis that large tetraxonid megascleres may be accomplished extracellularly and that both intracellular and extracellular megasclere secretion occurs in demosponges.

Other questions such as how silica polymerization proceeds in the spicules or part of spicules without any trace of axial filaments are partially resolved. Pisera (2003) has shown images that suggest the presence of organic molecules which would induce silica polymerization without forming discrete filaments. However, this issue remains to be verified by molecular techniques.

Several of the pending questions are to be fixed by ultrastructural studies. Yet, these studies face the technical difficulty of catching a dynamic complex process by means of microscale “snapshots”. What is the fate of sclerocytes once they have finished the spicule secretion? Is the same sclerocyte able to secrete several spicules or does it degenerate after the spicule secretion. How many sclerocytes participate in the complete spicule secretion when it is performed extracellularly?

The identification and quantification of the molecules that transport Si through the unit-type membranes, already identified for diatoms (Perry and Lu 1992), would help to confirm the role of membranes in defining the spicule shape, particularly in the zones devoid of axial filament.

There is still an important gap in the knowledge about how spicules are arranged to form the various skeletal frameworks that we can find in Demosponges. The video images recorded by Bond and Harris (1988), Bond (1992) and TokyoCinema (1996) gave precious information on continuous cell rearrangement in sponges that fixed SEM and TEM images cannot produce. Thanks to these videos, we know that sclerocytes are continuously moving while secreting spicules, following the general sponge cell flow, and that at least not all the cells that transport spicules are spongocytes as was previously assumed (Simpson 1984).

Morphogenesis, however, still remains enigmatic for spicule arrangement in the different zones of the sponge body. Several homeobox genes that confer positional information during development have been identified in demosponges (e.g., Coutinho et al. 1994; Seimyia et al. 1994; Degnan et al. 1995; Richelle-Maurer et al. 1998; Richelle-Maurer and Van der Vyver 1999). Possibly, they have a role in the differentiation and arrangement of sclerocytes and the various spicule types, but how they are expressed to form the sponge skeleton still requires a great deal of investigation.

The application of the panoply of new molecular and ultrastructural methods to old unresolved questions such as those outlined may produce a blossoming of new and interesting information about the complex process of silica deposition in sponges in the forthcoming years.

*Acknowledgements.* We acknowledge G. Agell and R. Coma from the CEAB for scanning the pictures. We particularly thank W. Müller for inviting us to participate in this volume. This research was funded by grants CICYT MAR98-1004-CO2 and REN2001-2312-CO3/MAR from the Spanish Government.

## References

Bavestrello G, Bonito M, Sarà M (1993) Influence of depth on the size of sponge spicules. In: Uriz MJ, Rützler K (eds) Recent advances in biology and systematics of sponges. Scientia Marina, Barcelona, pp 415–420

Bibiloni MA (1990) Fauna de Esponjas de las Islas Baleares. Variación cualitativa y cuantitativa de la Población de esponjas en un gradiente batimétrico. Comparación Baleares-Costa Catalana. PhD Thesis, Universidad de Barcelona

Bond C (1992) Continuous cell movements rearrange anatomical structures in intact sponges. J Exp Zool 263:284–302

Bond C, Harris AK (1988) Locomotion of sponges and its physical mechanism. J Exp Zool 246: 271–284

Boury-Esnault N (1971) Spongiaires de la zone rocheuse de Banyuls-sur-Mer. II. Systématique. Vie Milieu 22:287–350
Boury-Esnault N, Rützler K (eds) (1997) Thesaurus of sponge morphology. Smith Contrib Zool 596:1–55
Boury-Esnault N, Pansini M, Uriz MJ (1994) Spongiaires bathyaux de la mer d'Alboran et du golfe ibéro-marocain 160. Mém Mus Nat Hist Nat Zool:1–174
Bütschli O (1901) Einige Beobachtungen über Kiesel- und Kalknadeln von Spongien. Z Wiss Zool 69:235–286
Carballo JL, Uriz MJ (1998) *Guitarra flamenca* sp. nov. (Porifera: Poecilosclerida) with a SEM revision of the spiny isochelae and placochelae in the genus. J Mar Biol Assoc UK 78:807–819
Cha JN, Shimizu K, Zhou Y, Christiansen SC, Chmelka BF, Stucky GD, Morse DE (1999) Silicatein filaments and subunits from a marine sponge direct the polymerization of silica and silicones *in vitro*. Proc Natl Acad Sci USA 96:361–365
Cha JN, Stucky GD, Morse DE, Deming TJ (2000) Biomimetic synthesis of ordered silica structures mediated by block polypeptides. Nature 40:289–292
Corriero G, Scalera-Liaci L, Pronzatro R (1996) Two new species of *Dendroxea* Griessinger (Porifera: Demospongiae) from the Mediterranean Sea. Bull Inst Roy Sci Nat Belg Biol 66 [Suppl]:197–203
Coutinho CC, Vissers S, Van den Vyver G (1994) Evidence of homeobox genes in the fresh-water sponge *Ephydatia fluviatilis*. In: van Soest RWM, van Kemper TMG, Braekman J-C (eds) Sponges in time and space. Balkema, Rotterdam, pp 385–388
Cristobo FJ (1998) *Guitarra solorzanoi* (Porifera, Desmospongiae) a new species from the Galician Coast (northeast Atlantic). Ophelia 48:25–34
Degnan BM, Degnan SM, Giusti A, Morse DE (1995) A hox/hom homeobox gene in sponges. Gene 155:175–177
Dendy A (1921) The tetraxonid sponge-spicule: a study in evolution. Acta Zool Stockh 2:95–152
De Pomar B (1973) Acerca del canículo axial en espongolitos silíceos. Rev Assoc Cien Nat Lit 4: 167–176
Donadey C, Paris J, Vacelet J (1990) Occurrence and ultrastructure of microraphides in *Axinella polypoides*. In: Rützler K (ed) New perspectives in sponge biology. International Conference on the Biology of Sponges 1985. Smithsonian Institution Press, Washington, DC, pp 259–263
Drum RW (1968) Electron microscopy of siliceous spicules from the freshwater sponge *Heteromyenia*. J Ultrastruct Res 22:12–21
Elvin D (1971) Growth rates of the siliceous spicules of the fresh-water sponge *Ephydatia muelleri* (Lieberkühn). Trans Am Microsc Soc 90:219–224
Finks RM (1970) The evolution and ecological history of sponges during Paleozoic times. In: Fry WG (ed) The biology of Porifera. Symposia of the Zoological Society of London, 25, The Zoological Society of London, London, pp 3–22
Garrone R (1969) Collagène, spongine et squelette minéral chez l'éponge *Haliclona rosea* (Démosponge, Haploscléridа). J Microsc 8:581–598
Garrone R, Lethias C (1990) Freeze-fracture study of sponge cells. In: Rützler K (ed) New perspectives in sponge biology. International conference on the biology of sponges, 1985. Smithsonian Institution Press, Washington, DC, pp 121–128
Garrone R, Simpson TL, Pottu-Boumendil J (1981) Ultrastructure and deposition of silica in sponges. In: Simpson TL, Volcani BE (eds) Silicon and siliceous structures in biological systems. Springer, Berlin Heidelberg New York, pp 495–525
Hartman WD (1958) Natural history of the marine sponges of southern New England. Bull Peabody Mus Nat Hist 12:1–155
Hartman WD (1981) Form and distribution of silica in sponges. In: Simpson TL, Volcani BE (eds) Silicon and siliceous structures in biological systems. Springer, Berlin Heidelberg New York, pp 453–493

Hildebrand M, Volcani BE, Gassmann W, Schroeder JI (1997) A gene family silicon transporters. Nature 385:688–689
Hinde GJ, Holmes WM (1892) On the sponge remains in the Lower Tertiary Strata near Oamaru, New Zealand. J Linn Soc Zool 24:177–262
Holvoet S, van der Vyver G (1990) Effects of 2,2′-bipyridin on skeletogenesis of *Ephydatia fluviatilis*. In: Rützler K (ed) New perspectives in sponge biology. International conferences on the biology of sponges, third conference, 1985. Smithsonian Institution Press, Washington, DC, pp 206–210
Jorgensen CB (1944) On the spicule-formation of *Spongilla lacustris* (L.). 1. The dependence of the spicule-formation on the content of dissolved and solid silicic acid of the milieu. Det Kgl Danske Vidensks Selsk Biol Medd 19:2–45
Kraso A, Batel R, Schröder HC, Müller IM, Müller WEG (2000) Expression of silicatein and collagen genes in the marine sponge *Suberites domuncula* is controlled by silicate and myotrophin. Eur J Biochem 267:4878–4887
Kröger N, Deutzman R, Sumper M (1999) Polycationic peptides from diatom biosilica that direct silica nanosphere formation. Science 286:1129–1132
Le Pennec G, Perovic S, Shokry A, Ammar A, Grebenjuk VA, Steffen R, Brümmer F, Müller WEG (2002) Cultivation of primmorphs from the marine sponge *Suberites domuncula*, morphogenetic potential of silicon and iron. A review. J Biotechnol (in press)
Lévi C (1973) Systématique de la classe des Demospongiaria (Démosponges). In: Grassé P-P (ed) Traité de Zoologie, 3 Spongiaires. Masson, Paris pp 576–631
Lévi C (1989) A remarkably strong natural glassy rod: the anchoring spicule of the Monorhaphis sponge. J Mater Sci Lett 8:337–339
Maldonado M, Carmona MC, Uriz MJ, Cruzado A (1999) Decline in Mesozoic reef-building sponges explained by silicon limitation. Nature 401:785–788
Maldonado M, Uriz MJ (1996) Skeletal morphology of two controversial Poecilosclerid genera (Porifera, Demospongiae): Discorhabdella and Crambe. Helg Meeres 50:369–390
Müller W, Krasko A, Le Pennec G, Müller CI, Schröder HC (2003) Biochemistry and cell biology of silica formation in sponges. Micros Res Techn (in press)
Pé J (1973) Étude quantitative de la régulation du squelette chez une éponge d'eau douce. Arch Biol 84:147–173
Perry CC, Lu Y (1992) Preparation of silicas from silicon complexes: role of cellulose in polymerization and aggregation control. J Chem Soc Faraday Trans 88:2915–2921
Pisera A (2003) Some aspects of silica deposition in lithistid demosponge desmas. Micros Res Techn (in press)
Pisera A, Lévi C (2002) Lithistid demospongiae. In: Hooper JNA, Van Soest RWM (eds) Systema porifera 1. Kluger/Plenum, London
Richelle-Maurer E, Van der Vyver G (1999) Temporal and spatial expression of EmH-3, a homeobox-containing gene isolated from the freshwater sponge *Ephydatia muelleri*. Mech Ageing Dev 109:203–219
Richelle-Maurer E, Van der Vyver G, Vissers S, Coutihno CC (1998) Homeobox-containing genes in freshwater sponges: characterisation, expression and phylogeny. Progr Mol Subcell Biol 19: 157–175
Reiswig HM (1971) The axial symmetry of sponge spicules and its phylogenetic significance. Cah Biol Mar 12:505–514
Rosell D, Uriz MJ (2002) Excavating and endolithic sponge species (Porifera) from the Mediterranean: species descriptions and identification key. Org Divers Evol 2:55–86
Rützler K, Macyntyre (1978) Siliceous sponge spicules in coral reef sediments. Mar Biol 49: 147–159
Rützler K, Smith KP (1993) The genus Terpios (Suberitidae) and new species in the "Lobiceps" complex. In: Uriz MJ, Rützler K (eds) Recent advances in ecology and systematics of sponges. Scientia Marina, Barcelona, pp 381–393

Rützler K, Muzik K (1993) *Terpios hoshinota*, a new cyanobacteriosponge threatening Pacific reefs. In: Uriz MJ, Rützler K (eds) Recent advances in ecology and systematics of sponges. Sci Mar, Barcelona, pp 395–403
Sangster AG, Parry DW (1981) Ultrastructure of silica deposits in higher plants. In: Simpson TL, Volcani BE (eds) Silicon and siliceous structures in biological systems. Springer, Berlin Heidelberg New York, pp 384–407
Schönberg C (2001) New mechanisms in demosponge spicule formation. J Mar Biol Assoc UK 81: 345–346
Schröeder KN (1936) Beitrage zur Kenntnis der Spiculabbildung der Larven-spiculation und der Variationsbreite der Gerüstnadeln von Süβwasserschwammen. Z Morphol Oekol Tiere 31: 245–267
Schwab DW, Shore RE (1971a) Fine structure and composition of a siliceous sponge spicule. Biol Bull 140:125–136
Schwab DW, Shore RE (1971b) Mechanism of internal stratification of siliceous sponge spicules. Nature 232:501–502
Seimyia M, Ishiguro H, Miura K, Watanabe Y, Kurosawa Y (1994) Homeobox-containing genes in the most primitive metazoa, the sponges. Eur J Biochem 221:219–225
Seimyia M, Watanabe Y, Kurosawa Y (1997) Identification of POU-class homeobox genes in a freshwater sponge and the specific expression of these genes during differentiation. Eur J Biochem 243:27–31
Shimizu K, Cha JH, Stucky GD, Morse DE (1998) Silicatein alpha: Cathepsin L-like protein in sponge biosilica. Proc Natl Acad Sci USA 95:6234–6238
Shore RE (1972) Axial filament of siliceous sponge spicules, its organic component and synthesis. Biol Bull 143:689–698
Simpson TL (1978) The biology of the marine sponge *Microciona prolifera* (Ellis and Sollander). III Spicule secretion and the effects of temperature on spicule size. J Exp Mar Biol Ecol 35: 31–42
Simpson TL (1981) Effects of Germanium on silica deposition in sponges. In: Simpson TL, Volcani VE (eds) Silicon and siliceous structures in biological systems. Springer, Berlin Heidelberg New York, pp 527–550
Simpson TL (1984) The cell biology of sponges. Springer, Berlin Heidelberg New York
Simpson TL (1989) Silification processes in sponges: *Geodia* asters and the problem of morphogenesis of spicule shape. In: Crick RE (ed) Origin evolution, and modern aspects of biomineralization in plants and animals. Plenum Press, New York, pp 125–136
Simpson TL (1990) Recent data on patterns of silicification and the origin of monaxons from tetraxons. In: Rützler K (ed) New perspectives in sponge biology. International conference on the biology of sponges, 1985. Smithsonian Institution Press, Washington, DC, pp 264–270
Simpson TL, Vaccaro CA (1974) An ultrastructural study of silica deposition in the freshwater sponge *Spongilla lacustris*. J Ultrastruct Res 47:296–309
Simpson TL, Langenbruch PF, Scalera-Liaci L (1985) Silica spicules and axial filaments of the marine sponge *Stelletta grubii* (Porifera, Demospongiae). Zoomorphology 105:375–382
Thiele J (1899) Über *Crambe crambe* (O. Schmidt). Arch Naturges 87–94
TokyoCinema (1996) Life of the freshwater sponge. TokyoCinema Tokyo
Uriz MJ (1983) Contribución a la fauna de esponjas (Demospongiae) de Cataluña. Ann Sec Cienc Col Univ Gerona Monogr 1:1–220
Uriz MJ (1988) Deep-water sponges from the continental shelf and slope of Namibia (South-West Africa): classes hexactinellida and demospongia. Contributions to the marine fauna of Southern Africa, vol 3. Monografías de Zoología Marina, Scientia Marina, pp 9–157
Uriz MJ, Maldonado M (1993) Redescriptions of some rare sponge species in the western Mediterranean. In: Uriz MJ, Rützler K (eds) Recent advances in ecology and systematics of sponges. Sci Mar 57(4):81–94
Uriz MJ, Rützler K (eds) (1993) Recent advances in ecology and systematics of sponges. Sci Mar 57(4):273–432

Uriz MJ, Maldonado M (1995) A reconsideration of the relationship between polyaxonid and monaxonid spicules in Demospongiae: new data from the genera *Crambe* and *Discorhabdella* (Porifera). Biol J Linnean Soc 55:1–15

Uriz MJ, Turon X, Becerrro MA (2000) Silica deposition in demosponges: spiculogenesis in *Crambe crambe*. Cell Tissue Res 301:299–309

Uriz MJ, Turon X, Becerrro MA, Agell G (2003)Siliceous spicules of sponges: shapes and frameworks, biological functions and ultrastructural patterns. Micros Res Techn (in press)

Vacelet J, Boury-Esnault N (1995) Carnivorous sponges. Nature 373:333–335

Vacelet J, Boury-Esnault N (1996) A new species of carnivorous sponge (Demospongiae: Cladorhizidae) from a Mediterranean Cave. In: Willenz P (ed) Recent advances in sponge biodiversity inventory and documentation. Bull Inst R Sci Nat Belg Biol [Suppl] 66:109–115

Volcani BE (1981) Cell wall formation in diatoms: morphogenesis and biochemistry. In: Simpson TL, Volcani BE (eds) Silicon and siliceous structures in biological systems. Springer, Berlin Heidelberg New York, pp 157–200

Weissenfels N, Langenbruch PF (1985) Langzeitkulturen von Süßwasserschwämmen (Porifera, Spongillidae) unter Laborbedingungen. Zoomorphologie 105:12–15

Wiedenmayer F (1977) A monograph of the shallow-water sponges of the western Bahamas. Experientia supplementum 28. Birkhäuser, Basel

Wilkinson CR, Garrone R (1980) Ultrastructure of siliceous spicules and microsclerocytes in the marine sponge *Neofibularia irata* n. sp. J Morphol 166:51–64

Yourassowsky C, Rasmont R (1983) The differentiation of sclerocytes in fresh-water sponges grown in a silica-poor medium. Differentiation 25:5–9 [illegible] (1995) A reconsideration of the relationship between [illegible]

[illegible]

# Molecular Mechanism of Spicule Formation in the Demosponge *Suberites domuncula*: Silicatein – Collagen – Myotrophin

Werner E.G. Müller, Anatoli Krasko, Gaël Le Pennec, Renate Steffen, Matthias Wiens, Mohammed Shokry A. Ammar, Isabel M. Müller, and Heinz C. Schröder

## 1 Introduction

In living organisms four major groups of biominerals exist: (1) iron compounds, which are restricted primarily to Prokaryota; (2) calcium phosphates, found in Metazoa; (3) calcium carbonates, used by Prokaryota, Protozoa, Plantae, Fungi and Metazoa and (4) silica (opal) present in sponges and diatoms (reviewed in: Bengtson 1994; Baeuerlein 2000). It is surprising that the occurrence of silica as a major skeletal element is restricted to some Protozoa and to sponges (Porifera). The element silicon (Si) contributes to 28% of the earth crust and is – after oxygen – the second most abundant element on earth (Windholz 1983).

At present, it is not known why sponges have used silica as a biomineral to form one major innovation during the evolution from Protozoa to Metazoa, the hard skeleton. The transition from Protozoa to Metazoa dates back 600–1000 million years ago (reviewed in: Müller 1997; Nikoh et al. 1997). It has been proposed that oxygen level, temperature, and seawater chemistry played major roles in the evolution from Protozoa to Metazoa (Seilacher 1994). In particular, it was assumed that during the period of appearance of sponges the ocean was richer in sodium carbonate than in sodium chloride (Holland 1992); that such a "soda ocean" probably had a pH > 9. Under such conditions, the concentration of silica, the dioxide form of silicon, was presumably higher in seawater than today. Silica is the major constituent of sponge spicules (Demospongiae; Hexactinellida).

## 2 Sponges

It is now well established that the different sponge taxa are – together with other metazoan taxa – monophyletic and originate from the hypothetical ancestor, the Urmetazoa (see Müller 2001). The phylum Porifera (sponges) is

Institut für Physiologische Chemie, Abteilung Angewandte Molekularbiologie, Universität, Duesbergweg 6, 55099 Mainz, Germany

Progress in Molecular and Subcellular Biology, Vol. 33
W. E. G. Müller (Ed.)

subdivided into the following three classes, Demospongiae, Hexactinellida and Calcarea (Brien et al. 1973). Molecular data using nucleotide sequences from informative proteins resolved the problem about the phylogenetic relationships of the different classes of Porifera; the siliceous sponges, Hexactinellida and Demospongiae are closer related to the Urmetazoa, while the Calcarea share more similarities with the Cnidaria (Koziol et al. 1997; Kruse et al. 1998; Müller et al. 1998; Schütze et al. 1999). This positioning has recently been confirmed using sequence data of 18S rDNA (Borchiellini et al. 2001). It is documented that the major poriferan taxa, the Hexactinellida, Demospongiae and Calcarea, have existed since the Early Cambrian (Atdabatian); reviewed in Mehl et al. (1998). One specimen of the well-preserved hexactinellid sponge fossil, *Solactiniella plumata* is shown in Fig. 1A. This fossil record is from the Early Cambrian and has been found at the Sansha section in Hunan (China) in both the lower and upper levels of the Niutitang Formation (Steiner et al. 1993). A recent hexactinellid sponge, *Euplectella aspergillum*, is shown in Fig. 1B.

The spicules of sponges grouped in the two classes Demospongiae and Hexactinellida are composed of hydrated, amorphous, noncrystalline silica $[(SiO_2)_{2-5}\cdot H_2O]$ while those of the class Calcarea possess spicules formed from calcium carbonate (reviewed in: Simpson 1984). The Demospongiae and Hexactinellida are the only two taxa within the kingdom of Metazoa which use silica instead of calcium in their mineral skeleton; calcium is otherwise the dominant inorganic skeletal component. This is especially interesting in view of the fact that the concentration in seawater, the milieu in which most of the sponges live, is much higher for calcium ions (10 mM) than for silicon (100 µM) (Kennish 1994). The concentration of silicic acid in the surrounding milieu is crucial for the formation of spicules. The optimal concentration for the synthesis is between 5 and 100 µM (Jewell 1935); higher concentrations, e.g., 1.3 mM, are inhibitory (Jϕrgensen 1944). The secretion of spicules occurs in Demospongiae in specialized cells, the sclerocytes (Wilkinson and Garrone 1980); there silica is deposited around an organic filament (Garrone 1969). If the formation of siliceous spicules is inhibited, the sponge body collapses (Ankel et al. 1955). The synthesis of spicules is a rapid process; in *Ephydatia fluviatilis* a 100-µm-long megasclere is formed within 40 h (Weissenfels and Landschoff 1977).

Inhibition studies revealed that skeletogenesis of siliceous spicules is enzyme-mediated (Holvoet and Van de Vyver 1985a), more particularly, by an $Fe^{2+}$-dependent enzyme (Holvoet and Van de Vyver 1985b). If the freshwater sponge *Ephydatia fluviatilis* is grown in medium containing the $Fe^{2+}$ chelator 2,2′-bipyridin, skeletogenesis is inhibited (Holvoet and Van de Vyver 1985b). The siliceous spicules are known to contain a definite axial filament and their synthesis is under genetic control (Reiswig 1971). In Demospongiae the spicules are glued together by a collagenous "cement" made of microfibrils (Gross et al. 1956; Garrone 1969). Experimental evidence suggested that deposition of silica particles induces biological reactions, especially that of collagen fibrillogenesis (Garrone 1978).

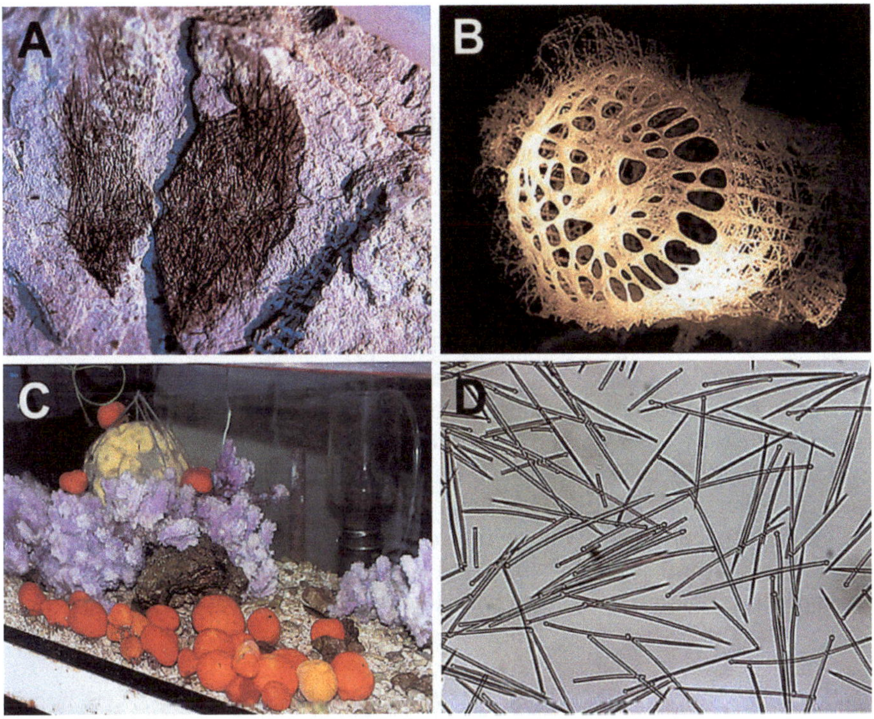


**Fig. 1A–D.** Siliceous sponges. **A**, **B** Specimens of the class Hexactinellida. **A** Earliest preserved sponges, *Solactiniella plumata* (Lowermost Cambrian Sansha section, Hunan, China; courtesy of Drs. B.D. Erdtmannn and M. Steiner); magnification ×0.6. **B** Spicule skeleton of the hexactinellid *Euplectella aspergillum*; ×0.5. **C**, **D** *Suberites domuncula*. **C** *S. domuncula* (*red*) specimens have been maintained in an aquarium for over 2 years together with two other demosponges *Geodia cydonium* (*yellow*) and *Dysidea avara* (*violet*); ×0.1; **D** spicules, oxeae and tylostyles from the species *S. domuncula*; ×100.(Partly taken from Müller 2001)

# 3 Spiculogenesis

As outlined, siliceous spicule formation, spiculogenesis, requires the synthesis of biosilica and collagen and takes place only if silicon and iron are present in the surrounding medium.

## 3.1 The Model Test System: Primmorphs

Spiculogenesis has been studied in vitro using the primmorph system established from the marine sponge *Suberites domuncula*. This sponge, which occurs in nature in red, orange, whitish-blue or as a mixture of these colors (Fig. 1C), possesses siliceous spicules, oxeae – styles – tylostyles with an average size of 150 μm (Fig. 1D).

Our group has developed the primmorph system (Müller 1998) and has focused especially on *S. domuncula* (Custodio et al. 1998; Müller et al. 1999). In brief, if tissue samples are transferred into $Ca^{2+}$ and $Mg^{2+}$-free artificial seawater containing EDTA, they dissociate into single cells (Fig. 2A) which form in $Ca^{2+}$ and $Mg^{2+}$-containing seawater aggregates (Fig. 2B). Usually after 5 days primmorphs, 1–5 mm in size, are formed (Fig. 2C) which are characterized by cells undergoing proliferation, as measured by BrdU (5-bromo-2′-deoxy-uridine)-labeling (Fig. 2D); the BrdU-positive cells are stained brownish in their nuclei. Cross sections through primmorphs revealed that their interior is surrounded by an almost complete single-cell layer of epithelial-like cells. The cells which compose the squamous epithelium of the primmorphs are pinacocytes. The cells inside the primmorphs are primarily spherulous cells, while the others may be called amoebocytes and archaeocytes.

The growth conditions for primmorphs could be optimized by supplementing the natural seawater with 0.2% RPMI 1640–medium and silicate. The optimal concentration of silicate was determined to be 60 μM (Krasko et al.

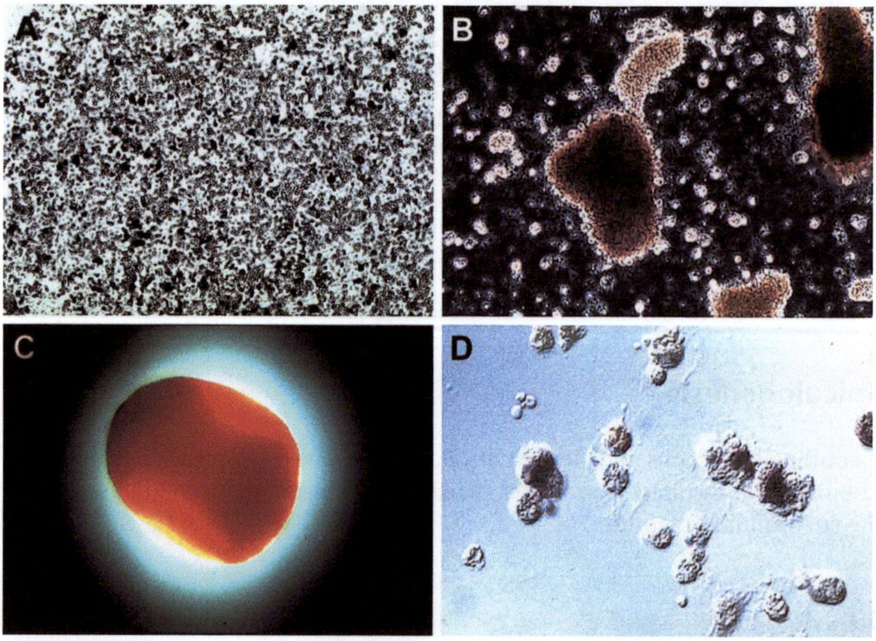


**Fig. 2A–D.** Primmorph formation from cells of the sponge *S. domuncula*. **A** Dissociated single cells; ×10. **B** Aggregates formed after their transfer into seawater after 6 h; ×20. **C** Primmorph formed after 5 days; ×8. **D** DNA synthesis in primmorphs was determined using the BrdU-labeling and detection assay; ×650

2000). Addition of silicate resulted in an increase in the size of the primmorphs (Fig. 3A). While the size of the primmorphs grown in the absence of additional silicate is ≈2 mm, that of primmorphs which grew in medium supplemented with 60 μM silicate reaches 6 mm; a further supplementation of the seawater medium containing 60 μM silicate with 30 μM Fe(3+) resulted in the formation of primmorphs larger than ≈10 mm (Fig. 3A).

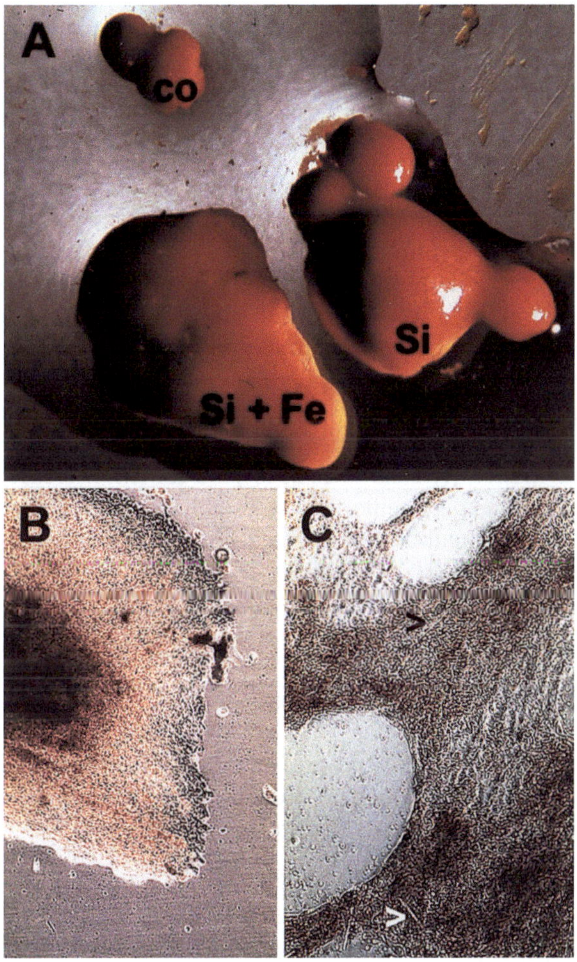


**Fig. 3.** **A** Effect of silicate and Fe(3+) on the size of the primmorphs. The primmorphs grown under different conditions are shown: primmorphs grown for 8 days in seawater medium (*co*), primmorphs grown for 8 days in seawater plus 60 μM silicate (*Si*) and primmorphs grown in seawater (60 μM silicate) together with 30 μM Fe(3+) (*Si + Fe*); ×3.5. **B**, **C** Effect of silicate on spicule formation in primmorphs. **B** Primmorphs grown in the absence of exogenous silicate for 5 days and **C** primmorphs incubated in medium containing 250 μM Na-silicate for the same period of time. Some spicules in **C** are marked; ×8

One proteinaceous growth-promoting protein has been isolated from *S. domuncula* which was shown to stimulate the proliferation of sponge cells; the myotrophin-like polypeptide (Schröder et al. 2000) that increases the size and influences the shape of the primmorphs.

## 3.2 Effect of Silicon on the Spicule Formation

In a semi-quantitative approach the effect of silicate on the spicule formation in vitro was determined (Krasko et al. 2000). If primmorphs, formed from single cells after 5 days were incubated further in the absence of exogenous silicate for another 5 days, no spicules could be seen by microscopic inspection (Fig. 3B). However, if the primmorphs were incubated in the presence of 250 μM exogenous silicate, bundles of spicules are present (Fig. 3C).

In a quantitative approach, the content of polymerized silica was determined in primmorphs in dependence on the presence of 60 μM exogenous silicate. Polymerized silica was determined as described (Cha et al. 1999). Primmorphs were incubated either in natural seawater containing 1 μM silicate, or the cultures were supplemented with extracellularly applied silicate. Only a low content of polymerized silica, 0.1 mg/0.4 g aliquot, was detected in primmorphs if they were cultivated in medium without additional silicate. If silicate is added at 60 μM, a concentration previously found to stimulate spicule formation (Krasko et al. 2000), the content increased to 4.4 mg. Based on these sets of experiments, it was established that the amount of spicules as well as the content of polymerized silica strongly depend on the level of exogenous silicate (Krasko et al. 2002).

## 3.3 Silicon-Responsive Genes

Until now it has been proven that the expression of the following three genes is upregulated in primmorphs of *S. domuncula* after exposure to silicon: silicatein, collagen and myotrophin.

### *3.3.1 Silicatein*

The first complete silicatein cDNA was cloned from *Tethya aurantia* (Shimizu et al. 1998) and later a second one from *S. domuncula* (Krasko et al. 2000). The complete *S. domuncula* silicatein cDNA is 1169 bp long (accession number AJ272013) and comprises a predicted translation product of 331 aa (amino acids), named SILICA_SUBDO with a calculated $M_r$ of 36.3 kDa. As outlined previously (Shimizu et al. 1998), silicatein is a member of the cathepsin subfamily. The *S. domuncula* silicatein shares 42% identical and 59% similar aa (identical plus physico-chemical related aa) with *G. cydonium* cathepsin (CAT-1_GEOCY) and 17% identical and 34% similar aa with the *S. domuncula* cathepsin (CAT-1_SUBDO). The three aa Cys, His and Asn, which form the catalytic triad of cysteine proteases (Rogers et al. 1985), are present in the sponge cathepsin at the characteristic sites: $Cys_{125}$, $His_{164}$ and $Asn_{184}$ (Krasko et al. 1997). However, in the sponge silicatein sequence the cysteine residue is replaced by serine. It is established that for the formation of active cathepsins, processing is required either by autolysis (Reilly et al. 1989) or by a second protease (Nishimura et al. 1988). Based on comparisons with cathepsins, the characteristic cleavage site in the sponge silicatein sequence is also present [Gln↓Asp $aa_{112}/aa_{113}$; Fig. 4A], suggesting a processing of the proenzyme of a $M_r$ of 36,306 to the mature form with a $M_r$ of 23,125. Also, the signal peptide is found in silicatein from $aa_1$ to $aa_{15}$ (Fig. 4A).

*Phylogenetic Analyses.* The databank search with the deduced *S. domuncula* silicatein revealed highest similarity with the *T. aurantia* silicatein and with the cathepsin L sequences (Fig. 4B). A rooted phylogenetic tree was constructed. The relationship showed that the *S. domuncula* cathepsin forms the basis of the sponge silicatein sequences as well as of the other cathepsin molecules from Metazoa, the human cathepsin L2, cathepsin L from *Drosophila melanogaster*, the probable cysteine proteinase from *Caenorhabditis elegans* (CYS-PRO_CAEEL; T24387) and also the *G. cydonium* cathepsin sequence. The probable cysteine proteinase from *Arabidopsis thaliana* was used to root the tree. The overall homology (identity and similarity on the protein level) between *S. domuncula* silicatein and *T. aurantia* silicatein is 79%. The homology to the cathepsin from *S. domuncula* is lower (34%; in prep.) while the relationship to the *G. cydonium* cathepsin (Krasko et al. 1997; Y10527), which forms the basis for the other most closely related cathepsins, is 59%; Fig. 4B.

### 3.3.2
### *Collagen*

Collagens constitute a superfamily of extracellular matrix proteins. All collagens have triple helical domains, with subunits (-chains) containing the (Gly–Xaa–Yaa)$_n$ repetitive sequence motif. Based on the configuration of the sequence domains, the collagens are classified into: (1) fibrillar collagens containing only one triple-helical domain which encompasses the entire molecule; (2) basement membrane collagens (type IV); (3) and Facit collagens - both comprising several shorter triple-helical domains which are separated by non-triple-helical segments; (4) the short-chain collagens as well as (5) some specialized collagen types (Bork 1992). In sponges, simple fibrillar collagens have been visualized in a series of species, e.g., *Chondrosia reniformis* (Garrone et al. 1975) or *G. cydonium* (Diehl-Seifert et al. 1985). The 20–25 nm (diameter) thin fibrils from *G. cydonium* display a 19.5-nm periodicity with one intrape-

**A**

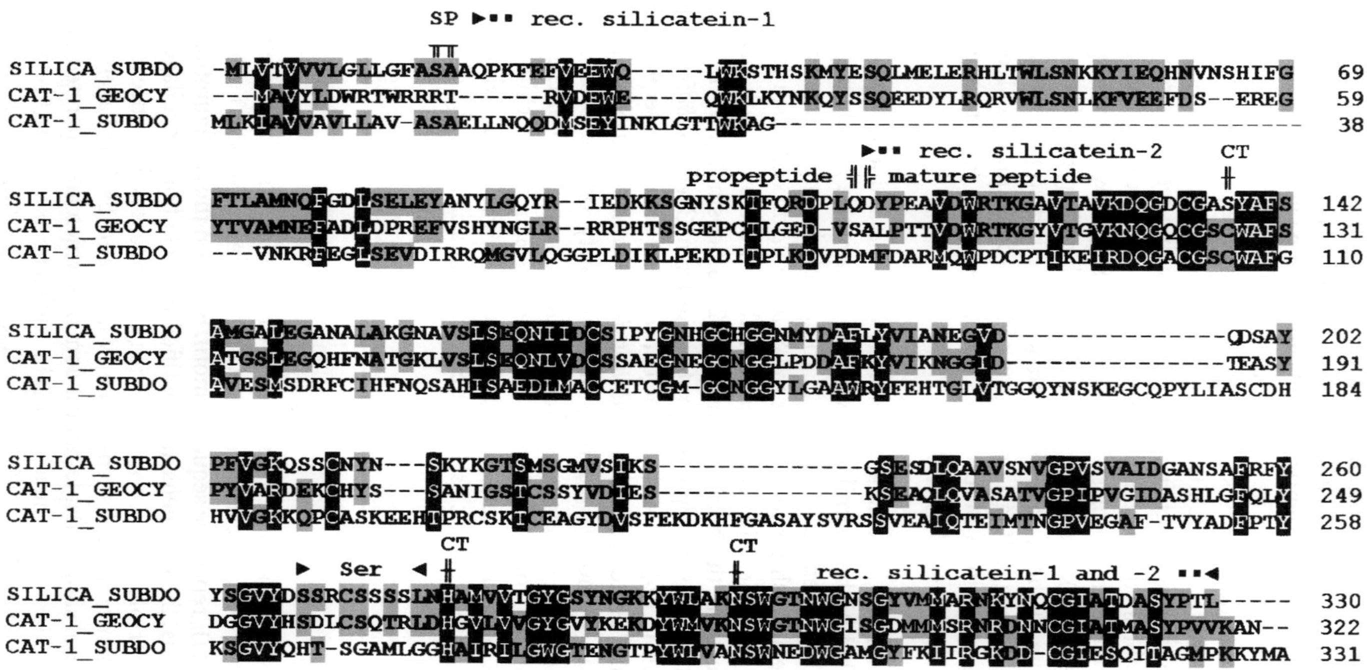


**Fig. 4A,B.** Silicon-responsive gene silicatein. A Silicatein sequence from *S. domuncula* (SILICA_SUBDO) is aligned with the sponge cathepsin sequences from *S. domuncula* (CAT-1_SUBDO; in prep.) and *G. cydonium* (CAT-1_GEOCY; Y10527). Residues conserved (similar or related with respect to their similar physico-chemical properties) in all sequences are shown in *white on black* and those in at least two sequences in *black on gray*. The characteristic sites within the silicatein sequences are indicated: the three aa Cys, His and Asn form the catalytic triad (*CT*), the 17-residue-long signal peptide (*SP*) and the processing site for the conversion of the proenzyme to the mature enzyme. The stretches of the polypeptides, r-silicatein-1 (long form) and -2 (short-form), which were expressed in *E. coli* are marked (▶ rec. silicatein-1 and -2 ◀).

B

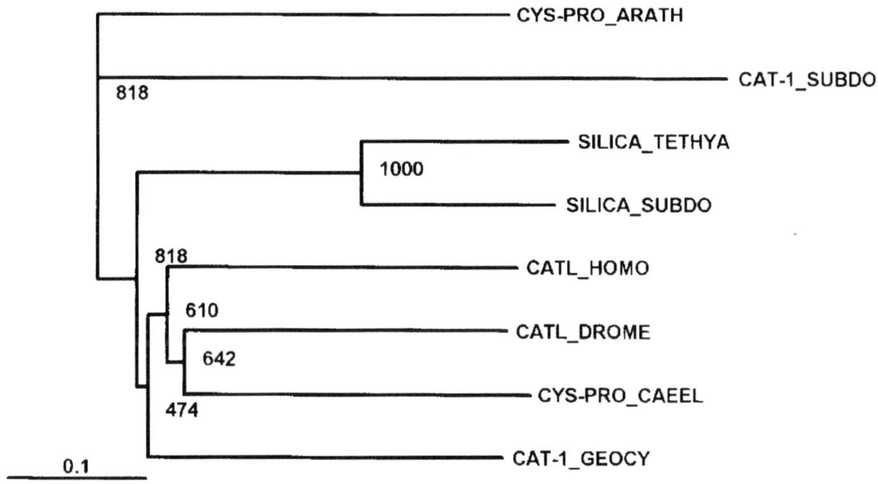


**Fig. 4.** B Phylogenetic relationships of the sponge silicatein sequence from *S. domuncula* (SILICA_SUBDO) with silicatein from *T. aurantia* (SILICA_TETHYA; AAC23951) as well as the cathepsin-like sequences, the human cathepsin L2 (CATL_HOMO; XP_005571.2), cathepsin L from *Drosophila melanogaster* (CATL_DROME; Q95029), the probable cysteine proteinase from *Caenorhabditis elegans* (CYS-PRO_CAEEL; T24387) and the probable cysteine proteinase from *Arabidopsis thaliana* (CYS-PRO_ARATH; P43297). The rooted tree (with respect to the *A. thaliana* sequence) was computed by neighbor-joining and distance matrix determinations. *Scale bar* indicates an evolutionary distance of 0.1 aa substitutions per position in the sequence

riod band. Exposito and Garrone (1990) succeeded for the first time to identify a collagen of that type on the cDNA level.

Focusing on *S. domuncula*, it has been demonstrated that cells of *S. domuncula* express a collagen gene in response to growth factors (Schröder et al. 2000). The cDNA for *S. domuncula* collagen was isolated. The *S. domuncula* collagen comprises three segments (Fig. 5A): (1) the noncollagenous N-terminal domain (NC1); (2) the collagenous internal domain (COL) and (3) the noncollagenous C-terminal domain (NC2). The collagenous internal domain in the *S. domuncula* molecule is unusually short and comprises only 24 G-x-y collagen triplets. In contrast, the deduced collagen polypeptide from *E. muelleri* (Exposito and Garrone 1990) is composed of two internal collagen domains (COL1 and COL2) with 79 G-x-y triplets (Fig. 5A).

### 3.3.3
### *Myotrophin*

It is known that myocardial hypertrophy is associated with a qualitative change in contractile protein composition in general and of collagen specifically

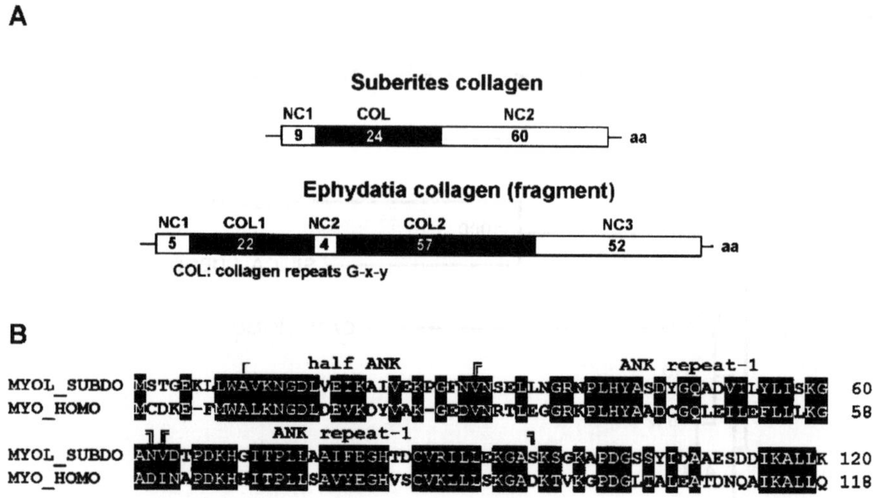


**Fig. 5A,B.** Silicon-responsive genes: collagen and myotrophin. **A** Collagen: schematic representation of the sponge collagen. Comparison of the *S. domuncula* collagen with the collagen isolated from *Ephydatia muelleri*. The *S. domuncula* collagen comprises three segments noncollagenous N-terminal domain [NC1], the collagenous internal domain [COL] and the noncollagenous C-terminal domain [NC2]. In contrast to the *E. muelleri* collagen, the collagenous internal domain of the *Suberites* collagen is unusually short and comprises only 24 G-x-y collagen triplets while the *E. muelleri* internal collagen domains [COL1 and COL2] are composed of 79 triplets. **B** Myotrophin: alignment of the deduced sponge myotrophin from *S. domuncula* (MYOL_SUBDO) with the related human myotrophin (MYO_HOMO; AAE54777 [US-patent: 6153423-A 3]). The two complete ankyrin motifs [ANK repeat-1 and -2] as well as the half repeat [half ANK] are indicated

(reviewed in: Brooks et al. 1997). Therefore, it was inviting to speculate that a sponge myotrophin-like molecule may cause an increased expression primarily of those genes which code for structural and morphogenetic proteins.

The cDNA encoding a myotrophin-like molecule from *S. domuncula*, SUBDOMYOL that comprises high sequence similarity to the mammalian cardiac myotrophin was cloned (Schröder et al. 2000). On the protein level, the sponge myotrophin shares identical/similar amino acids (%) with the following sequences: the human myotrophin (AAE54777) [47/70%], the hypothetical protein C43H6.3 from *Caenorhabditis elegans* (U51999) [45/68%], the CG7423 gene product from *Drosophila melanogaster* (AAF48915) [46/65%], the Yil112wp protein from *Saccharomyces cerevisiae* (NP_012154) [2/5%] and the SKOR polypeptide from *Arabidopsis thaliana* (CAA11280) [1/4%]. The sponge sequence shows the characteristic features known from the vertebrate myotrophins (Yang et al. 1997); one half of an ankyrin repeat is located at the $NH_2$ terminus of the protein ($aa_{10}$ to $aa_{28}$) followed by two complete repeats towards the COOH terminus ($aa_{30}$ to $aa_{62}$ and $aa_{63}$ to $aa_{95}$; Fig. 5B).

The sponge myotrophin was expressed and the protein was tested for biological activity. Incubation of cells for 5 days with recombinant myotrophin resulted in a change of the shape of the primmorphs formed. While those aggregates are round-shaped and ball-like in the absence of myotrophin, the majority of primmorphs which had been formed in the presence of myotrophin are elongated and oval-shaped assemblies (Schröder et al. 2000). From these data, it can be concluded that sponges are provided with factors, here myotrophin, which cause a comparable or even identical effect on cells and gene expression as the corresponding humoral stimuli in mammalian systems.

## 3.4 Effect of Silicon on Silicon-Responsive Genes

The primmorph system was used to study the effect of silicon on the expression of these three silicon-responsive genes (Krasko et al. 2000 and in prep.). Primmorphs were taken 5 days after reaggregation from single cells and incubated for different periods of time after addition of Na-hexafluorosilicate at a final silicate concentration of 60 μM.

The technique of Northern blotting was applied to estimate semi-quantitatively the steady-state expression level of the three silicon-responsive genes. The concentration of silicate in the natural seawater used for the experiments was determined to be ≈1 μM. Under this condition, the expression of *silicatein* in primmorphs is low and almost not detectable (Fig. 6A). If the concentration of silicate in the seawater was enhanced to 60 μM, the expression of silicatein already increased strongly after an incubation period of 1 day. After 5 days the expression reached a maximum; longer incubations for 10 days resulted in a decrease of the expression level by ≈25% (Krasko et al. 2000). The concentration of 60 μM silicate is the value which was previously found to be optimal for the spicule formation in *S. domuncula* (Simpson et al. 1985).

Likewise the gene encoding *collagen* was upregulated in primmorphs from *S. domuncula* (Krasko et al. 2000). Since this polypeptide is required for the formation of the functional skeleton in sponges (Hartman 1969), it was plausible to determine if the expression of this gene is also influenced by silicate. Once again, Northern blot experiments were performed which revealed a drastic increase in the expression of the collagen gene (Fig. 6B). At the natural silicate concentration of the seawater used (≈1 μM) the collagen expression was relatively low and amounted to ≈15% of that seen in the presence of 60 μM exogenous silicate.

The expression of the third silicon-responsive gene, *myotrophin*, was tested in parallel. Those Northern blot studies also revealed that the expression of this gene responded to silicon; in the presence of 60 μM of silicate the expression increased later – in comparison to silicatein or collagen; the first band was seen after 3 days and the intensity increased during the following 2 days of incubation (Fig. 6C).

While the inorganic silicon/silicate caused a strong effect on the expression of these selected genes, the silicon alkoxide tetraethoxysilane, added at a

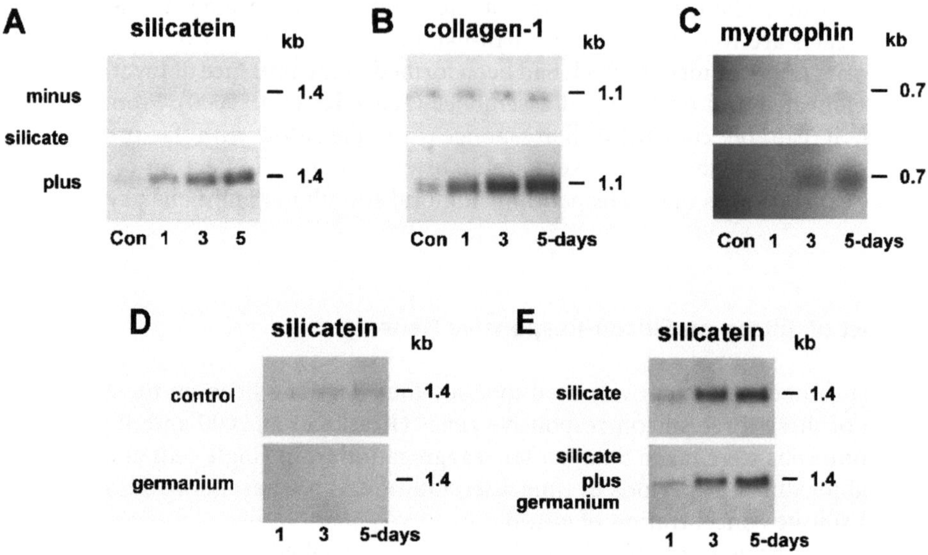


**Fig. 6A–E.** Effect of silicon on silicon-responsive genes. 5-day-old primmorphs were incubated for 0 (*Con* controls) to 5 days either in the absence of exogenous silicate (minus silicate) or presence of 60 μM Na-silicate (plus silicate). RNA was extracted and 5 g of total RNA was size-separated; after blot transfer, hybridization was performed either with the silicatein probe, *SUBDOSILICA* (accession number AJ272013) (**A**), the collagen probe, *SUBDOCOL1* (AJ252241) (**B**), or myotrophin *SUBDOMYOL* (AJ252240) (**C**), all isolated from *S. domuncula*. The relative degree of expression was correlated with that seen for the maximal expression (after 5 days in the presence of silicate). **D, E** Effect of germanium on the expression of silicatein. **D** Primmorphs were incubated in the absence of exogenous silicate or in the presence of 60 μM germanium for 5 days. **E** In parallel, primmorphs were incubated in the presence of exogenous silicate (60 μM) alone or with equimolar concentrations together with germanium. The expression of silicatein was determined using *SUBDOSILICA* as a probe

concentration of 60 μM displayed no effect on the expression of silicatein, collagen or myotrophin (Krasko et al. 2000); data not shown.

## 3.5 Inhibition of Biosilica Formation by Germanium

Interesting is the fact that the silicification machinery in sponges, as in diatoms, is effectively inhibited by the element germanium [Ge] (Elvin 1972a, b; Simpson et al. 1979). Germanium is a known competitive inhibitor of silicon transport (Azam et al. 1973). At a molar ratio of Ge/Si >1 complete inhibition of spicule formation occurs (Simpson et al. 1979). A detailed study has been performed with the sponge *S. domuncula* (Simpson et al. 1985). It was reported that at a Ge/Si ratio of 1 hatching of gemmules is inhibited in this species. Moreover, spicule formation is inhibited at a Ge/Si ratio of 0.1 (Simpson et al. 1985).

Germanium is also a known constituent of seawater at a concentration of 1 nM (Kennish 1994).

In our studies primmorphs were incubated with germanium for 5 days at a concentration of 60 μM. Under such conditions, as in the controls, no expression of the gene encoding silicatein could be detected (Fig. 6D). If exogenous silicate was added at a concentration of 60 μM, strong upregulation of the expression was measured again (Fig. 6E). If exogenous silicate (60 μM) was added together with an equimolar concentration of germanium, the expression of silicatein was the same as that seen in the absence of germanium (Fig. 6E).

## 3.6
## Proposed Pathway for Spicule Formation

The formation of siliceous spicules in sponges is genetically controlled (Simpson 1984) and can be assumed to initiate the *morphogenesis phase*. The data presented above (Krasko et al. 2000) demonstrate that at suitable concentrations silicate induces the genes encoding collagen, silicatein and myotrophin. However, the gene induction of silicatein appears to be a complex process, since the morphogen myotrophin induces only the collagen, but not the silicatein gene (Krasko et al. 2000; Schröder et al. 2000). This finding supports earlier observations indicating that the formation of collagen fibrils can proceed independently of spicule formation as well as closely connected with this process (reviewed in: Simpson 1984). Collagen fibrils are synthesized in several cell types (Simpson 1984) with the lophocytes as the prevailing cells (Garrone 1978); in addition, spongocytes have been identified which secrete spongin fibers that contribute to the synthesis of the horny skeleton (Garrone 1978). According to the chemical analysis, spongin should be considered as a collagen-like molecule (Junqua et al. 1974). Since only one gene coding for collagen has been identified in sponges, it remains open if spongin is only a processed form of collagen.

In contrast to the formation of collagen fibrils, disconnected from the synthesis of spicules, those collagen polypeptides which are formed in connection with the silicification are produced in sclerocytes and exopinacocytes (Weissenfels and Landschoff 1977). The sclerocytes secrete the axial filaments, while the exopinacocytes secrete the collagen-like spongin, which functions as a organic sheath around the spicules (Schröder 1936; Weissenfels 1989). Collagen is required for the formation of the functional skeleton in sponges (Hartman 1969). In this study it is reported that in newly formed tissue the spicules are not entrapped by these fibrils, while in older regions the spicules are encased in the organic matrix.

In a previous study (Krasko et al. 2000), the induction of silicatein gene expression could only be observed with silicate, but not with tetraethoxysilane. Therefore, it may be postulated that the organic silica substrate is formed intracellularly from silicate. Until now such a reaction has not been identified. In analogy with the synthesis of ether phospholipids, the following reaction can be proposed; (1) first an ester is formed using fatty acid coenzyme A as substrate and (2) the resulting carboxylate moiety is exchanged by ethanol. On the other side, the bond between silicate and the ethyl unit can be considered also as an "ester-like" ether bond, since the proton can dissociate from the hydroxyl group. The organosilicon acts now as substrate for the silicatein under formation of polymerized silica and/or silicone. According to this scheme, silicatein acts as an ether bond [or "ester-like" ether bond]-splitting enzyme (Zhou et al. 1999).

These data indicate that together with the induction of the gene expression for the axial filament/silicatein which leads to the initial process of spicule formation, the collagen gene is expressed in parallel and – as described (Weissenfels 1989) – in a second cell type. Based on inhibition studies, it could be demonstrated that the formation of siliceous spicules can occur in the absence of the synthesis of such collagen polypeptides which form the perispicular spongin sheath (Holvoet and Van de Vyver 1985a). This finding suggests that the expression of collagen is not directly connected with the formation of the siliceous spicules.

## 4 Expression of Silicatein in Primmorphs and in Sponge Tissue

To obtain further insight into the expression of silicatein in both primmorphs and tissue from *S. domuncula*, polyclonal antiserum against recombinant silicatein was raised in rabbits (see below; in prep.). After an initial injection of recombinant silicatein and three further boosts, antiserum with a titer of 1:5,000 (as determined in an ELISA assay) was obtained; it was termed anti-silicatein antiserum.

This serum was applied to stain sections through *S. domuncula* tissue. The immunofluorescence images revealed that the expression of silicatein on the protein level was highest in those regions where the spicules existed (Fig. 7). They had been dissolved with hydrofluoric acid prior to sectioning the specimens. At a lower magnification, the space where the spicules had been present stained brightly (Fig. 7A). In a control, where the section was not treated with the primary antibody, the anti-silicatein antiserum, no staining was visible (Fig. 7B). At higher magnification the staining of the canals, the empty space formed after desilicification, which represent the location of the formed spicules, strongly reacted with anti-silicatein antiserum and stained brightly (Fig. 7C, D). Based on these findings, it can be proposed that silicatein is present not only in the central axis of the spicules, as demonstrated earlier (Cha et al. 1999), but also on the surface, the rim of the spicules at a region where the expression of collagen has also been suspected (Weissenfels 1989).

After determining the tissue distribution of silicatein, it was interesting to see if the expression of silicatein in primmorphs could also be followed on the

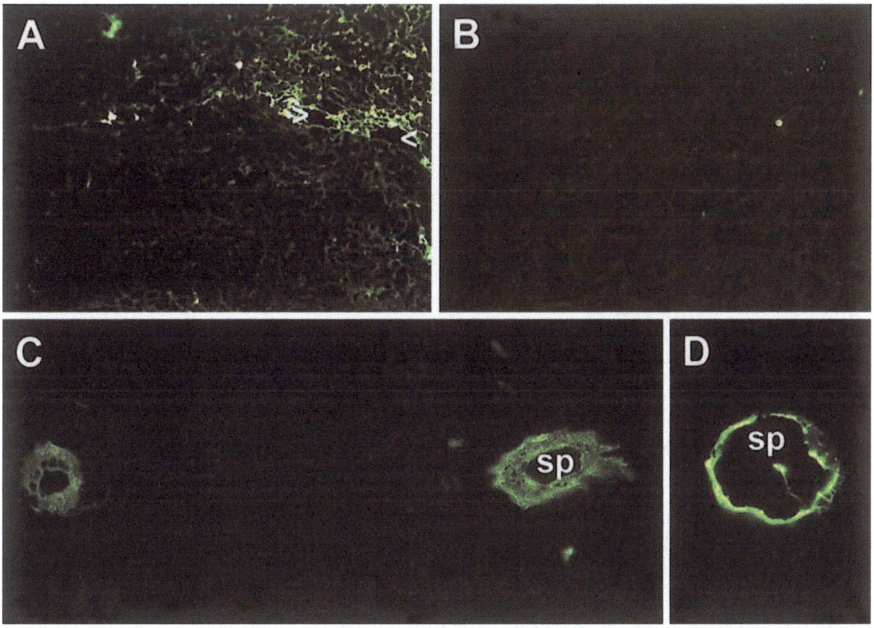


**Fig. 7A–D.** Immunohistological analysis of *S. domuncula* silicatein using antibodies raised against the recombinant silicatein. Tissue was treated with 6% Na-fluoride to dissolve the spicules, fixed with paraformaldehyde, dehydrated and embedded in Technovit 8100 (Wagner-Hülsmann et al. 1996). Sections of 2-μm thickness were prepared and reacted with polyclonal antisera raised in rabbits against recombinant silicatein (1:2000); **A, C, D.** After washing in PBS, the slides were incubated with FITC-conjugated anti-rabbit IgG (1:1000). In **B**, a control is shown; the section had been treated only with the labeled secondary antibody and not with anti-silicatein antiserum. Magnifications: **A** ×150; **B** to **C** ×500

protein level using this antibody. For this series of experiments primmorphs were treated with 60 μM of exogenous silicate. Protein was extracted both from tissue and primmorphs of *S. domuncula* and size-separated by $NaDodSO_4$/PAGE (Fig. 8A). After transfer to Immobilon sheets and subsequent treatment with primary (anti-silicatein antiserum) and secondary antibodies (Fig. 8B), one band became visible. The $M_r$ is 36 kDa, suggesting that only the signal peptide (calculated $M_r$ 1.4 kDa) has been cleaved off, while the postulated cleavage of the silicatein polypeptide (36 kDa) into the propeptide and "mature" peptide ($M_r$ 23 kDa) obviously does not occur. Furthermore, the Western blot data indicate that the level of protein expression is highest in sponge tissue (Fig. 8B, lane c); however, in primmorphs which had been incubated with 60 μm of silicate, the expression level reached approximately 50% of that seen in tissue (lane b). In contrast, no expression of silicatein could be seen in primmorphs incubated in the absence of silicate (lane a). These data confirm the Northern blot experiments which revealed that the steady-state level of transcripts for silicatein is absent in primmorphs incubated without silicate (Fig. 6A).

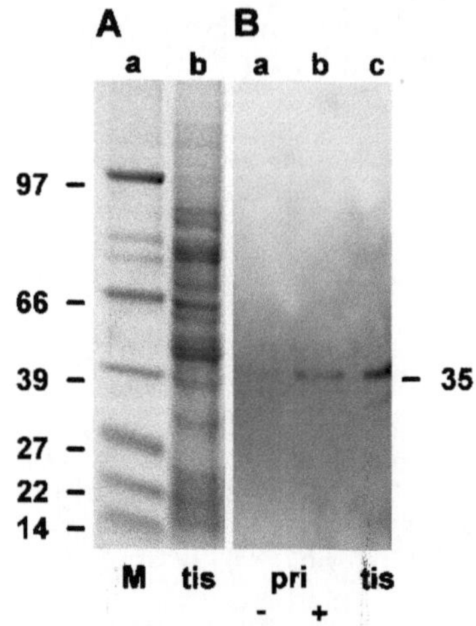


**Fig. 8.** Stimulation of silicatein expression in primmorphs after incubation with silicon. Primmorphs or tissue samples from *S. domuncula* were prepared and subjected to Western blotting after size separation by NaDodSO$_4$/PAGE. **A** In *lane b*, the protein pattern from tissue which was separated on a 15% gel and subsequently stained with Coomassie brilliant blue is given; 5 μg of protein was loaded onto the gel. In *lane a*, marker proteins were run in parallel. **B** Protein samples (5 μg of protein each) obtained from primmorphs cultivated for 5 days in the absence (*pri*: -; *lane a*) or presence of 60 μM silicate (*pri*: +; *lane b*) as well as from sponge tissue (*tis*; *lane c*) were size-separated using a 15% polyacrylamide gel, transferred to Immobilon sheets, and incubated with anti-silicatein antiserum; the immunocomplexes were visualized by alkaline phosphatase-conjugated secondary antibodies, followed by staining with 4-chloro-1-naphthol

# 5 Biosilica Formation

A major step forward to elucidate the formation of the siliceous spicules on the molecular level was the finding that what was formerly termed "axial organic filament" of siliceous spicules is in reality an enzyme, silicatein, which mediates the apposition of amorphous silica and hence, the formation of spicules (Shimizu et al. 1998; Cha et al. 1999); these studies have been performed with the demosponge *T. aurantia*. In a following study it was confirmed that in the marine demosponge *S. domuncula* the biosilica synthesis is also mediated by the silicatein.

## 5.1 Silicatein cDNA Expression

We succeeded in the preparation of recombinant silicatein. For the expression of silicatein in *E. coli*, two constructs were chosen: first, the sequence including the putative catalytic segment, from $aa_{17}$ to $aa_{330}$ ($nt_{49}$ to $nt_{993}$) [full-length protein without the terminal signal peptide, termed silicatein-long form], and second, the sequence comprising the active domain [mature protein, termed silicatein-short form], from $aa_{113}$ to $aa_{330}$ ($nt_{337}$ to $nt_{993}$), were used.

The cDNAs were inserted into the bacterial oligohistidine expression vector pQE-30 via the *Bam*HI (5′-end) and the *Hin*dIII (3′-end). Between the sequences encoding oligohistidine and silicatein, a nucleotide stretch corresponding to the enterokinase cleavage site was inserted (Fig. 9A). The *E. coli* strain Origami was transformed with this plasmid and expression of the fusion protein was induced with isopropyl 1-thio-β-D-galactopyranoside (IPTG; Ausubel et al. 1995). The fusion protein was extracted and purified on Ni-NTA-agarose resin, as described by Hochuli et al. (1987). The purified silicatein polypeptides, termed r-silicatein-long form and -short form, were subjected to digestion with recombinant enterokinase and analyzed on 15% polyacrylamide gels containing 0.1% $NaDodSO_4$ according to Laemmli (1970; Fig. 9B). The recombinant silicatein was cleaved from the His-tag by digestion with recombinant enterokinase; the purified long form of the r-silicatein preparations had a $M_r$ of 35 kDa (calculated size: 34,663; Fig. 9B-1, lane c) and the short form of $M_r$ 26 kDa (22,149) (Fig. 9B-2, lane a). These two recombinant r-silicatein preparations were used for analysis of enzymatic activity.

## 5.2 Silicatein Enzyme Assay

In a total volume of 500 µl, 100 µg of r-silicatein was added to the 25 mM Tris-HCl buffer (pH 7.2; 10 mM DL-dithiothreitol, 100 mM NaCl). As substrate, 4.5 mM tetraethoxysilane (Sigma) was used. Where indicated, 10 mM $MgSO_4$ and/or 10 mM (final concentration) of Fe(3+)citrate were added. The reaction was performed at 22 °C for 90 min and terminated by centrifugation (20,000 × g; 4 °C, 15 min). The sediment obtained was washed three times with ethanol (fraction: polymerized silica). To quantitatively measure the amount of polymerized silica, an aliquot of 250 µl was taken from the reaction mixture and the released silicic acid was determined (Cha et al. 1999) using the "Silicon Test" colorimetric assay kit (Merck; 1.14794). Based on the absorbance values at 810 nm, the absolute amounts of silicic acid were calculated after construction of a calibration curve using a silicon standard (Merck; 1.09947); the concentrations of the reaction products are given in ng of silicon/mg of protein.

The data show that (Table 1) r-silicatein causes a significant increase in the amount of polymerized silica from 2.2 to 5.9 ng/mg for the long form of the enzyme (2.4–6.1 nmol for the short form) if compared to the assays without r-silicatein. If ferric iron [Fe(3+)]was added to the reaction mixture, both in the assay composed of the long form of the enzyme as well as the one with the short form of r-silicatein, a threefold increase in the reaction product was measured. In contrast, the presence of Mg(2+) had no effect on the reaction.

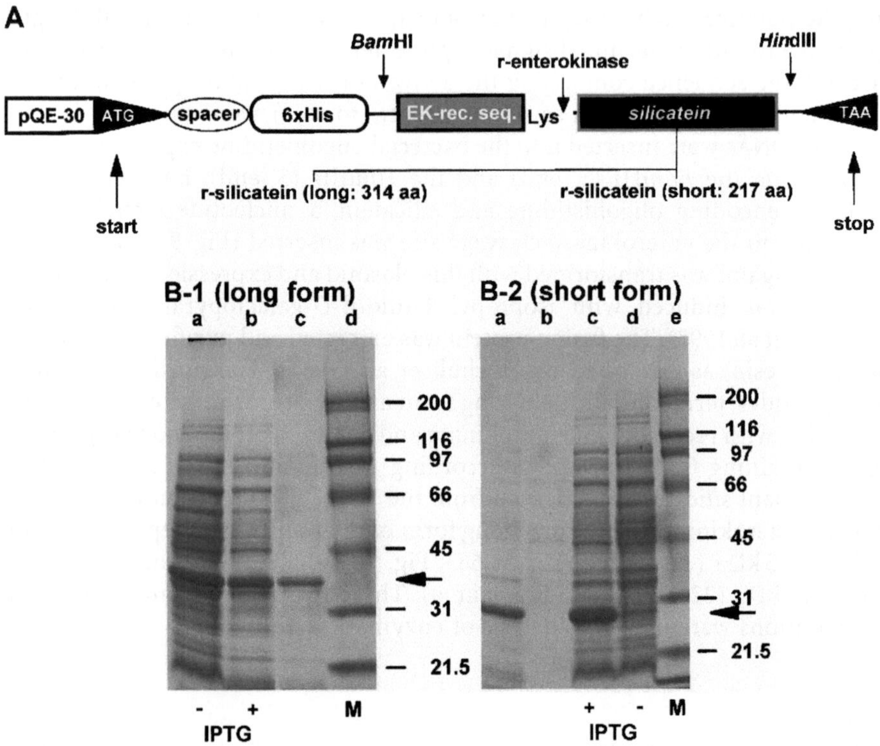


**Fig. 9A,B.** Expression of silicatein. **A** The *S. domuncula* silicatein (SILICA_SUBDO) was inserted into the expression vector pQE-30 via the *Bam*HI (5′-end) and the *Hin*dIII (3′-end). Between oligohistidine (*6xHis*) and *silicatein* a nucleotide stretch corresponding to the enterokinase recognition site [EK-rec. seq.] with the cleavage terminus [Lys↓] was inserted. The start ATG as well as the stop TAA codon for transcription are indicated. The position at which the enterokinase cleaved the recombinant fusion protein is indicated. Two constructs of silicatein cDNA have been inserted. For the preparation of the long form of silicatein (comprising $aa_{17}$ to $aa_{330}$) the stretch spanning $nt_{49}$ to $nt_{993}$ was selected, while for the short form of silicatein ($aa_{113}$ to $aa_{330}$) the $nt_{337}$ to $nt_{993}$ part of the sequence was used (Fig. 4A). **B** Expression of the His-silicatein fusion protein; **B-1** expression of the long form as well as the **B-2** short form of silicatein. Aliquots of lysates from bacteria, which remained noninduced (–IPTG; **B-1**, *lane a* and **B-2**, *lane d*) or had been induced with IPTG (+IPTG; **B-1**, *lane b* and **B-2**, *lane c*) as well as the purified fusion protein (**B-1**, *lane c* and **B-2**, *lane a*) were analyzed by $NaDodSO_4$/PAGE (15% gels) as described in the text. In **B-2** (*lane b*), the fraction which had not been adsorbed to the Ni-NTA-agarose resin was analyzed. The fusion protein was subjected to digestion with recombinant enterokinase. The *arrow* points to the band corresponding to the recombinant silicatein protein. Marker proteins (*M*) were run in parallel

**Table 1.** Stimulation of the activity of silicatein by Fe(3+). The assay was performed with the two r-silicatein preparations (long as well as short form) as described in the text using 4.5 mM tetraethoxysilane (TEOS) as substrate and, where indicated, $MgSO_4$ or Fe(3+) citrate. After incubation the amount of polymerized silica was quantified after hydrolyzation with NaOH and subsequent determination of the released silicic acid ($n = 5$; ±SD). The values are given in ng of silicon (polymerized), and if enzyme was added, correlated with 1 mg of protein; means of five separate experiments are presented

| r-Silicatein | TEOS | Addition of: Fe(3+)citrate (mM) | Addition of: $MgSO_4$ (mM) | Polymerized silica (ng silicon) |
|---|---|---|---|---|
| Long form | – | – | – | 2.21 ± 0.51 |
| | + | – | – | 5.90 ± 1.59 |
| | + | – | 10 | 6.25 ± 1.43 |
| | + | 10 | – | 18.72 ± 5.48 |
| Short form | – | – | – | 2.36 ± 0.69 |
| | + | – | – | 6.09 ± 1.81 |
| | + | – | 10 | 8.94 ± 2.48 |
| | + | 10 | – | 20.70 ± 5.62 |

## 6
## Effect of Iron

Recently, we could establish that in addition to silicon, iron has a key role in the synthesis of spicules (Krasko et al., in prep.). We demonstrated that after incubation of primmorphs with Fe(3+) at concentrations above 10 µM, a significant increase in spicule formation is observed; in parallel a Fe(3+)-mediated stimulation of DNA and protein synthesis occurs.

This process is paralleled by a strong upregulation of the gene expressions of the iron storage protein ferritin (reviewed in: Chasteen and Harrison 1999) as well as of a putative septin polypeptide.

The complete cDNAs from ferritin, termed *SUBDOFTN1* and 2 *SUBDOFTN2* as well as from septin were cloned from *S. domuncula* (accession numbers AJ306615 and AJ306614). As a marker for cell proliferation, the cDNA encoding septin, *SUBDOSEPTL*, was isolated (AJ310805). Septin molecules are conserved from yeast, *Saccharomyces cerevisiae* (Longtine et al. 1996) to mammals (Sanders and Field 1994) and are known to be involved in the completion of cytokinesis (reviewed in: Xie et al. 1999) and hence included in the elucidation of those genes which respond to iron. Finally, since it is known that the majority of spicules in Demospongiae such as *S. domuncula* are formed extracellularly (Uriz et al. 2000), it must be assumed that cell surface receptors are involved in the growth guidance/control of the spicules, composed of silica. It was found that the expression of the sponge integrin receptor does not depend on the presence of extracellular iron, while a considerable upregulation of a scavenger receptor is observed after addition of iron to primmorph cultures. This scavenger receptor, *SDSRCR*, was also cloned (AJ306613).

### 6.1 Effect of Iron on the Expression of Ferritin, Septin and Scavenger Receptor in Primmorphs

After formation of primmorphs, these aggregates were transferred into seawater which had been supplemented in addition to 60 μM silicate with 30 μM ferric citrate [Fe(3+)]. In one series the primmorphs remained in the culture in the absence of Fe(3+). Subsequently, the primmorphs were either immediately (time zero: controls Fig. 10) used for the preparation of total RNA or were incubated for 1–5 days before the RNA was isolated. Equal amounts of RNA were analyzed by Northern blotting.

If the RNA – after size separation and blot transfer – was hybridized with the ferritin probe, *SUBDOFTN2* (cDNA-encoding ferritin), it became obvious that in the absence of additional Fe(3+) only a slight increase of the expression of the 0.9 kb ferritin band is seen after 5 days. However, if RNA has been taken from primmorphs which had been cultivated in the presence of Fe(+++), a strong increase was already seen after 3 days; this increased during the 5-day incubation period (Fig. 10A).

In parallel, samples were subjected to hybridization to the septin probe, *SUBDOSEPTL*. Again, it was measured that the expression of the septin gene strongly increased in cultures treated with Fe(3+), while in the absence of iron the expression level is much lower (Fig. 10B).

There was a drastic effect measured with the probe of the scavenger receptor, *SDSRCR*. Expression is only seen in the presence of Fe(3+) after 3 and 5 days, respectively, while no transcripts can be traced in RNA isolated from primmorphs kept in the absence of Fe(3+) (Fig. 10C).

Two further series of hybridization experiments have been performed with the probe measuring the expression of the β-subunit of integrin, *SDINTB*, as well as with the probe for silicatein, *SUBDOSILICA*. Both probes already gave signals at time zero of the incubation period, and the expression of both the 2.9-kb signal for the integrin receptor (Wimmer et al. 1999) and the 1.2-kb signal (Krasko et al. 2000) for silicatein remained unchanged irrespective of whether the primmorphs were incubated with or without Fe(3+).

From these data we conclude that iron is a metal ion essential for proliferation and morphogenesis of cells in the sponge primmorph system. In the initial phase of reorganization of cells in the primmorphs, iron causes an enhanced proliferation of cells and in a later phase, a morphogenetic effect via its accelerative and supportive influence on spicule synthesis, through the activation of silicatein, and on size/shape, likely via the scavenger receptor.

## 7 Conclusion

In addition to silicon, iron is one of the major elements found in sponge spicules (Vinogradov 1953). Based on the studies performed with *T. aurantia*

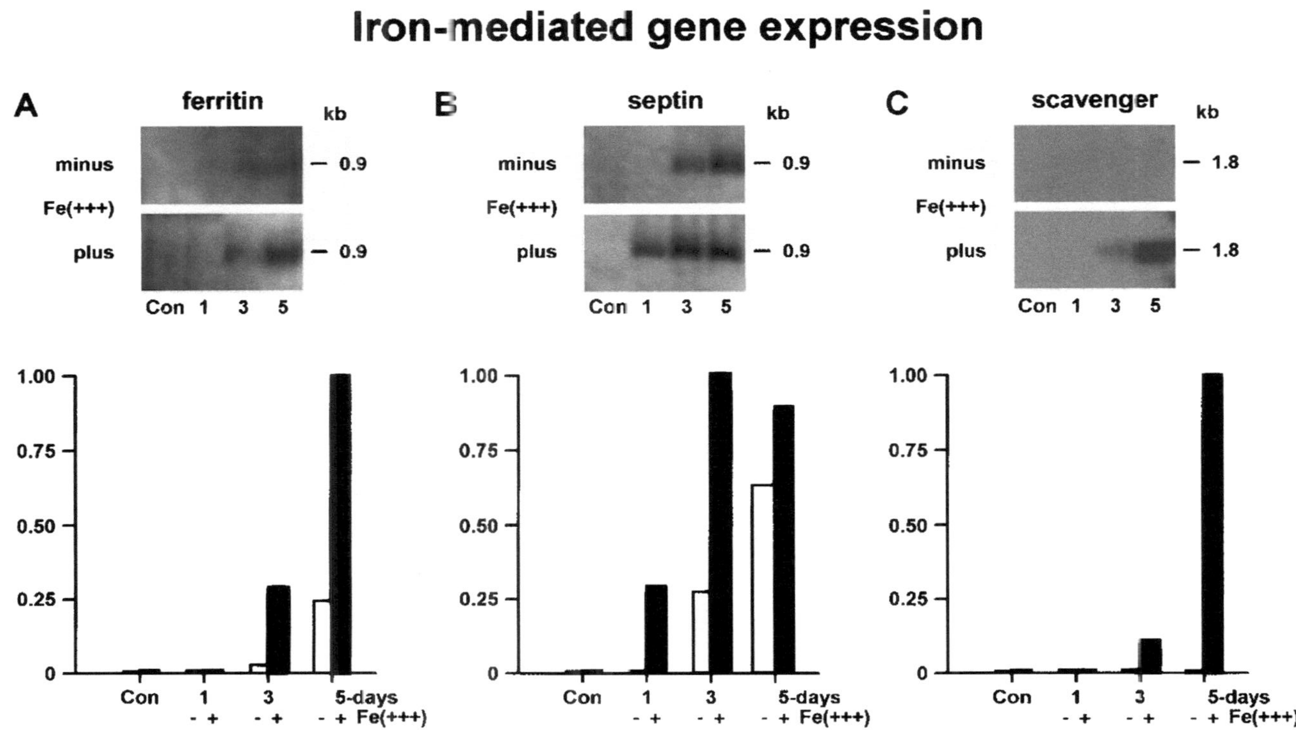


**Fig. 10A–C.** Iron-mediated gene expression. Primmorphs were incubated for 5 days in seawater, supplemented with 60 µM silicate. The primmorphs were then transferred into seawater with silicate which had been complemented with 30 µM Fe(3+); the primmorphs were incubated from 0 (*Con* controls) to 5 days in the absence (*minus* [–] Fe(3+); *open bars*) or presence of 30 µM Fe (*plus* [+] Fe(3+); *closed bars*). Then Northern blotting was performed with *SUBDOFTN2* to determine the expression of ferritin (**A**). The intensities of the bands were quantified, as described in the text. The relative degree of expression was correlated with that seen for the maximal expression [after 5 days in the presence of Fe(3+)]. In parallel, RNA samples were hybridized with the probe *SUBDOSEPTL* to determine the expression of septin (**B**) as well as with *SDSRCR* to measure the expression of the sponge scavenger receptor (**C**)

(Cha et al. 1999) and *S. domuncula* (Krasko et al. 2000), it appears to be well-established that silicatein is the major enzyme involved in the spicule formation. However, the bulk of information is lacking which addresses the questions of the initiation of spiculogenesis as well as the factors which control the size and shape of the spicules.

The experiments using the primmorph system from *S. domuncula* suggest that besides silicon, iron ions play an important role in the initiation of spicule formation. In the soluble form, iron ions as Fe(2+) are better absorbed than as Fe(3+) (Crichton and Ward 1998; Fig. 11). In seawater, the concentration of iron is 40 ng/l (≈1 nM; Kennish 1994), mainly present as Fe(2+). Experimental data indicate that silica uptake into sponges follows the Michaelis-Menten kinetics (Reincke and Barthel 1997; Fröhlich and Barthel 1997), suggesting that a silica transporter might modulate the uptake of silicon into the cells.

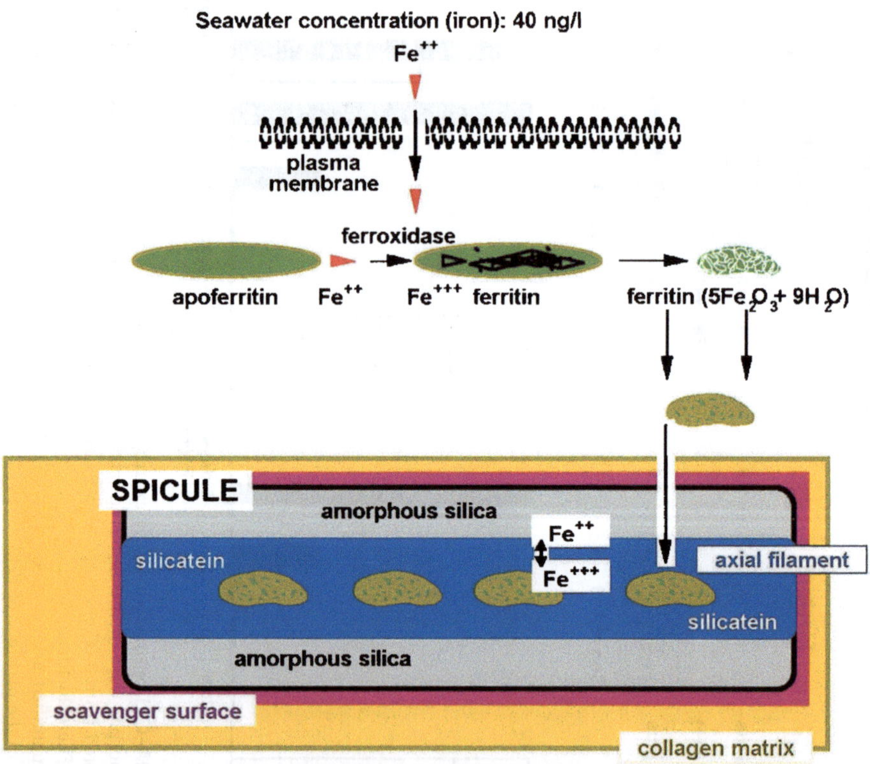


**Fig. 11.** Proposed pathway for the synthesis of spicules in Demospongiae. It is suggested that Fe(2+) is taken up by sponge cells and undergoes oxidation to Fe(3+) via the ferroxidase of ferritin. Ferritin/Fe(3+) is translocated to the axial filament where it becomes surrounded by silicatein. This enzyme, very likely activated by iron, forms amorphous silica and finally the spicule. This skeletal element becomes surrounded by the scavenger receptor finally embedded into the collagen matrix

The data gathered in our group indicate that iron causes an upregulation of ferritin in sponge cells which very likely binds the intracellular Fe(2+). In the next step, Fe(2+) is oxidized to Fe(3+) under the consumption of molecular dioxygen (Ha et al. 1999), a process which is mediated by the ferroxidase site of ferritin. By this step, apoferritin is converted to ferritin. Finally, iron is deposited in ferritin as $5Fe_2O_3{\cdot}9H_2O$ and then as the stable mineral $\alpha$-$Fe_2O_3$ (see: Chasteen and Harrison 1999). The deposited, mineralized iron, the ferric oxide nanoparticles remain surrounded by the ferritin protein shell. In the iron-free state the ferritin molecules (apoferritin) comprise hollow spheres with an outer diameter of 12 nm and an inner diameter of 8 nm (Lawson et al. 1991). Hence, ferritin forms a shell that causes both the oxidation of iron ions and controls the growth of the iron mineral.

In situ studies with *S. domuncula* revealed that ferritin is abundantly present in cells surrounding the canals within the sponge (in prep.). The function of ferritin in those cells remains to be established. Based on the analytical (Vinogradov 1953) as well as functional data (Krasko et al., in prep.), it can be proposed that the deposition of iron in ferritin spheres is involved also in the initiation of spicule formation (Fig. 11). Since the diameter of the axial filament, which comprises also the silicatein, measures $\approx 1\ \mu m$ (Uriz et al. 2000), there is ample space for the ferritin shell. It is suggested that these spheres become covered by silicatein, a process which results in the initiation of spiculogenesis. Silicatein is finally activated by iron, either as Fe(2+) or Fe(3+). This assumption is supported by our recent data which indicate that Fe(3+) stimulates the activity of silicatein. It is well established that iron within the ferritin shell becomes mobilized after reduction (Chasteen 1998).

After or concomitantly with the synthesis of amorphous silica, the growing spicules are surrounded with the scavenger receptor which is assumed to be the docking molecule for the collagen matrix into which the spicules are embedded (Fig. 11).

*Acknowledgements.* This work was supported by grants from the Bundesministerium für Bildung und Forschung [project: Center of Excellence BIOTEC-MARIN] Germany, the European Commission [project: SPONGE] and the International Human Frontier Science Program [RG-333/96-M]. *Disclosure statement*: A patent application entitled "Silicatein-vermittelte Synthese von amorphen Silikaten und Siloxanen und ihre Verwendung" (DE 100 40 033.7; inventors: W.E.G. Müller, B. Lorenz, A. Krasko and H.C. Schröder authors: 2000) has been submitted.

## References

Ankel WE, Wintermann-Kilian G, Kilian EF (1955) Fehlbildungen bei *Ephydatia fluviatilis* und ihre Bedeutung für das Verständnis der normalen Korrelationen. Verh Dtsch Zool Ges Tübingen 1954:361–371

Ausubel FM, Brent R, Kingston RE, Moore DD, Smith JA, Seidmann JG, Struhl, K (1995) Current protocols in molecular biology. Wiley, New York

Azam F, Hemmingsen BB, Volcani BE (1973) Germanium incorporation into silica of the diatom cell walls. Arch Mikrobiol 92:11–20

Baeuerlein E (2000) Biomineralization. Wiley-VCH, Weinheim
Bengtson S (1994) The advent of animal skeletons. In: Bengtson S (ed) Early life on earth. Columbia Univ Press, New York, pp 412–425
Borchiellini C, Manuel M, Alivon E, Boury-Esnault N, Vacelet J, Le Parco Y (2001) Sponge paraphyly and the origin of Metazoa. J Evol Biol 14:171–179
Bork P (1992) The molecular architecture of vertebrate collagens. FEBS Lett 307:49–54
Brien P, Levi C, Sarà M, Tuzet O, Vacelet J (1973) Spongiaire, traité de zoologie. In: Grassé PP (ed) Anatomie, systématique, biologie, vol 3. Masson, Paris
Brooks WW, Bing OH, Conrad CH, O'Neill L, Crow MT, Lakatta EG, Dostal DE, Baker KM, Boluyt MO (1997) Captopril modifies gene expression in hypertrophied and failing hearts of aged spontaneously hypertensive rats. Hypertension 30:1362–1368
Cha JN, Shimizu K, Zhou Y, Christianssen SC, Chmelka BF, Stucky GD, Morse DE (1999) Silicatein filaments and subunits from a marine sponge direct the polymerization of silica and silicones in vitro. Proc Natl Acad Sci USA 96:361–365
Chasteen ND (1998) Ferritin: uptake, storage, and release of iron. In: Sigel A, Sigel H (eds) Metal ions in biological systems. Dekker, New York, pp 479–514
Chasteen ND, Harrison PM (1999) Mineralization in ferritin: an efficient means of iron storage. J Struct Biol 126:182–194
Crichton RR, Ward RJ (1998) Iron homeostasis. In: Sigel A, Sigel H (eds) Metal ions in biological systems. Dekker, New York, pp 633–665
Custodio MR, Prokic I, Steffen R, Koziol C, Borojevic R, Brümmer F, Nickel M, Müller WEG (1998) Primmorphs generated from dissociated cells of the sponge *Suberites domuncula*: a model system for studies of cell proliferation and cell death. Mech Ageing Dev 105:45–59
Diehl-Seifert B, Kurelec B, Zahn RK, Dorn A, Jerecevic B, Uhlenbruck G, Müller WEG (1985) Attachment of sponge cells to collagen substrata: effect of a collagen assembly factor. J Cell Sci 79:271–285
Elvin DW (1972a) Effect of germanium upon development of siliceous spicules of some freshwater sponges. Exp Cell Res 72:551–553
Elvin DW (1972b) Growth rates of the siliceous spicules of the freshwater sponge *Ephydatia fluviatilis* (Lieberkühn). Trans Am Microsc Soc 90:219–224
Exposito JY, Garrone R (1990) Characterization of a fibrillar collagen gene in sponges reveals the early evolutionary appearance of two collagen families. Proc Natl Acad Sci USA 87:6669–6673
Fröhlich H, Barthel D (1997) Silica uptake of the marine sponge *Halichondria panicea* in Kiel Bight. Mar Biol 128:115–125
Garrone R (1969) Collagène, spongine et squelette minéral chez l'éponge *Haliclona rosea*. J Microscop 8:581–598
Garrone R (1978) Phylogenesis of connective tissue. Karger Press, Basel
Garrone R (1998) Evolution of metazoan collagens. Progr Mol Subcell Biol 21:191–139
Garrone R, Huc A, Junqua S (1975) Fine structure and physicochemical studies on the collagen of the marine sponge *Chondrosia reniformis* Nardo. J Ultrastruct Res 52:261–275
Gross J, Soka, Z, Rougvie A (1956) Structural and chemical studies on the connective tissue of marine sponges. J Histochem Cytochem 4:227–246
Ha Y, Shi D, Small GW, Theil EC, Allewell NM (1999) Crystal structure of bullfrog M ferritin at 2.8 Å resolution: analysis of subunit interactions and the binuclear metal center. J Biol Inorg Chem 4:243–256
Hartman WD (1969) New genera and species of coralline sponges (Porifera) from Jamaica. Postilla (Yale Univ) 137:1–39
Hochuli E, Döbeli H, Schacher A (1987) New metal chelate adsorbent selective for proteins and peptides containing neighbouring histidine residues. J Chromatogr 411:177–184
Holland HD (1992) Chemistry and evolution of the Proterozoic ocean. In: Schopf JW, Klein C (eds) The proterozoic biosphere. Cambridge Univ Press, Cambridge, pp 169–172

Holvoet S, Van de Vyver G (1985a) Skeletogenesis in *Ephydatia fluviatilis* grown in the presence of puromycin and hydroxyurea. In: Rützler K (ed) New perspectives in sponge biology. Smithsonian Institution Press, Washington, DC, pp 200–205

Holvoet S, Van de Vyver G (1985b) Effects of 2,2′-bipyridin on skeletogenesis of *Ephydatia fluviatilis*. In: Rützler K (ed) New perspectives in sponge biology. Smithsonian Institution Press, Washington, DC, pp 206–210

Jewell ME (1935) An ecological study of the fresh-water sponge of northern Wisconsin. Ecol Monogr 5:461–504

Jϕrgensen CB (1944) On the spicule-formation of *Spongilla lacustris* (L.). 1. The dependence of the spicule-formation on the content of dissolved and solid silicic acid of the milieu. Det Kgl Danske Viensks Selsk Biol Med 19:2–45

Junqua S, Robert L, Garrone R, Pavans de Ceccatty M, Vacelet J (1974) Biochemical and morphological studies on collagens of horny sponges. *Ircinia* filaments compared to spongines. Connect Tissue Res 2:193–203

Kennish MJ (1994) Practical handbook of marine science. CRC Press, Boca Raton

Koziol C, Leys SP, Müller IM, Müller WEG (1997) Cloning of Hsp70 genes from the marine sponges *Sycon raphanus* (Calcarea) and *Rhabdocalyptus dawsoni* (Hexactinellida). An approach to solve the phylogeny of sponges. Biol J Linnean Soc 62:581–592

Krasko A, Gamulin V, Seack J, Steffen R, Schröder HC, Müller WEG (1997) Cathepsin, a major protease of the marine sponge *Geodia cydonium*: purification of the enzyme and molecular cloning of cDNA. Mol Mar Biol Biotechnol 6:296–307

Krasko A, Batel R, Schröder HC, Müller IM, Müller WEG (2000) Expression of silicatein and collagen genes in the marine sponge *Suberites domuncula* is controlled by silicate and myotrophin. Eur J Biochem 267:4878–4887

Krasko A, Schröder HC, Batel R, Grebenjuk VA, Steffen R, Müller IM, Müller WEG (2002) Iron induces proliferation and morphogenesis in primmorphs from the marine sponge *Suberites domuncula*. DNA Cell Biol 21:67–80

Kruse M, Leys SP, Müller IM, Müller WEG (1998) Phylogenetic position of the Hexactinellida within the phylum Porifera based on amino acid sequence of the protein kinase C from *Rhabdocalyptus dawsoni*. J Mol Evol 46:721–728

Laemmli UK (1970) Cleavage of structural proteins during the assembly of the head of bacteriophage T4. Nature 227:680–685

Lawson DM, Artymiuk PJ, Yewdall SJ, Smith JMA, Livingstone JC, Treffry A, Luzzago A, Levi S, Arosio P, Cesareni G, Thomas CD, Shaw WV, Harrison PM (1991) Solving the structure of human H-ferritin by genetically engineering intermolecular crystal contacts. Nature 349: 541–544

Longtine MS, DeMarini DJ, Valencik ML, Al-Awar OS, Fares H, Virgilio CD, Pringle JR (1996) The septins: roles in cytokinesis and other processes. Curr Opin Cell Biol 8:106–119

Mehl D, Müller I, Müller WEG (1998) Molecular biological and palaeontological evidence that Eumetazoa, including Porifera (sponges), are of monophyletic origin. In: Watanabe Y, Fusetani N (eds) Sponge science – multidisciplinary perspectives. Springer, Berlin Heidelberg New York, pp 133–156

Müller WEG (1997) Origin of metazoan adhesion molecules and adhesion receptors as deduced from their cDNA analyses from the marine sponge *Geodia cydonium*. Cell Tissue Res 289: 383–395

Müller WEG (1998) Herstellung von Primmorphen aus dissoziierten Zellen von Schwämmen, Korallen und weiteren Invertebraten: Verfahren zur Kultivierung von Zellen von Schwämmen und weiteren Invertebraten zur Produktion und Detektion von bioaktiven Substanzen, zur Detektion von Umweltgiften und zur Kultivierung dieser Tiere in Aquarien und im Freiland. German, European and international patent application (DE 198 24 384)

Müller WEG (2001) How was metazoan threshold crossed: the hypothetical Urmetazoa. Comp Biochem Physiol [A] 129:433–460

Müller WEG, Kruse M, Koziol C, Leys SP (1998) Evolution of early Metazoa: Phylogenetic status of the Hexactinellida within the phylum of Porifera [sponges]. Progr Mol Subcell Biol 21: 141–156

Müller WEG, Wiens M, Batel R, Steffen R, Borojevic R, Custodio MR (1999) Establishment of a primary cell culture from a sponge: Primmorphs from *Suberites domuncula*. Marine Ecol Progr Ser 178:205–219

Nikoh N, Iawabe N, Kuma KI, Ohno M, Sugiyama T, Watanabe Y, Yasui K, Shi-Cui Z, Hori K, Shimura Y, Miyata T (1997) An estimate of divergence time of Parazoa and Eumetazoa and that of Cephalochordata and Vertebrata by aldolase and triose phosphate isomerase clocks. J Mol Biol 45:97–106

Nishimura Y, Kawabata T, Kato K (1988) Identification of latent procathepsins B and L in microsomal lumen: characterization of enzymatic activation and proteolytic processing in vitro. Arch Biochem Biophys 261:64–71

Reilly JJ, Mason RW, Chen P, Joseph LJ, Sukhatme VP, Yee R, Chapman HA (1989) Synthesis and processing of cathepsin L, an elastase, by human alveolar macrophages. Biochem J 257: 493–498

Reincke T, Barthel D (1997) Silica uptake kinetics of *Halichondria panicea* in Kiel Bight. Mar Biol 129:591–593

Reiswig H (1971) The axial symmetry of sponge spicules and its phylogenetic significance. Cah Biol Mar 12:505–514

Rogers JC, Dean D, Heck GR (1985) Aleurain: a barley thiol protease closely related to mammalian cathepsin H. Proc Natl Acad Sci USA 82:6512–6516

Sanders SL, Field CM (1994) Septins in common? Curr Biol 4:907–910

Schröder HC, Krasko A, Batel R, Skorokhod A, Pahler S, Kruse M, Müller IM, Müller WEG (2000) Stimulation of protein (collagen) synthesis in sponge cells by a cardiac myotrophin-related molecule from *Suberites domuncula*. FASEB J 14:2022–2031

Schröder K (1936) Beiträge zur Kenntnis der Spiculabildung der Larven-Spiculation und Variationsbreite der Gerüstnadeln von Süßwasserschwämmen. Z Morph Ökol Tiere 31:245–267

Schütze J, Reis Custodio M, Efremova SM, Müller IM, Müller WEG (1999) Evolutionary relationship of Metazoa within the eukaryotes based on molecular data from Porifera. Proc R Soc Lond B 266:63–73

Seilacher A (1994) Early multicellular life: late Proterozoic fossils and the Cambrian explosion. In: Bengtson S (ed) Early life on earth. Columbia Univ Press, New York, pp 389–400

Shimizu K, Cha J, Stucky GD, Morse DE (1998) Silicatein alpha: cathepsin L-like protein in sponge biosilica. Proc Natl Acad Sci USA 95:6234–6238

Simpson TL (1984) The cell biology of sponges. Springer, Berlin Heidelberg New York

Simpson TL, Refolo LM, Kaby M (1979) Effects of germanium on the morphology of silica deposition in a freshwater sponge. J Morphol 159:343–354

Simpson TL, Gil M, Connes R, Diaz JP, Paris J (1985) Effects of germanium (Ge) on the silica spicules of the marine sponge *Suberites domuncula*: transformation of the spicule type. J Morphol 183:117–128

Steiner M, Mehl D, Reitner J, Erdtmann BD (1993) Oldest entirely preserved sponges and other fossils from the Lowermost Cambrian and a new facies reconstruction of the Yangtze Platform (China). Berliner Geowiss Abh (E) 9:293–329

Uriz MJ, Turon X, Becerro MA (2000) Silica deposition in Demospongiae: spiculogenesis in *Crambe crambe*. Cell Tissue Res 301:299–309

Vinogradov AP (1953) The elementary chemical composition of marine organisms. Sears Foundation, New Haven

Wagner-Hülsmann C, Bachinski N, Diehl-Seifert B, Blumbach B, Steffen R, Pancer Z, Müller WEG (1996) A galectin links the aggregation factor to cells in the sponge [*Geodia cydonium*] system. Glycobiol 6:785–793

Weissenfels N (1989) Biologie und Mikroskopische Anatomie der Süßwasserschwämme (Spongillidae). Fischer, Stuttgart

Weissenfels N, Landschoff HW (1977) Bau und Funktion des Süßwasserschwammes *Ephydatia fluviatilis* L. (Porifera). IV. Die Entwicklung der monaxialen $SiO_2$-Nadeln in Sandwich-Kulturen. Zool Jahrb Abt Anat 98:355–371

Wilkinson CR, Garrone R (1980) Ultrastructure of siliceous spicules and microsclerocytes in the marine sponge *Neofibularia irata*. Span J Morphol 166:51–64

Wimmer W, Perovic S, Kruse M, Krasko A, Batel R, Müller WEG (1999) Origin of the integrin-mediated signal transduction: functional studies with cell cultures from the sponge *Suberites domuncula*. Eur J Biochem 178:156–165

Windholz M (1983) The Merck index. Merck, Rahway

Xie H, Surka M, Howard J, Trimble WS (1999) Characterization of the mammalian septin H5: distinct patterns of cytoskeletal and membrane association from other species. Cell Motil Cytoskel 43:52–62

Yang Y, Rao NS, Walker E, Sen S, Qin J (1997) Nuclear magnetic resonance assignment and secondary structure of an ankyrin-like repeat-bearing protein: myotrophin. Protein Sci 6: 1347–1351

Zhou Y, Shimizu K, Cha JN, Stucky GD, Morse DE (1999) Efficient catalysis of polysilane synthesis by silicatein α requires specific hydroxy and imidazole functionalities. Angew Chem [Int Edit] 38:780–782

[illegible]

# Biotechnology

# Biotechnological Advances in Biosilicification

Jan L. Sumerel and Daniel E. Morse

## 1 Introduction

In nature, enormous quantities of silica are produced by a diverse group of eukaryotes. Remarkably, this production of biosilica occurs under ambient conditions while at the same time obeying physical and chemical laws to build exquisite nanoscale architectures that in many cases cannot be reproduced in vitro. Because these organisms are able to sequester and condense silicon-based monomeric compounds into patterned morphologies unparalleled in complexity compared to anthropogenic materials, understanding the molecular mechanisms and the proteins that mediate these syntheses has come to center stage.

In most eukaryotes, pattern formation is largely governed by genetic control and manifests itself primarily in two ways, transcriptionally (at the level of gene expression) and translationally (at the level of protein synthesis and/or protein modification). Promoters, genetic elements at the beginning of each gene, are platforms for the assembly of initiators and repressors that control messenger RNA (mRNA) synthesis. This transcriptional regulation allows cells to become phenotypically distinct from neighboring cells since the combinatorial protein assembly on the promoter will determine whether or not a specific gene is transcribed. After transcription, additional proteins control mRNA half-lives and localization. During translation, ribosome-associated proteins regulate which mRNAs are translated into their corresponding polypeptides at discrete times. Because pattern formation is controlled genetically at so many stages, it often is difficult to determine at the molecular level how a specific pattern emerges.

Fluctuating resources can cause changes in gene expression resulting in phenotypic plasticity. Carnivore jaw morphologies are induced in tadpoles by the presence of specific prey (fairy shrimp) and in their absence, omnivore jaw morphologies dominate (Pfennig and Murphy 2002). Likewise, caterpillar larvae express different body styles depending on whether they hatch in the

Department of Molecular, Cellular and Developmental Biology, Materials Research Laboratory and California NanoSystems Institute, University of California at Santa Barbara, Santa Barbara, California 93106, USA

Progress in Molecular and Subcellular Biology, Vol. 33
W. E. G. Müller (Ed.)

spring or summer, and this phenotypic plasticity is governed by their diet (Greene 1989). In the spring, larvae feed on oak catkins and develop to look like catkins, whereas in the summer the catkins have fallen, and larvae develop to look like oak twigs and develop larger jaws so that they can successfully feed on oak leaves.

This propensity for adaptively induced morphological change coupled to biosilicification was demonstrated in *Crambe crambe* sponge populations (Maldonado et al. 1999). In these experiments, it was shown that the morphology of both sponges and their supporting skeletal elements (silica spicules) could be modulated as a function of dissolved silicic acid concentration. The sponges were cultured in the following silicic acid ($Si(OH)_4$) concentrations; 0.7 μM (the silicic acid concentration most typical of their native habitat), 30 and 100 μM $Si(OH)_4$. As the availability of silicic acid increased, the sponge's silicic acid uptake also increased proportionally. At 0.7 μM $Si(OH)_4$, styles (needle-like spicules) were the only spicule type produced. However, when the concentration was increased to 30 μM, there was a major change in spiculogenesis. Not only were large and small styles produced, but isochelae (symmetrical claw-like spicules) also were made. At 100 μM $Si(OH)_4$, the sponge produced four spicule types, large and small styles, isochelae and aster-like desmas (branched spicules). This study clearly demonstrated phenotypic plasticity of biosilica morphologies governed by silicic acid concentrations in the environment. However, it does not demonstrate the molecular basis for this response; we do not yet know the specific element that responds to this switch, or the molecular mechanism of gene regulation that mediates the observed changes in biosilicification.

It is our goal in this review to share with the reader how researchers studying biosilicification have employed the tools of molecular biology and biochemistry to further understand the underlying molecular mechanisms and their genetic control. We describe not only the discoveries, but also focus on the "silicon biotechnology" (Morse 1999, 2001; Cha et al. 2000a; Shimizu and Morse 2000) utilized to facilitate these discoveries. Many standard biochemical and molecular biological tools have been employed, including the isolation of proteins and sequencing of their amino acids, the construction and probing of DNA libraries for the isolation of genes and coding sequences of interest, the cloning and genetic engineering of recombinant proteins, and combinatorial bacteriophage display technologies, all allowing the elucidation of gene products that were not previously accessible. Also utilized are in vitro "directed evolution" techniques to produce novel "unnatural" or synthetic gene products that can recognize inorganic substrates and/or direct inorganic materials synthesis. These experiments have captured the interest of the materials science community, creating a new discovery path that we will call "inorganic biotechnology" that merges biological principles with materials science in order to produce advanced inorganic/organic hybrid materials with novel properties.

## 2 Silicon Transport in Diatoms

Diatoms are single-celled organisms in the plant class Bacilariophyceae characterized by silica walls that usually are beautifully sculpted and ornamented. In diatoms, biosilicification commences with the uptake of dissolved silicon in the form of silicic acid from surrounding waters. Silicic acid is pumped into these single-celled organisms against a concentration gradient through the plasma membrane into the cytosol. A method in biotechnology called subtractive hybridization has been elegantly used to isolate silicon-responsive genes from diatoms (Hildebrand et al. 1993). Briefly, two separate diatom populations were cultured; both populations were silicon-starved, but one population had silicon replenished in the form of silicic acid after starvation. mRNAs were isolated from both populations. The mRNAs were copied enzymatically to yield the complimentary DNAs (cDNAs) from the silicon-replenished cells, which were then hybridized with mRNA from silicon-starved cells. cDNAs specific for silicon uptake were identified and isolated by their lack of binding targets in the original silicon-starved mRNA population. A subpopulation of these silicon-uptake-specific cDNAs was found to encode the multipass transmembrane silicic acid transporters (Hildebrand et al. 1993). One of these transporter cDNAs was sequenced, and its corresponding amino acid sequence was shown to represent an integral cell membrane protein (Hildebrand et al. 1997). The predicted structure of this protein has ten right-handed $\alpha$-helices that traverse the membrane and pack cooperatively, presumably forming a barrel-shaped molecule with an interior channel constituting a hydrophilic environment for silicic acid movement through the hydrophobic plasma membrane (Fig. 1A).

Dramatic proof that the cloned DNA does in fact encode a silicon transporter was obtained when its corresponding mRNA was microinjected into *Xenopus laevis* oocytes, frog eggs that normally do not transport silicon (Fig. 1B). After microinjection of the mRNA, radioisotopically labeled germanic acid (often used as a surrogate for silicic acid) was transported into the oocytes from the surrounding medium. This specific uptake required sodium, adenosine triphosphate (ATP; an energy source) and protein synthesis. These requirements mimicked those requirements already demonstrated for silicic acid uptake in diatoms (Sullivan 1976), suggesting that the silicon transporter that had been cloned is the molecule responsible for silicic acid uptake in the diatom.

The intracellular form(s) of silicon and the molecular and biochemical pathways involved in biosilicification have not been fully elucidated, although it is hypothesized that silicic acid may be either conjugated or sequestered after import into the cytosol. The resulting silicon-containing precursors are then delivered preferentially to a membrane-bound organelle called the silicon deposition vesicle (SDV) where silica condensation occurs (Drum and

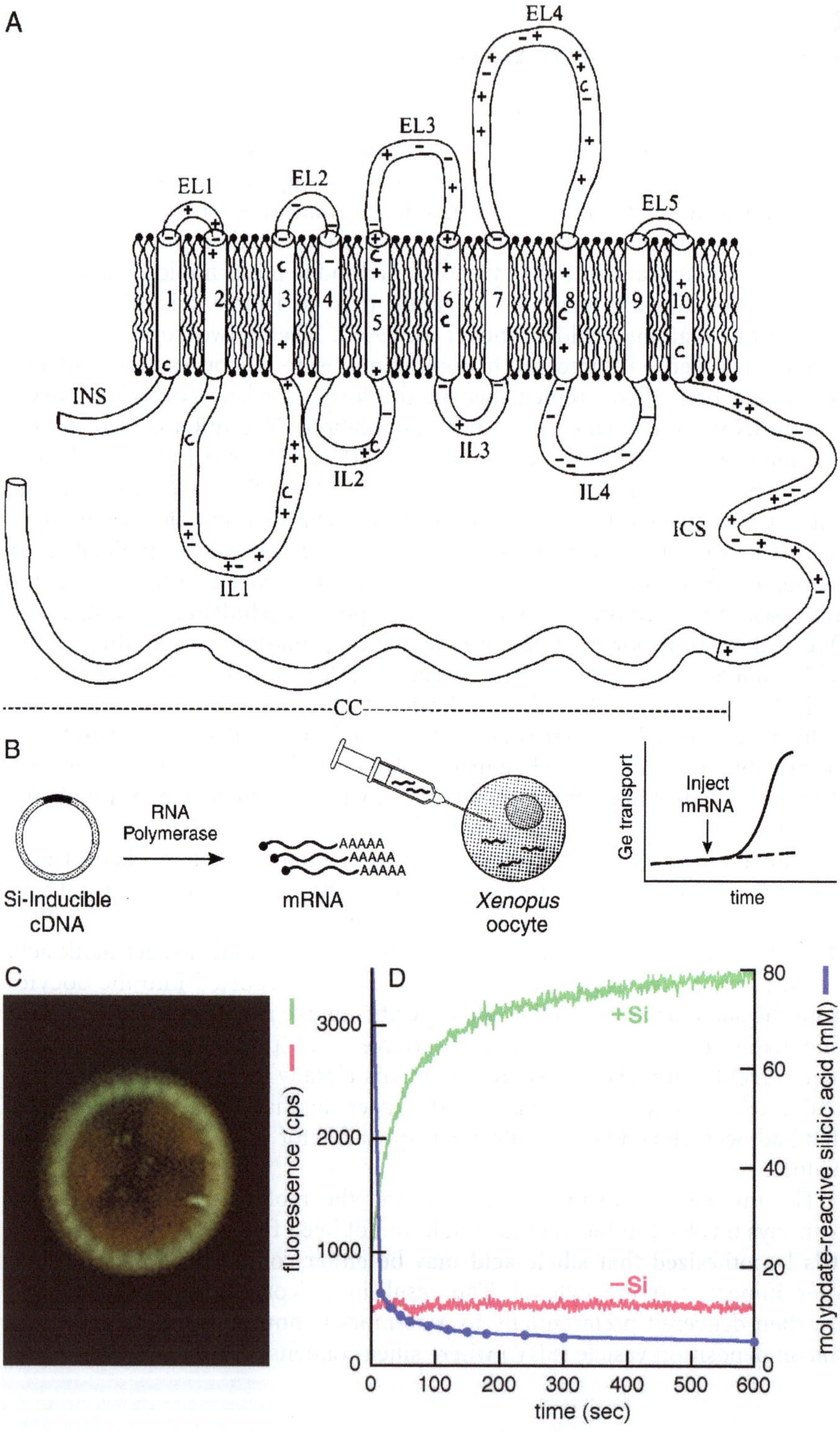
A
EL4
EL3
EL1
EL2
EL5
1
2
3
4
5
6
7
8
9
10
INS
IL1
IL2
IL3
IL4
ICS
CC
B
RNA Polymerase
AAAAA
Si-Inducible cDNA
mRNA
Xenopus oocyte
Inject mRNA
Ge transport
time
C
D
+Si
−Si
fluorescence (cps)
molybdate-reactive silicic acid (mM)
time (sec)
0
1000
2000
3000
100
200
300
400
500
600
20
40
60
80

Pankratz 1964; Reimann et al. 1966). Although these vesicles have not been isolated biochemically, it has been suggested that SDVs may contain silicon transporter proteins (Hildebrand et al. 1998). The resulting accumulation of silicic acid produces concentrations within the SDV that far exceed the threshold for spontaneous condensation (Sullivan 1986). This finding supports the suggestion of sequestration or chemical conjugation of the silicic acid, to prevent premature condensation of the silica in the cytosol. Spatial control of the site of silicification may require the action of proteins; first to conjugate the silicic acid to some protective moiety, and then to hydrolyze the conjugate (like the silicateins in the sponge; see below).

In order to resolve some of these unanswered questions, the import of silicic acid to form condensed silica has been studied in vivo using fluorescence microscopy and in vitro using fluorimetery (Shimizu et al. 2001). These studies made use of PDMPO [(2-(4-pyridyl)-5-((4-(2-dimethylaminoethylamino-carbamoyl)methoxy)-phenyl)oxazole], an acidotropic probe with a low $pK_a$ that has been shown to accumulate in acidic organelles including lysosomes (Diwu et al. 1999), In its acidotrophic behavior, PDMPO behaves like another probe (DAMP; 3-(2,4-dinitroanilino)-3′-amino-*N*-methylpropoylamine) used to investigate the acidic nature of the SDVs (Vrieling and Gieskes 1999). It was shown that PDMPO quickly entered diatom cells and was specifically incorporated into newly synthesized frustules (the silica walls of these cells; Fig. 1C; Shimizu et al. 2001). PDMPO also can be incorporated into newly synthesized spicules of cultured sponges (Weaver and Morse, unpubl. observ.). It was also demonstrated that PDMPO provides a convenient fluorimetric measure of silica formation in vitro (Fig. 1D). While these experiments and methods made it possible to observe silica formation in biological systems, characterization of the proteins that mediate these processes is needed to help us understand the underlying molecular mechanisms.

**Fig. 1.** A Silicic acid transporter protein. Topological model displaying how the protein may reside in the cellular membrane with ten lipid bilayer spanning regions, five extracellular loops (*EL*), four intracellular loops (*IL*), the amino terminal part of the protein (*INS*) and the intracellular carboxyl terminus (*ICS*) residing in the intracellular space. and the coil-coil region (*CC*) (reproduced with permission from Hildebrand 2000). **B** Diagram depicting microinjection experiments from Hildebrand et al. (1997). mRNA was synthesized from a cDNA template and microinjected into *Xenopus laevis* oocytes. After microinjection of the mRNA encoding the silicon transporter, the oocyte took up germanic acid, whereas control oocytes did not. **C** Fluorescence micrograph of a living diatom cell. The diatom had been grown in the presence of PDMPO (described in text), and the girdle bands show fluorescently labeled newly synthesized silica (*green*) (reproduced with permission from Shimizu et al. 2001). **D** Effect of silica condensation on the fluorescence emission intensity of PDMPO at 510 nm (excitation at 338 nm). Silica is produced from silicic acid, and the emission intensity increases by more than sixfold. Residual silicic acid was quantified with the molybdate assay (*blue*). *Pink* Fluorescence in the absence of added silicic acid; *green* fluorescence after the addition of silicic acid. (Reproduced with permission from Shimizu et al. 2001)

# 3
# Proteins Closely Associated with the Silica Wall of Diatoms

Thus far, understanding biosilicification has largely been accomplished by purifying proteins closely associated with silica structures. The diatom, *Cylindrotheca fusiformis* has a silica cell wall that encases and protects the single-celled organism's cellular contents. This cell wall has two predominant components, the smaller hypotheca overlapped by the larger epitheca at the cingulum. Close to the cingulum's center on the epitheca is an organic cell wall component called a girdle band. As a diatom divides into two new cells, the epitheca and hypotheca become the progenitor's epithecae, such that two hypothecae are synthesized for the progeny cells. Frustulins, glycoproteins that make up the outermost protein coat, appear to be cross-linked to the wall by calcium bridges (Kroger et al. 1994, 1996) and may play a structural role (van de Poll et al. 1999). These proteins are not occluded within the silica wall; thus their relationship to the dynamic process of silica formation remains unclear.

In order to study the occluded proteins, the silica wall was separated from the rest of the diatom, and proteins within the wall were isolated by dissolution of the silica (Kroger et al. 1994, 1997, 1999). Polyanionic proteins called pleuralins (formally known as HEPs), were found encased within the silica wall and could only be extracted from the wall by treatment with hydrofluoric acid (HF; Kroger et al. 1997). In order to isolate one of these proteins, pleuralin-1, the HF-soluble protein mixtures were neutralized, bound to hydroxyapatite, and differentially eluted. A high molecular weight fraction was separated that was substantially enriched with pleuralin-1; this protein was isolated and digested with trypsin, and the resulting peptides were sequenced. Sequence-specific degenerate DNA oligonucleotides were employed in the polymerase chain reaction (PCR) to generate a short double-stranded DNA probe. The resulting 35-bp DNA fragment was used in a rapid amplification of cDNA ends (RACE) PCR reaction to generate a 392-bp DNA fragment that was subsequently used to screen a cDNA library to isolate the full-length clone. From this screening, a full-length open reading frame (ORF) was isolated that represented a good pleuralin gene candidate. The deduced amino acid sequence corresponding to this full-length gene did not give any clues to its function. To determine the cellular location of this protein in the diatom, a truncated clone encoding the C-terminal fragment of the molecule was overexpressed in bacteria, and used to generate a polyclonal antibody (Kroger and Wetherbee 2000). This antibody ($\alpha$-pleuralin-1) was used in immunolabeling and electron microscopy experiments to determine pleuralin-1 localization through the diatom's cycle of cell division (Kroger and Weatherbee 2000). Whole mount diatoms were fixed with glutaraldehyde, mounted and thin-sectioned. Ultrathin sections were blocked with nonspecific protein to reduce background and then incubated in $\alpha$-pleuralin-1. After incubation, sections were washed and incubated in the presence of an α-rabbit gold-labeled secondary antibody revealing pleuralin-1 by the presence of electron-dense areas in an electron micrograph.

These analyses revealed that pleuralin-1 is localized to the terminal girdle bands on the epithecal inner surface at the cingulum in diatoms that have not entered cell division. During cytokinesis (the physical splitting of the cell during cell division) pleuralin-1 remains associated with the girdle bands, but is also detected in the cleavage furrow. It is not clear whether pleuralin-1 moves during cytokinesis or whether newly synthesized protein appears in the cleavage furrow. These studies did not elucidate the role pleuralin-1 plays, but its localization suggested the possibility that pleuralin-1 may be involved in both triggering the separation of the two daughter cells and the conversion of a hypotheca to an epitheca. Furthermore, these studies indicate that although many proteins are closely associated with or occluded within biosilica, not all of these proteins will have a direct role in biosilicification; some may in fact couple biosilicification to other processes such as cell division.

## 4 Polycationic Peptides and Polyamines Accelerate Silica Condensation

With the discovery of the dynamic role of the structure-directing silicatein proteins occluded in the sponge biosilica (Shimizu et al. 1998; Cha et al. 1999; see below), attention in the diatoms intensified in the search for mediators of silicification among the silica-occluded molecules. These efforts were rewarded with the discovery of the silaffins, low molecular weight polypeptides occluded within the silica that initially copurify with the pleuralins (Kroger et al. 1999). In order to isolate these proteins, instead of absorbing the HF-extracted proteins to hydroxyapatite (see above), the extract was fractionated by size-exclusion chromatography. The fractions containing proteins between 4 and 17 kDa (kilodaltons) were pooled, and three separate proteins, silaffin 1A (4 kDa), silaffin 1B (8 kDa) and silaffin 2 (17 kDa), were resolved by sodium dodecyl sulfate polyacrylamide gel electrophoresis (SDS-PAGE), a technique that separates proteins based on their Stokes' radii. Amino acid sequence information was obtained for silaffin 1B, and a double-stranded DNA probe was generated from this information to screen a phage cDNA library. Surprisingly, the isolated cDNA represented a gene encoding a protein of approximately 265 amino acid residues (~28 kDa), a protein much larger than the protein from which the probe was generated (Fig. 2A) and shared no sequence homology to any proteins in the database. The first 19 residues represent a signal peptide sequence, presumably responsible for targeting to the SDV. Although residues 20–107 are highly enriched in serine and glutamic acid, the role of this domain in the full-length peptide is unknown. The rest of the sequence is composed

A

| | | |
|---|---|---|
| | *MKLTAIFPLLFT* | 12 |
| | *AVGYCAAQ*SIADLAAANLS | 31 |
| | TEDSKSAQLISADSSDDAS | 50 |
| | DSSVESVDAASSDVSGSSV | 69 |
| | ESVDVSGSSLESVDVSGSS | 88 |
| | LESVDDSSEDSEEEELRIL | 107 |
| R1 | SSKKSGSYYSYGTKK | 122 |
| | SGSYSGYSTKKSASRRIL | 140 |
| R2 | SSKKSGSYSGYSTKKSGSRRIL | 162 |
| R3 | SSKKSGSYSGSKGSKRRIL | 181 |
| R4 | SSKKSGSYSGSKGSKRRNL | 200 |
| R5 | SSKKSGSYSGSKGSKRRIL | 219 |
| R5 | SSKKSGSYSGSKGSKRRNL | 238 |
| R7 | SSKKSGSYSGSKGSKRRIL | 257 |
| | SGGLRGSM | 265 |

B

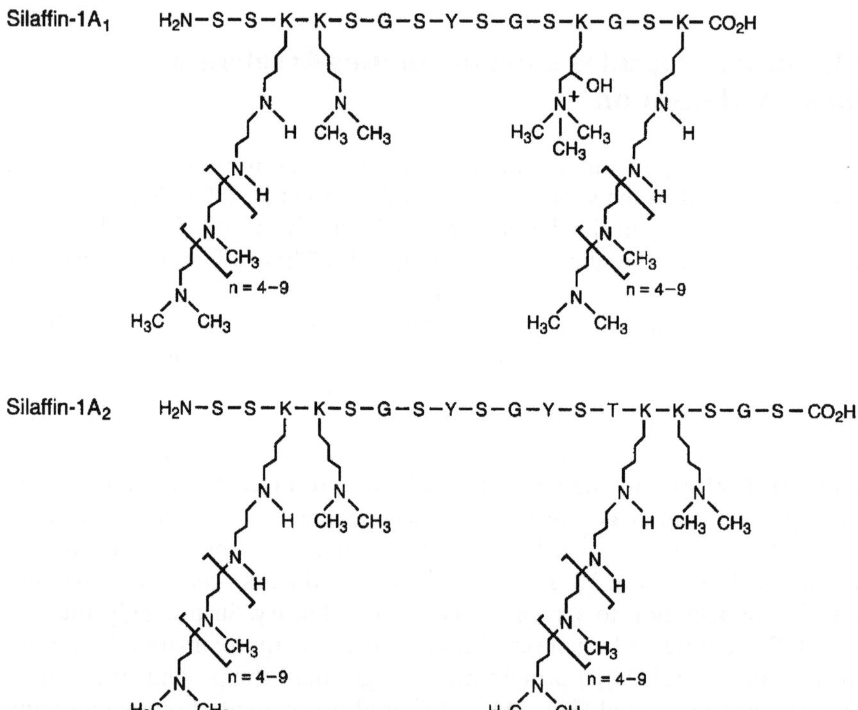


**Fig. 2.** A Amino acid sequence of *silp*, the gene isolated using the silaffin 1B DNA sequence as a probe. *Italicized residues 1–19* Signal sequence; *R1–R7* repeating peptides 1–7. One-letter code for amino acids used: *A* Ala, *C* Cys, *D* Asp, *E* Glu, *F* Phe, *G* Gly, *I* Ile, *K* Lys, *L* Leu, *M* Met, *N* Asn, *Q* Gln, *R* Arg, *S* Ser, *T* Thr, *V* Val, and *Y* Tyr (reproduced with permission from Kroger et al. 1999). B Comparison of silaffin $1A_1$ and silaffin $1A_2$ peptides as in A, illustrating the structures of the modified lysine residues. (Reproduced with permission from Kroger et al. 2001)

of seven repeating peptide sequences enriched in lysine residues (named R1–R7) followed by a short 8-amino acid C-terminal domain. It has not been determined how the gene or mRNA encoding the full-length 28-kDa silaffin are processed to yield the resulting low molecular weight silaffins, but it was suggested that silaffin 1A and 1B peptides are generated by endoproteolytic cleavages (Kroger et al. 1999). The alternative possibility that cleavage of acid-labile bonds during HF treatment could also generate these peptides was not initially excluded, but further experiments demonstrated that the sites of cleavage are highly specific rather than random, and that bacterially expressed recombinant full-length silaffin-1 is not degraded by HF (Kroger, pers. comm.). Peptide sequencing revealed that silaffin 1B contains the R1 peptide sequence, residues 102–120, and silaffin 1A is represented by the first 11 residues in the identical peptide repeating sequences R3–R7 (see Fig. 2A; Kroger et al. 1999). While it was determined that silaffin 1A and 1B are integral components of this large open reading frame (Kroger et al. 1999), the possibility that the 17-kDa protein, silaffin 2, could also be represented by this clone was not discussed even though the molecular weight of the protein represented by the full-length clone could encompass all three proteins isolated using this procedure.

Silaffin 1A and silaffin 1B peptides were found to contain multiple lysine residues (each with an n-butylamine side chain) making these peptides polycationic. The silaffin 1A fraction consisted of two structurally distinct peptides derived from the parent protein (silaffin $1A_1$ and silaffin $1A_2$; Fig. 2B; Kroger et al. 2001), both containing post-translational modifications of lysine. High-pressure liquid chromatography (HPLC) was employed to separate the two peptides; silaffin $1A_2$ eluted after silaffin $1A_1$ on a reverse phase C18 column (Kroger et al. 2001). Electrospray ionization mass spectrometry (ESI-MS) revealed subpopulations of the two peptides that differed in mass by multiples of 71 Da. Moreover, when silaffin $1A_1$ was digested with chymotrypsin, peptides resolved using reverse phase HPLC also differed in mass by multiples of 71 Da. It was determined that these differences correspond to a variations in long-chain polyamine modification (*N*-methylpropylamine$_n$) on lysine residues that with variable numbers of building blocks covalently linked and terminated with a dimethyl amine (Kroger et al. 1999, 2001). These long-chain derivatives were found at residue positions 3 and 5 on silaffin $1A_1$ and positions 3 and 14 on silaffin $1A_2$. Another lysine derivative, ε-*N*,*N*-dimethyllysine is at residue 4 in silaffin-$1A_1$ and residues 4 and 15 in silaffin-$1A_2$. It was also determined that silaffin-$1A_1$ has an additional ε-*N*,*N*,*N*-trimethyl-δ-hydroxylysine that is not present on silaffin-$1A_2$ at residue position 12.

The silaffin peptides were found to accelerate the precipitation of silica from silicic acid as verified by scanning electron microscopy (SEM) and quantified with molybdate (Iler 1979; Cha et al. 1999; see below). These peptides accelerated silica condensation from silicic acid, and silica/peptide spheres were produced. The size of spheres depended on the protein source; large spheres with 500–700 nm diameters were produced by silaffin 1A, and spheres whose diameters were less than 50 nm were produced by the protein fraction from the size exclusion column containing all the silaffins (Kroger et al. 1999). However, despite the ε-*N,N,N*-trimethyl-δ-hydroxylysine post-translational modification that distinguished the two silaffin 1A peptides (see above), there was no morphological distinction between the composite silica spheres produced from these two peptides (Kroger et al. 2001). Indeed, the activity of the silaffins resembles that of other polyamines that accelerate the condensation of silicic acid (Mizutami et al. 1998).

While characterizing the silaffins from different species of diatoms, Kroger and his colleagues discovered that the biosilica from the diatom *Navicula angularis* also contains a significant quantity of low molecular weight polyamines not detected in the original studies with *C. fusiformis* (Kroger et al. 2000). The range of molecular masses observed is similar to the mass differences from the polyamine adducts added post-translationally in a stepwise manner to the silaffins (see above). In addition, some of the *C. fusiformis* polyamines differ in mass by multiples of 57 Da, consistent with the molecular weight of a nonmethylated propylamine monomer, whereas the free polyamines from *N. angularis* exhibit multiple masses differing by 71 Da, consistent with *N*-methyl propylamine monomers. Isolated subpopulations of *N. angularis* polyamines were found to condense silica from silicic acid and the size of the resulting composite silica spheres depended on the size of the polyamines. Polyamines with molecular masses ranging from 1000 to 1250 Da produced predominantly particles within the 800-nm range, whereas polyamines with molecular masses between 600 and 750 Da produced particles of approximately 150 nm. In addition, using the entire mass range of polyamines (600–1250 Da), the size of silica particles decreased as the pH increased, reflecting the fact that pH controls the aggregation of condensing silica (Iler 1979). While these results demonstrate that polyamines isolated from *N. angularis* biosilica promote silicic acid polycondensation, it has also been postulated that phase-separation of these long-chain amines is involved in pattern formation (Sumper 2002). In this model, as the polyamine phase separates into smaller and smaller aqueous droplets, the silica condenses around these droplets, thereby generating regular patterns. It was further suggested that producing asymmetric silica structures could require the coupled action of the polyamines and silaffins (Kroger et al. 2000). While such a model remains to be validated, these studies provide important links between the size of the silica particles produced, the occluded polyamines and silaffins, and the local pH environment.

## 5 Employing Silica-Condensing Peptides to Fabricate Nanostructured Devices

A silaffin peptide without any post-translational modifications derived from the R5 peptide sequence (see Fig. 2A) has been used to develop the first proof-of-principle that silica-condensing peptides can be harnessed for advanced technological applications (Brott et al. 2001). Innovative use of a photolithographic process was made to develop a defined spatial pattern of silaffin molecules. A dispersion of acrylate monomers was seeded with silaffins and a photocatalyst. Laser-irradiation through a reticular grating induced photopolymerization to form regular rows of highly cross-linked polyacrylate, forcing a phase-separation of the silaffin-containing aqueous solution to the spaces between the rows of the cross-linked polymer. The result was a holographic "image" of the original grating, as rows of silaffin peptide-rich domains lying in valleys between highly cross-linked polyacrylate ridges. When this array was incubated with silicic acid, the peptides accelerated the in situ polycondensation of silicic acid to form regular rows of silica particles with an approximate spherical diameter of about 450 nm along the peptide valleys (Fig. 3). Because these particles have a dimension in the order of magnitude of the wavelength of light, the regular array behaves as a "photonic crystal". While the optical performance of this device is limited by the polydispersity of the silica particles and slight deviations from perfectly regular spacing, it provides a clear demonstration of the potential utility of silica-condensing peptides for the development of sophisticated applications.

However, these experiments also raise an important question: Why is it that peptides with polyamine linkages condense silica at pH 5 (the pH used in vitro to mimic the pH environment of the SDV; see Vrieling and Gieskes 1999), but the peptides without any post-translational modifications condense silica just as quickly at pH 8.0? How would both sets of in vitro experiments turn out had they been done at neutral pH? Even with these questions, these sophisticated

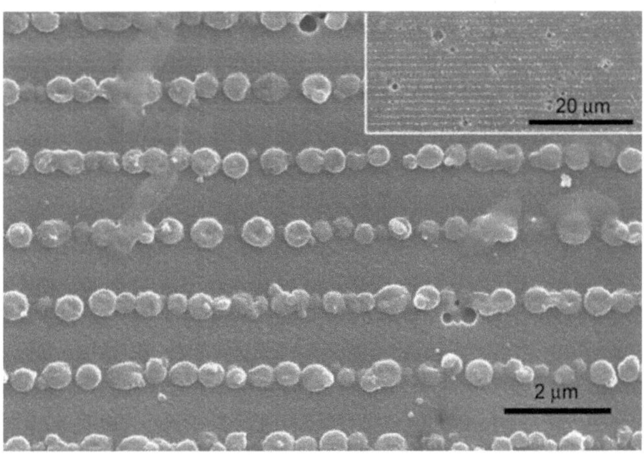


**Fig. 3.** Photonic crystal composed of silica spheres made from a holographically positioned array of silaffin peptides. Details as described in text. (Reproduced with permission from Brott et al. 2001)

experiments (Kroger et al. 2000; Brott et al. 2001) are an apparent leap of technology that interfaces the genetic information uncovered through molecular biology with the production of organic/inorganic hybrid materials that are of great interest in materials science.

## 6 Polycondensation-Catalyzing, Structure-Directing Catalytic Proteins from Sponge Biosilica

Biosilica also is condensed into both microscopic and macroscopic structures in sponges, and the molecular basis of this process has been investigated in the orange puffball sponge, *Tethya aurantia* (Shimizu et al. 1998; Cha et al. 1999). The sponge is skeletally supported and defended against predators by spicules, predominantly glassy needles that are approximately 2 mm in length and 30 μm in diameter. The protein axial filaments occluded within the spicules have been isolated by dissolving the silica in HF, yielding 1.7-mm protein threads that are approximately 2 μm in diameter (Shimizu et al. 1998). The protein filaments are predominantly composed of one protein family, named silicateins (for *silica* prot*eins*). Two subunits are present, silicateins α and β in the approximate ratio of 2:1; these are very similar in SDS-PAGE electrophoretic mobility and amino acid composition (Shimizu et al. 1998; Sumerel and Morse, unpubl. observ.). Silicateins α and β have been cloned, sequenced and characterized (Shimizu et al. 1998; Cao and Morse, unpubl. observ.). Using the BLOSUM62 matrix from the National Center for Biotechnology Information, protein sequence alignment of silicatein α indicates a 45% identity and 57% overall similarity to cathepsin L, a member of the family of lysosomal "catalytic triad" proteases . Using the same matrix, protein sequence alignment of silicatein β indicates a 40% identity and 56% overall similarity to Cathepsin L (Fig. 4A). Significantly, it was found that essential features of the catalytic active site and the position of the six cysteines (forming intramolecular disulfides) are highly conserved in the silicateins.

Cathepsin L and its family members are proteases that generally require a reducing, slightly acidic environment for their optimal activity, and their primary function is to degrade proteins inside lysosomes (Turk et al. 2000). Some members are selectively expressed in specific cell types. For example, cathepsin K is exclusively expressed in osteoclasts and plays a major role in bone resorption; its role in bone remodeling is to degrade collagen in the surrounding extracellular matrix (Bossard et al. 1996; Drake et al. 1996). These proteins are relatively small with molecular weights ranging from 20 to 35 kDa. As shown by X-ray crystallography, they have a heart-shaped structure composed of two domains (designated left and right) which interact with each other through a large planar interface maintained by three intramolecular disulphide cross-links (Turk et al. 2000). This structure is representative of many members of the larger superfamily to which these enzymes belong, the

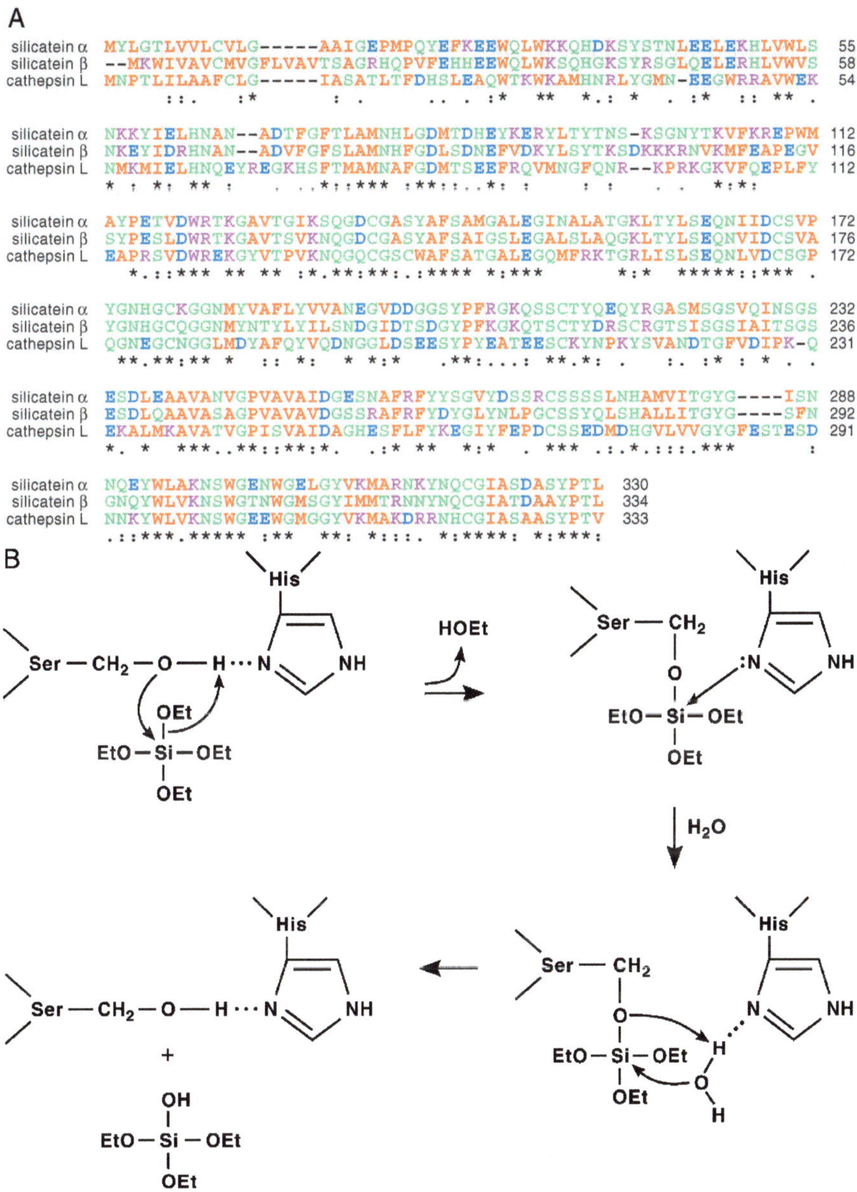


**Fig. 4.** A Sequence alignment of silicatein α, silicatein β and human cathepsin L. One-letter amino acid code used. *Red* Small and hydrophobic residues *A, V, P, M, I, L,* and *W*. *Blue* Acidic residues *D* and *E*. *Magenta* Basic residues *R, H,* and *K*. *Green* Hydroxyl, amide, imidazole, aromatic with hydroxyl *S, T, Y, H, C, N, G, Q*. *Asterisk* indicates identical or conserved residues in all sequences in alignment, *colon* indicates conserved substitutions, and *dot* indicates semi-conserved substitutions. Alignment was performed using the ClustalW program from the European Molecular Biology Laboratory, European Bioinformatics Institute. **B** Proposed mechanism of catalysis of the rate-limiting step in tetraethoxysilane (TEOS) hydrolysis leading to polycondensation catalyzed by silicatein. Mechanism confirmed by site-directed mutagenesis; see text for explanation. (Reproduced with permission from Cha et al. 1999)

catalytic triad hydrolases, in which the active site is characterized by a catalytic triad (Dodson and Wlodawer 1998). Molecular modeling of silicatein α shows that the catalytic triad active site is found in a V-shaped cleft between the two domains with two of the three presumptive active site residues, serine and histidine, in the middle of the cleft and the third amino acid, asparagine, immediately below these two residues further down in the cleft. The cysteine is donated from the left domain, and the histidine and asparagine are donated from the right domain.

Whereas most lysosomal cysteine protease members are endopeptidases that accomplish their task by nucleophilic attack on an amide bond at the carbonyl carbon, other members use the same nucleophilic attack on a carbonyl carbon to cleave ester bonds. The discovery that the silicateins are closely related to members of a family of enzymes suggested for the first time that proteins involved in biosilicification might act catalytically, rather than as passive templates (Shimizu et al. 1998). Testing this hypothesis, it was verified that the silicatein filaments and protein produced from recombinant DNA templates cloned in bacteria, do in fact catalyze the hydrolysis and subsequent polycondensation of silicon alkoxide substrates at neutral pH in a reaction analogous to those reactions by the catalytic triad hydrolases (Cha et al. 1999). When TEOS (tetraethoxysilane) was used as a substrate, silica was formed; when the methyl or phenyl triethoxysilanes were used as substrates, the corresponding methyl- and phenyl-silsesquioxanes (silicones) were produced. These products were quantified and characterized both by hydrolysis and colorimetric reaction with molybdate (Iler 1979), and by magic angle spinning solid-state nuclear magnetic resonance (NMR; Cha et al. 1999).

A mechanism for hydrolysis and polycondensation of silicon alkoxide substrates at the active site was proposed (Fig. 4B), based on the well-characterized mechanism of action of the catalytic-triad hydrolases with various carbon-based substrates (Cha et al. 1999). In this model, the active site functions as a general acid-base catalyst in the rate-limiting step during silicon alkoxide hydrolysis at neutral pH, closely paralleling the mechanism of catalysis of peptide bond or ester hydrolysis by other members of the enzyme family. It is suggested that hydrogen bonding between the nitrogen in histidine's imidazole ring and serine's hydroxyl increases the nucleophilicity of the serine oxygen, facilitating its attack on the silicon atom. Nucleophilic attack on the silicon atom would then displace ethanol in the case of TEOS, with the formation of a transitory covalent protein-*O*-Si intermediate. The addition of water would then complete the hydrolysis of the first alkoxide bond, with condensation then initiated by nucleophilic attack of the released Si-OH on a second substrate molecule (Cha et al. 1999).

Site-directed mutagenesis experiments performed with recombinant silicatein α confirmed the requirement for histidine and serine in the active site (Zhou et al. 1999), thus providing validation to this molecular model of hydrolysis. Similar experiments yielded identical confirmation of this mechanism for silicatein β (Cao and Morse, unpubl. observ.). When either the hydroxyl-containing serine or the imidazole-containing histidine residues in the active site were replaced with alanine, a dramatic decrease in silicatein-mediated catalysis resulted (Zhou et al. 1999; Cao and Morse, unpubl. observ.). The requirement for histidine was further substantiated by partially rescuing the activity of purified protein from the histidine mutant protein by exogenous imidazole (Merget, Sumerel and Morse, unpubl. observ.). These observations further defined the structural determinants required to catalyze the hydrolysis of a precursor silicon alkoxide.

# 7 Structure-Directing Polycondensation-Catalyzing Diblock Copolypeptides

As further confirmation that the interaction between the nucleophilic serine residue and the hydrogen-bonding with the nitrogen in the imidazole of the histidine at the active site are required for silicatein $\alpha$ to catalyze the hydrolysis and condensation of silicon alkoxides at neutral pH, diblock copolypeptides incorporating similar features were synthesized and found capable of structure-directing catalysis (Cha et al. 2000b). In the presence of TEOS, serine and histidine oligopeptides or mixtures of the two oligopeptides did not produce silica at neutral pH and ambient temperatures, suggesting that most simple oligopeptides lack the structural complexity of polypeptides required for substrate hydrolysis and/or condensation of silicon alkoxides at neutral pH. However, when a cysteine oligopeptide was tested under nitrogen (preventing oxidation of the sulfur side chains), it produced unstructured silica, indicating that reduced sulphydryl groups can play a role in the hydrolysis of TEOS. To test this hypothesis, using organonickel-catalyzed ring-opening polymerization of amino acid *N*-carboxyanhydrides (Deming 1997), diblock copolypeptides were synthesized containing precisely controlled blocks of defined features. Using this systematic approach, it was demonstrated that every coblock polypeptide that contained lysine had some activity in silica formation, and that these activities increased with the nucleophilicity of the side chains contributed by the second type of residue in the diblock copolypeptides. Based on these observations, cysteine-lysine diblock copolypeptides were generated. In a reducing environment under nitrogen, these diblock copolypeptides catalyze the formation of silica microspheres when reacted with TEOS at neutral pH (Fig. 5A). When identical diblock copolypeptides are exposed to air, disulphide bridges form between some of the cysteine residues, transforming the diblock copolypeptides to a self-assembled supramolecular concatenate. These concatenated peptides catalyze the synthesis of ordered fibrillar silica bundles from TEOS (Fig. 5B). These synthetic peptides thus mimic both activities of silicatein, serving both as catalysts and as structure-directing agents for the synthesis of silica from TEOS at neutral pH. These

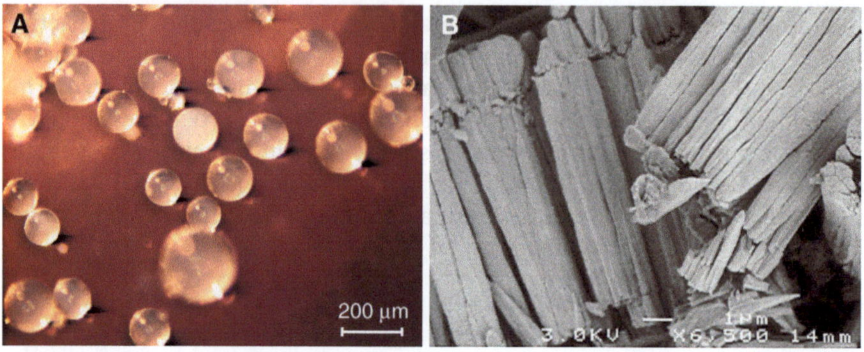


**Fig. 5A,B.** Structure-directed synthesis of silica catalyzed by biomimetic diblock copolypeptide from TEOS at neutral pH. Details of these experiments with self-assembling peptide mimics of silicatein are described in text (reproduced with permission from Cha et al. 2000b). **A** Silica spheres produced when concatenation of the peptide was blocked (optical micrograph). **B** Fibrillar bundles of silica produced when concatenation of the structure-directing catalytic peptides was promoted by oxidative cross-linking (electron micrograph)

experiments strongly suggest that the three-dimensional architecture resulting from self-assembly of the block copolypeptides plays a critical role in determining the resulting three-dimensional structure of the condensed silica.

# 8 Gene Expression During Sponge Development

Sponges, representing the simplest level of metazoan organization provide a uniquely tractable model system for the analysis of genetic mechanisms underlying cellular differentiation and the regulation of silicatein gene expression, and biosilicification is starting to play an important role in these studies. Sponges have been discovered to express a number of the *homeobox* family genes (Kruse et al. 1994; Seimiya et al. 1994; Degnan et al. 1995; Coutinho et al. 1998) that regulate pattern development and tissue-specific gene expression in higher organisms (Ferrier and Holland 2001), and sponge cells and tissues exhibit a remarkable developmental plasticity resembling that of the stem cells in higher organisms.

Recent work in *Suberites domuncula* has shown that silicate induces the expression of both silicatein and myotrophin genes, and that myotrophin subsequently induces the expression of collagen genes (Krasko et al. 2000). It is thought that the parallel induction of silicatein and collagen genes allows the parallel production of the skeletal spicule framework (in the case of silicateins) and spongin (in the case of collagen). Further work has shown that iron induces the expression of the ferritin gene which also stimulates the production of the skeletal spicule framework (Krasko et al. 2002). Furthermore, in vitro experiments showed that when iron was added to a reaction with recombinant silicatein derived from the silicatein gene from *S. domuncula* and TEOS, there was a threefold increase in silica polycondensation, but such a requirement for the silicateins from *T. aurantia* has not been found (Sumerel and Morse, unpubl. observ.). Moreover, the spicules from *T. aurantia* do not contain significant levels of iron as shown by inductively coupled plasma atomic emission spectroscopy (ICP-AES; Sumerel and Morse, unpubl. observ.). Studies in other species have begun (Weaver and Morse 2002, Weaver, Murr and Morse, unpubl. observ.) to analyze the complex spatial and temporal regulation of biosilicification responsible for the complex structures produced.

## 9
## The Biological Precursor for Silica Synthesis

The results with silicatein proteins gave independent experimental credence to the earlier suggestion that in some biological systems, the proximate precursor for biosilicification may consist of an organic conjugate of silicic acid. Further support for this suggestion was provided recently by NMR analysis showing that silicic acid forms higher valence conjugates of simple sugars at near-physiological pH and that these organosilicon compounds form easily between silicate anions and aliphatic carbohydrates (Kinrade et al. 1999). Even more recent evidence demonstrated a unique NMR chemical shift of $^{29}Si$ at −131.1 ppm after incubation of spin-labeled silicic acid with diatoms in culture. This signal was suggested to be attributable to an organic pentaoxo-azo-silicon complex (Kinrade et al. 2002), although this finding awaits further confirmation. In addition to catalysis, the silicatein proteins also were seen to direct the structures of the resulting polysiloxanes (Cha et al. 1999). Moreover, both silicatein $\alpha$ and $\beta$ showed good Michaelis-Menton kinetics, indicating that these proteins truly do behave as enzymes with the siloxane substrates (Zhou, Cao, and Morse, unpubl. observ.). Thermal denaturation of the silicateins from both the sponge and recombinant sources abolishes their catalytic activity, demonstrating that catalysis is dependent on the precise three-dimensional structure of the native protein (Cha et al. 1999; Cao and Morse, unpubl. observ.). These results demonstrated for the first time in any system that proteins involved in biosilicification might act dynamically as catalysts rather than simply provide a template for the condensing silica.

## 10
## Recognition of Inorganic Compounds Using Phage Display

Traditionally, bacteriophage (phage) peptide libraries that utilize the ability of viruses to infect bacterial cells and faithfully multiply themselves many-fold in the bacteria have been employed to elucidate protein–protein or drug–protein binding interactions (Kay et al. 2001). Using this system to find peptides that bind inorganic substrates has just added a new twist, and it has become a fundamental and powerful technology to engineer peptides capable of directing inorganic materials synthesis. The M13 filamentous bacteriophage viral coat contains the surface-displayed protein pIII in five copies per virion; genetic fusion of combinatorially generated peptides with pIII thus creates a means to display a wide variety of peptide sequences on the surface of the recombinant viruses. Millions of unique sequences of peptide-encoding DNA molecules can be generated and inserted into the pIII reading frame, thereby producing phage particles that differ from one another only in the 7-amino acid inserted peptide sequence expressed on the surface of the phage. Selection begins with libraries having up to $1 \times 10^{15}$ unique peptide-containing phage, and peptides with specific binding properties are "panned" by incubating the phage library with a target molecule. Unbound phage are removed by washing to eliminate peptides that do not recognize the target molecule. The bound phage are then eluted with high salt concentrations, amplified by infection of new host cells, and taken through additional panning cycles to successively enrich the phage pool with peptides that have the highest affinity for the target. After several rounds, individual phage clones are isolated and their sequences determined using the pIII flanking sequences as the priming sites for PCR amplification and DNA sequencing. The ability to rapidly identify the DNA that encodes the target peptide sequence in this "directed evolution" process is unparalleled by traditional genetic screening procedures (which are more cumbersome and time-consuming). Another advantage of phage display is that it provides a powerful means to select and recognize high-affinity peptide interactions that would otherwise be genetically elusive.

Using phage display, it was shown that silica recognition and binding by phage peptides can be used to enrich for a subset of peptides that accelerate the condensation of silica from silicic acid (Naik et al. 2002). Even though binding does not by itself guarantee that a peptide will condense silicic acid to silica, it is a prerequisite. Thus, the strategy of these experiments was to isolate specific phage that display peptides capable of binding silica and then assay those phage for the ability to precipitate silica from silicic acid. To isolate these silica-precipitating peptides, silica particles were incubated with phage, and the unbound phage were removed by washing. In order to 'find' specific silica particles with phage attached, a rabbit polyclonal antibody generated against the pIII protein was reacted with the phage-silica mixture. This mixture was further reacted with a fluorescent secondary antibody against rabbit antibodies, and positive phage were selected using immunofluorescence microscopy. Phage populations that showed fluorescence in this test were amplified and reselected using the same procedure, further purifying the phage with each panning cycle. The ability of each phage population to condense silica was tested and quantified using antibodies to capture the phage attached to silica, followed by measurement of affinity using the molybdate assay for

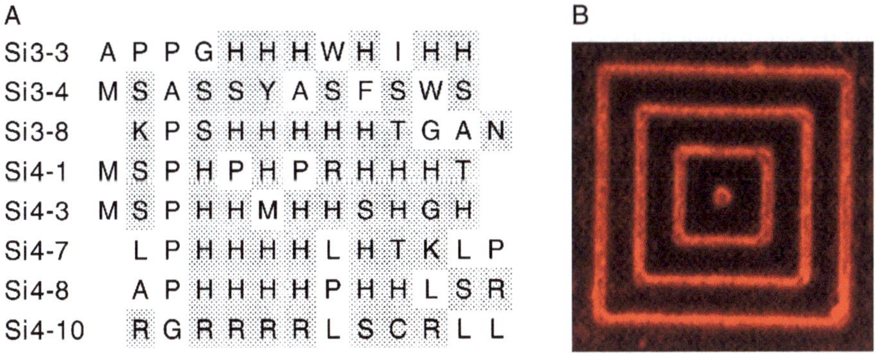


**Fig. 6.** **A** Multiple sequence alignment of silica-binding peptides obtained after several rounds of phage display and selection. One-letter code for amino acids used. *Gray shading* represents amino acid residues that have functional side chains that may interact with silica (reproduced with permission from Naik et al. 2002). **B** Phage recognition of GaAs semiconductor materials. The binding-selected phage was interacted with a GaAs patterned substrate. After interaction, the phage particles that bound to the substrate were reacted with an antibody against a phage coat protein and revealed by fluorescent labeling of the antibody with rhodamine-labeled secondary antibody (*red*). (Reproduced with permission from Whaley et al. 2000)

silica (see above, Cha et al. 1999 adapted from Iler 1979). Several peptides with affinities for silica were isolated in this screen (Fig. 6A). One of the peptides, Si4–1, has the highest ability to both recognize silica and condense silicic acid. In contrast, peptide Si4–10 has the second highest affinity for silica particles, but only marginal activity for condensation. Conversely, Si3–8 has the second highest activity in condensation, but its ability to recognize and bind silica is relatively low. Thus, from this phage display screening method, subpopulations can be further categorized based on two different properties, binding to condensed silica and condensation of a silica precursor. It is significant to note that the catalytic peptides identified in these studies are uniquely rich in those residues identified by site-directed mutagenesis as being essential for catalysis by silicatein (see above; Zhou et al. 1999).

Peptide phage display had been extended previously to find peptides that recognize semiconductor materials (Whaley et al. 2000), thus establishing a link between these inorganic materials and biology. Substrates for binding phage were prepared by etching Si (100) wafers in dilute hydrofluoric acid and by etching GaAs substrates in ammonium hydroxide. These prepared surfaces were incubated with phage, and those that bound to the surfaces were isolated, amplified and reselected with increasing detergent concentrations for higher binding stringency. It is interesting to note that as the purity and strength of binding of the selected phage increased, the overall amino acid composition of the peptides progressively converged toward a high proportion of uncharged polar functional groups and Lewis base groups. These results suggest that these two functionalities are most frequently required in a peptide for recognition of a target semiconductor material. Experiments such as these have selected peptides that can distinguish between different semiconductor materials, and one particular phage, G12–3, selectively recognizes GaAs (Fig. 6B). It was further shown that this recognition is specific to the composition of the semiconductor material, as the G12–3 clone is also able to distinguish GaAs from AlGaAs layers. These experiments further suggest that peptides which recognize specific inorganic surfaces could be used to fabricate novel electronic structures by providing a coordinating interface between the inorganic and organic surfaces with nanoscale precision and selectivity.

## 11 Future Prospects

The experiments discussed in this review have focused on the uses of silicon biotechnology and inorganic biotechnology both to elucidate the molecular mechanisms of biosilicification and to develop novel and environmentally benign new routes to inorganic material synthesis. These experiments have merged two fields, molecular biology and materials science. The search for silicon-processing molecules and organic molecules that recognize inorganic substrates will continue. The results are anticipated to provide further insight into the molecular mechanisms of biosilicification and bioinorganic synthesis. Moreover, they may facilitate the production of novel devices that will improve interfacial compatibility between cells and inorganic materials, such as medical prostheses and indwelling sensors. If the intimacy of such interfacial contacts can be improved, then it is also possible that biologically inspired recognition elements can be engineered on surfaces or incorporated into inorganic substrates for a wide range of biomedical applications. Further, these discoveries may lead scientists to a general route that could be used for many inorganic syntheses, thus allowing the production of new materials with enhanced electrical, optical and other physico-chemical properties.

*Acknowledgements.* We would like to thank our colleagues who helped make this work possible: Jennifer Cha, Katsuhiko Shimizu, Yan Zhou, Jixiang Cao, James Weaver, Markus Merget, Galen Stucky, Mark Brzezinski, Yolando Del Amo, Timothy Deming, Brad Chmelka and Sean Christianson. We also thank Herbert Waite, Mark Brzezinski, Mark Hildebrand, and Morley O. Stone for insightful discussions, and Meredith Murr and Germaine Fu for critical suggestions. This work was supported in part by grants from NASA (through the University Research, Engineering and Technology Institute on Bio-Inspired Materials under award No. NCC-1-02037, and through grant NAG1-01-003), the US Department of Energy (DEFG03-02ER46006), the US Army Research Office (DAAH04–96–1-0443 and DAA19-00-1-0564), the NOAA National Sea Grant Program, US Department of Commerce, under grant NA36RG0536, Project E/G-2 of the California Sea Grant College System, and the MRSEC Program of the National Science Foundation under award number DMR-96–32716 to the UCSB Materials Research Laboratory.

## References


Bossard MJ, Tomaszek TA, Thompson SK, Amegadzie BY, Hanning CR, Jones C, Kurdyla JT, McNulty DE, Drake FH, Gowen M, Levy MA (1996) Proteolytic activity of human osteoclast cathepsin K. Expression, purification, activation, and substrate identification. J Biol Chem 271: 12517–12524

Brott LL, Naik RR, Pikas DJ, Kirkpatrick SM, Tomlin DW, Whitlock PW, Clarson SJ, Stone MO (2001) Ultrafast holographic nanopatterning of biocatalytically formed silica. Nature 413: 291–293

Cha JN, Shimizu K, Zhou Y, Christiansen SC, Chmelka BF, Stucky GD, Morse DE (1999) Silicatein filaments and subunits from a marine sponge direct the polymerization of silica and silicones in vitro. Proc Natl Acad Sci USA 96:361–365

Cha JN, Shimizu K, Zhou Y, Christiansen SC, Chmelka BF, Deming TJ, Stucky GD, Morse DE (2000a) Learning from biological systems: novel routes to biomimetic synthesis of ordered silica structures. Mater Res Symp Proc 599:239–247

Cha JN, Stucky GD, Morse DE, Deming TJ (2000b) Biomimetic synthesis of ordered silica structures mediated by block copolypeptides. Nature 403:289–292

Coutinho CC, Seack J, Van de Vyver G, Borojevic R, Muller WE (1998) Origin of the metazoan bodyplan: characterization and functional testing of the promoter of the homeobox gene EmH-3 from the freshwater sponge *Ephydatia muelleri* in mouse 3T3 cells. Biol Chem 379: 1243–1251

Degnan BM, Degnan SM, Giusti A, Morse DE (1995) A *hox/hom* homeobox gene in sponges. Gene 155:175–177

Deming TJ (1997) Facile synthesis of block copolypeptides of defined architecture. Nature 390: 386–389

Diwu Z, Chen C-S, Zhang C, Klaubert DH, Haugland RP (1999) A novel acidotropic pH indicator and its potential application in labeling acidic organelles of live cells. Chem Biol 2:411–418

Dodson G, Wlodawer A (1998) Catalytic triads and their relatives. Trends Biochem Sci 23:347–352

Drake FH, Dodds RA, James IE, Connor JR, Debouck C, Richardson S, Lee-Rykaczewski E, Coleman L, Rieman D, Barthlow R, Hastings G, Gowen M (1996) Cathepsin K, but not cathepsins B, L, or S, is abundantly expressed in human osteoclasts. J Biol Chem 271:12511–12516

Drum RW, Pankratz HS (1964) Post-mitotic fine structure of *Gomphonema parvulum*. J Ultrastruct Res 10:217–233

Ferrier DE, Holland PW (2001) Ancient origin of the *Hox* gene cluster. Nat Rev Genet 2:33–38

Greene E (1989) A diet-induced developmental polymorphism in a caterpillar. Science 243: 643–646

Hildebrand M (2000) Silicic acid transport and its control during cell wall silicification in diatoms. In: Baeurlein E (ed) Biomineralization: from biology to biotechnology and medical application. Wiley-VCH, Weinheim, pp 171–188

Hildebrand M, Higgins DR, Busser K, Volcani BE. (1993) Silicon-responsive cDNA clones isolated from the marine diatom *Cylindrotheca fusiformis*. Gene 15:213–218

Hildebrand M, Volcani BE, Gassman W, Schroeder JI (1997) A gene family of silicon transporters. Nature 385:688–689

Hildebrand M, Dahlin K, Volcani BE (1998) Characterization of a silicon transporter gene family in *Cylindrotheca fusiformis*: sequences, expression analysis, and identification of homologs in other diatoms. Mol Gen Genet 260:480–486

Iler RK (1979) The chemistry of silica. Wiley, New York, pp 195–205

Kay BK, Kasanov J, Yamabhai M (2001) Screening phage-displayed combinatorial peptide libraries. Methods 24:240–246

Kinradi SD, Del Nin JW, Schach AS, Sloan TA, Wilson KL, Knight CTG (1999) Stable five- and six-coordinated silicate anions in aqueous solution. Science 285:1542–1545

Kinradi SD, Gillson A-ME, Knight CTG (2002) Silicon-29 NMR evidence of a transient hexavalent silicon complex in the daitom *Navicula peliculosa*. J Chem Soc Dalton Trans 3:307–309

Krasko A, Lorenz B, Batel R, Schroder HC, Muller IM, Muller WE (2000) Expression of silicatein and collagen genes in the marine sponge *Suberites domuncula* is controlled by silicate and myotrophin. Eur J Biochem 267:4878–4887

Krasko A, Schroder HC, Batel R, Grebenjuk VA, Steffen R, Muller IM, Muller WE (2002) Iron induces proliferation and morphogenesis in primmorphs from the marine sponge *Suberites domuncula.* DNA Cell Biol 21:67–80

Kroger N, Wetherbee R (2000) Pleuralins are involved in theca differentiation in the diatom *Cylindrotheca fusiformis.* Protist 151:263–273

Kroger N, Bergsdorf C, Sumper M (1994) A new calcium binding glycoprotein family constitutes a major diatom cell wall component. EMBO J 13:4676–4683

Kroger N, Bergsdorf C, Sumper M (1996) Frustulins: domain conservation in a protein family associated with diatom cell walls. Eur J Biochem 239:259–264

Kroger N, Lehmann G, Rachel R, Sumper M (1997) Characterization of a 200-kDa protein that is specifically associated with a silica-based substructure of the cell wall. Eur J Biochem 250: 99–105

Kroger N, Deutxmann R, Stumper M (1999) Polycationic peptides from diatom biosilica that direct silica nanosphere formation. Science 286:1129–1132

Kroger N, Deutzmann R, Bergsdorf C, Sumper M (2000) Species-specific polyamines from diatoms control silica morphology. Proc Natl Acad Sci USA 97:14133–14138

Kroger N, Deutzmann R, Sumper M (2001) Silica-precipitating peptides from diatoms: the chemical structure of silaffin-1A from *Cylindrotheca fusiformis.* J Biol Chem 276:26066–26070

Kruse M, Mikoc A, Cetkovic H, Gamulin V, Rinkevich B, Muller IM, Muller WE (1994) Molecular evidence for the presence of a developmental gene in the lowest animals: identification of a homeobox-like gene in the marine sponge *Geodia cydonium.* Mech Ageing Dev 77:43–54

Maldonado M, Carmona MC, Uriz MJ, Cruzado A (1999) Decline in Mesozoic reef-building sponges explained by silicon limitation. Nature 401:785–788

Mizutani T, Nagase H, Fujiwara N, Ogoshi H (1998) Silicic acid polymerization catalyzed by amines and polyamines. Bull Chem Soc Jpn 71:2017–2022

Morse DE (1999) Silicon biotechnology: harnessing biological silica production to construct new materials. Trends Biotech 17:217–259

Morse DE (2001) Biotechnology reveals new routes to synthesis and structural control of silica and polysilsesquioxanes. In: Rappoport Z, Apeloig Y (eds) The chemistry of organic silicon compounds, vol 3. Wiley, New York pp 805–819

Naik RR, Brott LL, Clarson SJ, Stone MO (2002) Silica-precipitating peptides isolated from a combinatorial phage display peptide library. J Nanosci Nanotech 2:95–100

Pfennig DW, Murphy PJ (2002) How fluctuating competition and phenotypic plasticity mediate species divergence. Evolution 56:1217–1228

Reimann BEF, Lewin JC, Volcani BE (1966) Studies on the biochemistry and fine structure of silica shell formation in diatoms. II. The structure of the cell wall of *Navicula pelliculosa.* J Phycol 2:74–84

Seimiya M, Ishiguro H, Miura K, Watanabe Y, Kurosawa Y (1994) Homeobox-containing genes in the most primitive metazoa, the sponges. Eur J Biochem 221:219–225

Shimizu K, Morse DE (2000) The biological and biomimetic synthesis of silica and other polysiloxanes. In: Baeurlein E (ed) Biomineralization: from biology to biotechnology and medical application. Wiley-VCH, Weinheim, pp 171–188

Shimizu K, Cha J, Stucky GD, Morse DE (1998) Silicatein α: cathepsin L-like protein in sponge biosilica. Proc Natl Acad Sci USA 95:6234–6238

Shimizu K, Del Amo Y, Brzezinski MA, Stucky GD, Morse DE (2001) A novel fluorescent silica tracer for biological silicification studies. Chem Biol 8:1051–1060

Stumper M (2002) A phase separation model for the nanopatterning of biosilica. Science 295: 2430–2433

Sullivan CW (1976) Diatom mineralization of silicic acid. I. $Si(OH)_4$ transport characteristics of *Navicula pelliculosa.* J Phycol 12:390–396

Sullivan CW (1986) Silicification by diatoms. In: Evered D, O'Conner MO (eds) Silicon biochemistry. Ciba Foundation Symposium 121. Wiley, Chichester

Turk B, Turk D, Turk V (2000) Lysosomal cysteine proteases: more than scavengers. Biochim Biophys Acta 1477:98–111

van de Poll WH, Vrieling EG, Gieskes WWC (1999) Location and expression of frustulins in the pennate diatoms *Cylindrotheca fusiformis*, *Navicula pelliculosa*, and *Navicula salinarum* (Bacillariophyceae). J Phycol 35:1044–1053

Vrieling EG, Gieskes WWC (1999) Silicon deposition in diatoms: control by the pH inside the silicon deposition vesicle. J Phycol 35:548–559

Weaver JC, Morse DE (2002) Molecular biology of Demosponge axial filaments and their roles in biosilicification. In: Uriz MJ (ed) Biology of silica deposition in sponges. Wiley, New York (in press)

Whaley SR, English DS, Hu EL, Barbara PF, Belcher AM (2000) Selection of peptides with semiconductor binding specificity for directed nanocrystal assembly. Nature 405:665–668

Zhou Y, Shimizu K, Cha JN, Stucky GD, Morse DE (1999) Efficient catalysis of polysiloxane synthesis by silicatein $\alpha$ requires specific hydroxyl and imidazole functionalities. Angew Chem Int Ed 38:780–782

# Silicase, an Enzyme Which Degrades Biogenous Amorphous Silica: Contribution to the Metabolism of Silica Deposition in the Demosponge *Suberites domuncula*

Heinz C. Schröder, Anatoli Krasko, Gaël Le Pennec, Teresa Adell, Matthias Wiens, Hamdy Hassanein, Isabel M. Müller, and Werner E.G. Müller

## 1 Introduction

Two classes of the phylum Porifera, the Demospongiae and the Hexactinellida, produce hydrated, amorphous, and noncrystalline silica which they deposit in their spicules. In contrast, the third class of Porifera, the Calcarea, have a skeleton which is composed exclusively of calcite spicules. In Demospongiae the amorphous hydrated silica, similar to opal, is formed primarily intracellularly in a special type of cells, the sclerocytes (see Uriz et al. 2000).

Sponges represent the phylogenetically oldest organisms which produce opal; they diverged approximately 800 million years ago (Ma; Müller 1997) from the common ancestor of the metazoans, the Urmetazoa (Müller 2001). Molecular sequence data revealed that among the Porifera the Hexactinellida and the Demospongiae which possess a siliceous skeleton are the oldest taxa, while the Calcarea (calcareous sponges) can be considered as a sister group to the Cnidaria and to the other metazoan taxa (Kruse et al. 1998; reviewed in Müller 1997). Other opal-producing organisms appeared after sponges, first the radiolarians and since the Cretaceous the diatom algae (see Racki 1999). Even though opaline silica is susceptible to dissolution, sponge siliceous spicules from hexactinellids have been preserved since the Ediacara, 550 Ma (Brasier et al. 1997). The earliest sponges (*Solactiniella plumata*; Hexactinellida) in body preservation are known from the Lowermost Cambrian Sansha section (Hunan, China), ≈530 Ma (see Steiner et al. 1993; Mehl et al. 1998), found at the Sansha section in both lower and upper levels of the Niutitang Formation (Fig. 1A). The Archaeocyatha, an extinct Porifera-related taxon, from the Flinders Ranges (South Australia), which had presumably a carbonate skeleton, are from the mid-Early Cambrian age (James and Gravestock 1990), 520 Ma (Fig. 1B). One of the oldest freshwater sponges was identified in oil shales of the Middle Eocene of Messel and was named *Spongilla gutenbergiana*

Institut für Physiologische Chemie, Abteilung Angewandte Molekularbiologie, Universität, Duesbergweg 6, 55099 Mainz; Germany

Progress in Molecular and Subcellular Biology, Vol. 33
W. E. G. Müller (Ed.)

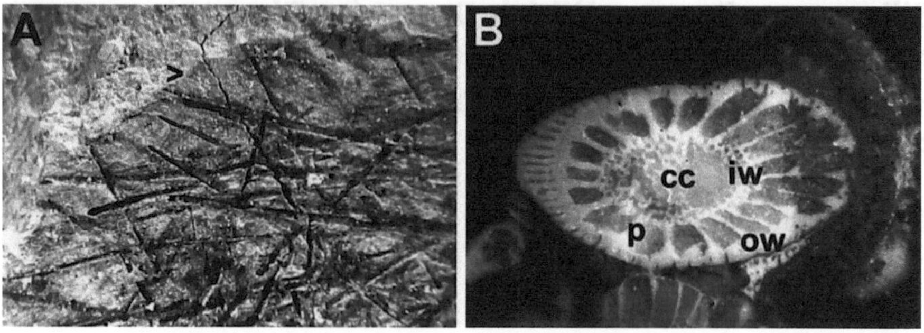


**Fig. 1A,B.** Fossil sponges. **A** Earliest sponges in body preservation with diactines, *Solactiniella plumata* (lowermost Cambrian Sansha section, Hunan, China). The border of the body preservation is indicated (>); magnification: ×5 (by courtesy of Drs. B.D. Erdtmann and M. Steiner). **B** Thin section through an archaeocyath from Flinders Ranges (South Australia); magnification: ×10. The characteristic structures are indicated: *cc* central cavity, *iw* and *ow* inner and outer wall, as well as *p* pariety

(Müller et al. 1982). Huge spongillites formed from spicules of freshwater sponges during the Holocene that have been found in Brazil (Volkmer-Ribeiro and Motta 1995), further support for the fact that - under suitable conditions - amorphous silica of spicules is provided with considerable stability.

It is striking that among the invertebrate (metazoan) groups, taxa which utilize silicon as the primary element to form a mineral skeleton never evolved again. This finding has been explained by the assumption that at the Precambrian/Cambrian boundary the coastal waters were silicate-rich (Brasier 1992). However, even though a huge reserve of silicate exists at present on earth, dissolved silicate is a limiting factor for the production of spicules due to progressive eutrophication (Racki 1999). Currently, seawater contains <3–150 μM of dissolved silicon (silicate; Kennish 1994). The surface water especially is poor with respect to silicon; the concentration is often below 3 μM (Nelson et al. 1995). In comparison, the level of silicon in human blood is ~140 μM (Geigy Scientific Tables 1984).

As comprehensively outlined (see Uriz et al. and Müller et al., this Vol.), the spicule formation in demosponges starts around an axial filament that also contains an enzyme termed silicatein, which has been attributed to the synthesis of amorphous hydrated silica (Cha et al. 1999; Krasko et al. 2000; Müller et al. 2002b). In contrast to the anabolic pathway which results in silica deposition, the catabolism of the siliceous spicules is only insufficiently known.

## 2 Siliceous Spicule Turnover

The formation of the spicules is a rapid process; under optimal conditions the 100–300 μm long spicules are synthesized within 40 h in the freshwater sponge

*Ephydatia fluviatilis* (Weissenfels 1989). The rate of appearance of new spicules is correlated with the growth of the oscular tube (Jones 1965), implying that an efficient relocation system must occur to place spicules at their functional destination. In addition, re- and disorganization processes in sponges, like bud formation, require an efficient machinery to replace spicules. To the best of our knowledge, no information is available on the postulated mechanism of spicule dissolution. However, one report exists which showed that sponges, with the example of the demosponge *Chondrosia reniformis*, have the capacity to etch and dissolve siliceous materials, e.g., marine sand (Bavestrello et al. 1995). In this context, another experiment of those authors is important. They could demonstrate that opaline spicules can only be dissolved in vitro after removal of the organic coat. In an earlier contribution, Schwab and Shore (1971) described that in mature spicules, the silica which is directly connected to the axial filament is more susceptible to hydrofluoric acid than the more external silica.

Opals are amorphous and contain water varying from 0 to 35%; if the water concentration falls to 7% [$4(SiO_2)\cdot H_2O$], opals become hard (Vinogradov 1953). The opaline silica in sponge spicules contains 6–13% water, yielding an approximate formula of $(SiO_2)_{2-5}\cdot H_2O$ (Schwab and Shore 1971). It has been established that the solubility of amorphous silica is strongly dependent on the pH value in the medium; the degree of polymerization of silica increases as (1) silicon concentration is increased, (2) pH is decreased or (3) temperature is decreased (Knight and Kinrade 2001).

# 3 Screening for Silica Degrading Enzymes

The enzyme very likely involved in catabolism of amorphous silica in sponges, here with *Suberites domuncula* (Demospongiae; Tetractinomopha; Hadromerida) as a model, was detected by application of the technique of "differential display" of mRNA using the in vitro primmorph cell culture system.

## 3.1 The Model Test System: Primmorphs

The cell culture system, the primmorphs, is described elsewhere (see Müller et al., this Vol.; Müller and Brümmer 1998; Müller et al. 1999). The primmorphs were cultivated for 8 days in seawater, supplemented with 30 µM $Fe(3+)$ [added as ferric citrate] and 1–10% of RPMI-1640 medium (natural seawater; Sigma; Krasko et al. 2002). The ambient silicon concentration in the medium was 5 µM. After 8 days some of the primmorphs were further incubated in this medium while others were transferred into a medium containing 60 µM of silicon (added as Na-hexafluorosilicate) for 1 or 3 days.

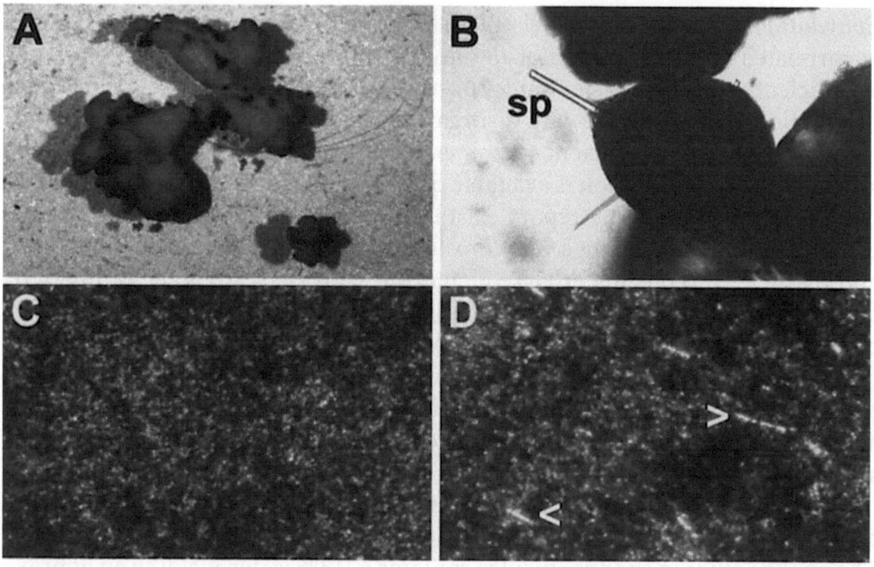


**Fig. 2A–D.** Effect of silicon on the formation of spicules in primmorphs. For the formation of primmorphs, dissociated cells were incubated in seawater supplemented with 10% RPMI-1640 medium and 30 µM Fe(3+). Then the primmorphs were transferred for 3 days into medium [RPMI-1640, Fe(3+)] which was additionally enriched with 60 µM silicon. **A** Primmorphs incubated in medium plus silicon; magnification: ×3. **B** In a few cases, the primmorphs start to synthesize spicules (*sp*); ×5. For a semiquantitative measurement, the primmorphs were squeezed between two cover slides. **C** Primmorphs which were incubated in the absence of silicon were almost completely devoid of spicules, while those which were cultured in silicon contained newly formed spicules (>) as shown in **D**; magnification: ×100

If the primmorphs were cultivated in the absence of additional silicon most of them had a round, spherical shape (see Müller et al., this Vol.). However, after cultivation in the presence of 60 µM silicon for an additional 3 days, most of them became oval-shaped (Fig. 2A). As already reported, primmorphs start to form spicules in the presence of silicon (Krasko et al. 2000). In some cases, synthesis of long (>100 µm) spicules was found in primmorphs (Fig. 2B), but usually only small 30 µm primordial spicules are seen (Fig. 2D). In the absence of any silicon, no spicules are present (Fig. 2C).

## 3.2 "Differential Display" of Transcripts

The detailed technique of "differential display" of mRNA/transcripts was described earlier (Müller et al. 2002a). Briefly, total RNA from control cultures (kept at the ambient silicon concentration of 5 µM) as well as from cells taken from treated samples (60 µM) was isolated using the TRIzol Reagent (Gibco

BRL). The first strand cDNA synthesis was performed with the anchored oligo(dT) primers and the AMV reverse transcriptase (Promega) according to the manufacturer's instructions. After first strand synthesis, the resulting cDNA preparation was diluted to one tenth with $H_2O$ and an aliquot of 2 µl was subjected to polymerase chain reaction (PCR). The reaction was performed in a volume of 20 µl using either the arbitrary primer 1 (5′-GTGATCGCAG-3′) or primer 2 (5′-CTTGATTGCC-3′), and 2 µM dNTP, $T_{11}$GC and 5 units of BioThem polymerase (Genecraft) together with [α-$^{32}$P] dATP. PCR reaction conditions were an initial denaturation at 95 °C for 5 min, then 40 amplification cycles each at 95 °C for 20 s, 42 °C for 120 s, 72 °C for 30 s, followed by a final incubation of 10 min at 72 °C. Samples were then subjected to a 5% polyacrylamide gel (in 1 × TBE). After the run the gel was dried and exposed to an X-ray film for 4 days. The gel sections corresponding to bands (autoradiograph) of interest were excised, boiled for 15 min in 200 µl $H_2O$, cooled on ice, and centrifuged for 10 min (14000 × g). The resulting supernatants were adjusted with equal volumes of 10 M ammonium acetate, 20 µg/ml tRNA and precipitated with 2.5 volumes of ethanol (–80 °C; overnight). The cDNA precipitates were washed three times in 75% ethanol and dissolved in 20 µl $H_2O$.

Approximately 2 µl of the eluted DNA was reamplified in 50 µl reactions using the same conditions and primers (see above), subcloned in *pGEM-T* vector (Promega) and sequenced.

Differentially expressed transcripts were analyzed here; their lengths varied between 240 and 600 bp. Only those transcripts were selected which became visible in the gels from products obtained with RNA isolated from cells, treated with 60 µM silica (Fig. 3). The cDNAs/transcripts identified were compared with the sequences listed in the BLAST data base (human). The following most related molecules were: calcium/calmodulin-dependent protein kinase (CaM kinase) II gamma (XM_044349, "expect value" [*E*] (Colligan et al. 2000) 1e $^{-16}$); hypothetical protein (XP_101359, *E* 1.6); MUC3B mucin (AJ291390, *E* 0.20); hypothetical protein (XP_067115, *E* 5.9); hypothetical protein (XP_090138, *E* 2.9); ATP-binding cassette, subfamily A member 4 (XM_001290, *E* 1.6); polypeptide similar to the zinc finger protein 91 (XM_091947, *E* 3.1); hypothetical protein (XP_104250, *E* 0.48), hypothetical protein (XP_169372, *E* 8.6); hypothetical protein (XP_104250, *E* 4.1), hypothetical protein (XP_098020, *E* 3.3) and hypothetical protein (XP_169372, *E* 8.6).

In addition to these sequences, two transcripts have been identified, which were analyzed here in greater detail: those which encode the silicase and the alpha subunit of the isocitrate dehydrogenase 3.

## 3.3
## Cloning of the Gene Encoding the Silicase

Applying the technique of "differential display" of mRNA, one cDNA was identified which codes for an enzyme with sequence similarity to carbonic

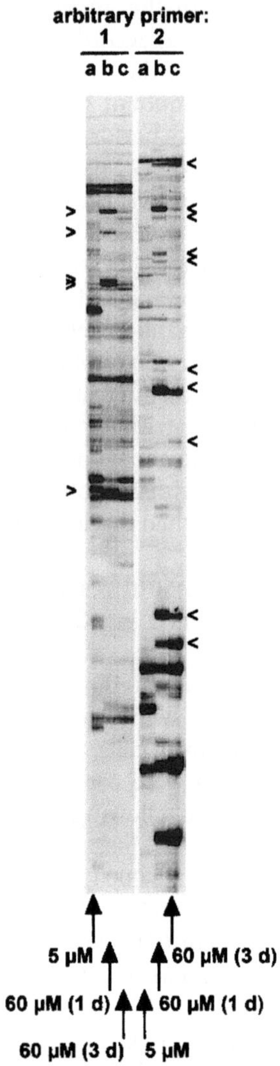


**Fig. 3.** Identification of transcripts in primmorphs which became upregulated after incubation in 60 μM silicon for 1 or 3 days by the technique of differential display. The primmorphs remained either at the ambient silicon concentration of 5 μM (*lanes a*) or were incubated with 60 μM silicon (*lanes b* and *c*). RNA was extracted and used for the analysis. For the amplification of the transcripts two different arbitrary primers (*1* and *2*) were used. Further details for cDNA preparation and amplification by PCR reaction are given in the text. Those transcripts which appeared only at higher silicon concentration (*lanes b* and *c*) and had been analyzed are indicated (>)

anhydrase, an enzyme which is involved in pH regulation, $HCO_3^-$ reabsorption and $CO_2$ expiration (see Sun and Alkon 2002; Deitmer and Rose 1996).

### *3.3.1*
### *Silicase*

The complete cDNA for the putative enzyme, termed *SDSIA* is 1395 nt long and comprises one open reading frame from $nt_{122}$–$nt_{124}$ (start methionine) to

$nt_{1259}$–$nt_{1261}$ (stop codon). The deduced polypeptide, SIA_SUBDO, of 379 aa has a putative size ($M_r$) of 43131 and an estimated isoelectric point (pI) of 6.5 (PC/GENE 1995; Fig. 4A). The deduced protein from *SDSIA* was termed silicase, SIA_SUBDO, since it is shown in the following that the recombinant silicase has the capacity to dissolve amorphous silica (see below). Northern blot analysis performed with the sponge *SDSIA* clone as a probe yielded one prominent band of ~1.5 kb, indicating that the full-length cDNA was isolated (see below).

The deduced sponge protein shares highest similarity to the family of carbonic anhydrases. Up to now more than seven isoenzymes of carbonic anhydrases have been identified in humans (Sun and Alkon 2002). The *S. domuncula* silicase has a similarity score "expect value" [*E*] of $2e^{-29}$ to human carbonic anhydrases II (CAH2_HUMAN; P00918). The eukaryotic-type carbonic anhydrase domain (PFAM00194, www.ncbi.nlm.nig.gov) is found in the sponge silicase between $aa_{87}$ to $aa_{335}$ (Fig. 4A). An alignment of the sponge silicase with the human carbonic anhydrase II shows that most of the characteristic amino acids forming the eukaryotic-type carbonic anhydrase signature are also present in the sponge silicase (Fujikawa-Adachi et al. 1999; Okamoto et al. 2001); however, in the sponge sequence, the residues 192 [alanine], 205 [phenylalanine] and 207 [again phenylalanine] are replaced (Fig. 4A).

The carbonic anhydrases are a family of zinc metal enzymes that are involved in the reversible hydration of $CO_2$ (Sly and Hu 1995). The three conserved histidine residues are found in the silicase at $aa_{181}$, $aa_{183}$ and $aa_{206}$ (Fig. 4A).

### 3.3.2 *Phylogenetic Analysis of Silicase*

A phylogenetic tree was constructed to determine the position of the sponge silicase among the different, selected members of the carbonic anhydrase family (Fig. 4B). After alignment, the tree was calculated and rooted with the bacterial carbonic anhydrase sequence from *Neisseria gonorrhoeae*. It becomes obvious that the sponge silicase forms the basis for the other metazoan carbonic anhydrases with the carbonic anhydrase from *Caenorhabditis elegans*. The metazoan enzymes are separated from the plant carbonic anhydrases and also from the bacterial enzymes; in yeast no related carbonic anhydrase has yet been identified.

## 4 Cloning of a Marker Gene of the Intermediary Metabolism

Another cDNA of *S. domuncula* identified by "differential display" was found to be highly related to the α-subunit of the $NAD^+$-dependent isocitrate

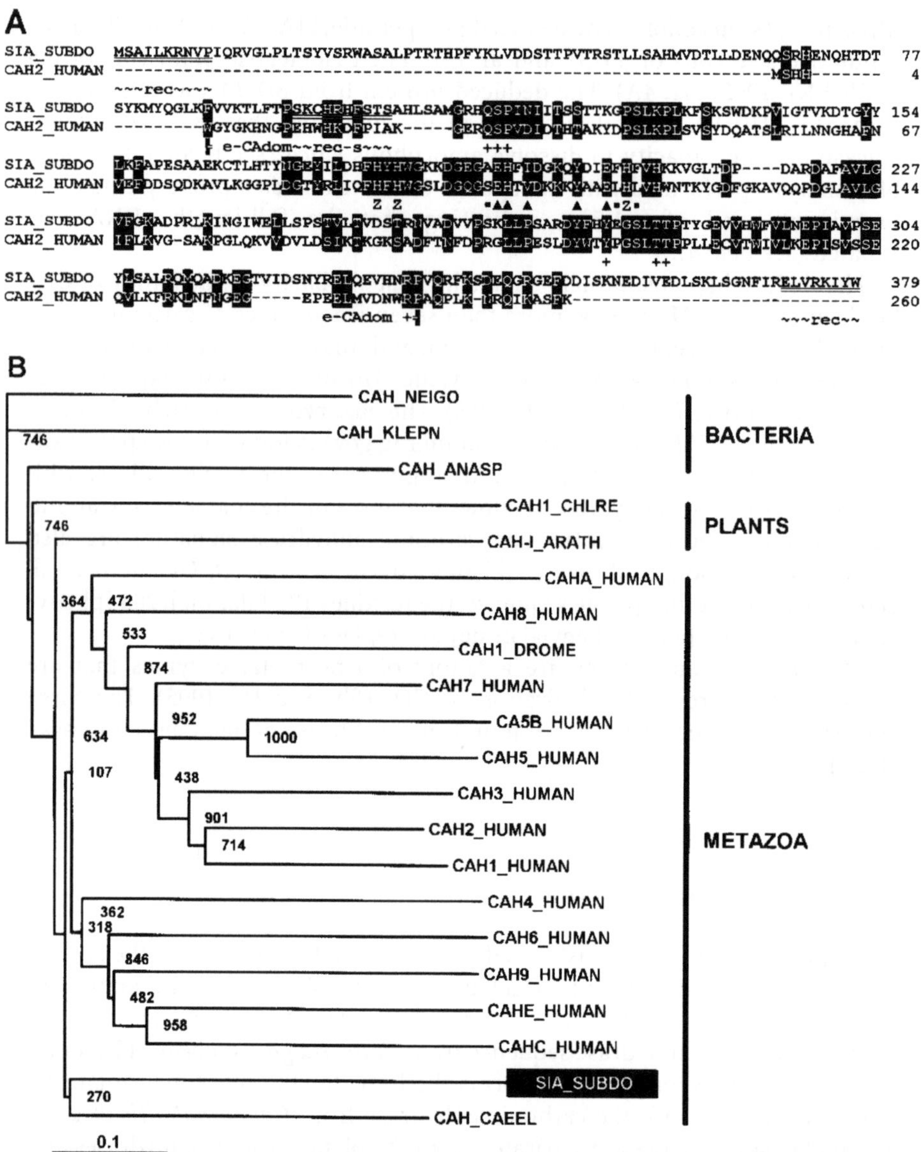


**Fig. 4A,B.** Sponge silicase. A Alignment of the sponge silicase (carbonic anhydrase-like) from *S. domuncula* (SIA_SUBDO) with the human carbonic anhydrase II (carbonate dehydratase II) (CAH2_HUMAN; P00918). The carbonic anhydrase domain is framed (╟ e-CAdom ╢). The characteristic amino acids forming the eukaryotic-type carbonic anhydrase signature are indicated [*filled triangles* (found in both sequences); *filled squares* (present only in the carbonic anhydrases, but not in the silicase)]; the additional *plus signs* (+) indicate the residues that form the active-site hydrogen network. The three zinc-binding histidine residues are indicated (*Z*). The similar amino acid residues within the two sequences are *highlighted*. The borders of the large (~rec~ to ~rec~) as well as of the short recombinant silicase (~rec-s~ to ~rec~) are *marked* and *double underlined*. B Rooted phylogenetic tree constructed with the sponge silicase

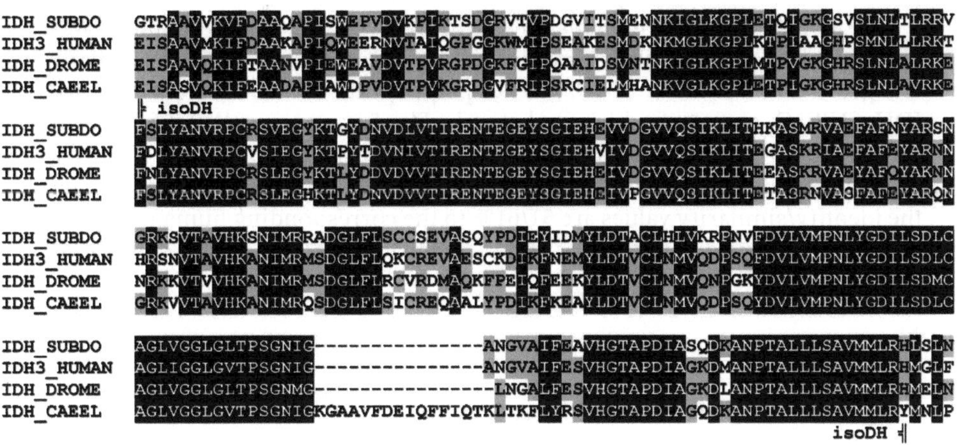


**Fig. 5.** Sponge isocitrate dehydrogenase 3 [(NAD$^+$) alpha subunit]. The segment of the *S. domuncula* isocitrate dehydrogenase (IDH_SUBDO) was aligned with the corresponding sequences from human (IDH3_HUMAN, $aa_{45}$ to $aa_{321}$; NP_005521), *D. melanogaster* (IDH_DROME, $aa_{61}$ to $aa_{336}$; AAF48965) and *C. elegans* (IDH_CAEEL, $aa_{38}$ to $aa_{329}$; NP_492330). Similar/identical amino acids are in *white on black*; residues conserved in at least three sequences are *shaded*. The conserved isocitrate dehydrogenase motif is marked (╟ isoDH ╢)

dehydrogenase, and was termed *SDIDH*; the deduced protein was named IDH_SUBDO (Fig. 5).

The NAD$^+$-dependent isocitrate dehydrogenase catalyzes the rate-limiting step of the citric acid cycle, the conversion of isocitrate to α-ketoglutarate; this reaction is coupled to the production of NADH (see Cupp and McAlister-Henn 1991). The enzyme is present in all eukaryotic cells. The enzyme from *Saccharomyces cerevisiae* has been found to be an octamer composed of two non-identical subunits –1 and –2; in mammals, the isocitrate dehydrogenase exists as a heterotetramer, composed of 2α, 1β, and 1γ subunits (see Kim et al. 1999).

**Fig. 4A,B.** *Continued* and the following related enzymes: the human carbonic anhydrases I (carbonate dehydratase I) (CAH1_HUMAN; P00915), II (CAH2_HUMAN), III (CAH3_HUMAN; P07451), IV (CAH4_HUMAN; P22748), VI (CAH6_HUMAN; P23280), VII (CAH7_HUMAN; P43166), VIII (CAH8_HUMAN; P35219), IX (CAH9_HUMAN; Q16790), X (CAHA_HUMAN; Q9NS85), VA (CAH5_HUMAN; P35218), VB (CA5B_HUMAN; Q9Y2D0), XII (CAHC_HUMAN; O43570), XIV (CAHE_HUMAN; Q9ULX7), the carbonic anhydrase from *Caenorhabditis elegans* (CAH_CAEEL; NP_510674.1), from *Drosophila melanogaster* (CAH1_DROME; NP_523561.1), from the plants *Arabidopsis thaliana* (CAH-l_ARATH; NP_196038.1) and *Chlamydomonas reinhardtii* (carbonate dehydratase 1) (CAH1_CHLRE; P20507) and also the bacterial carbonic anhydrases from *Neisseria gonorrhoeae* (CAH_NEIGO; Q50940), *Klebsiella pneumoniae* (CAH_KLEPN; O52535) and the cyanobacteria *Nostoc* sp. *PCC 7120* (CAH_ANASP; P94170). *Scale bar* indicates an evolutionary distance of 0.1 aa substitutions per position in the sequence. The phylogenetic tree was constructed by neighbour-joining, as implemented in the "Neighbor" program. (Felsenstein 1993)

The cDNA of the *S. domuncula* isocitrate dehydrogenase α-subunit was identified by "differential display" technique; the sequence was subsequently elongated and termed *SDIDH*. The polypeptide deduced from *SDIDH*, IDH_SUBDO, a part of which is shown in Fig. 5, comprises the strictly conserved isocitrate dehydrogenase motif (PFAM00180). The sponge sequence is highly related to the metazoan sequences; within the conserved domain the identity/similarity values are 51/61% to the corresponding human enzyme, 64/77% to *D. melanogaster* isocitrate dehydrogenase, and 63/72% to the *C. elegans* molecule (Fig. 5).

## 5
## Preparation of Recombinant Silicase

The goal was to prepare the recombinant *S. domuncula* silicase (rSIA_SUBDO) to elucidate its enzymatic activity. Expression of the *SDSIA* gene was performed in *Escherichia coli* using the "GST (glutathione-*S*-transferase) Fusion" system (Amersham) as described (Ausubel et al. 1995; Coligan et al. 2000) following the instructions of the manufacturer. Two inserts were used in order to eliminate potential effects of signal peptides during expression; one insert comprised the total deduced protein (long form; from $aa_1$ to $aa_{379}$), while the short insert had only $aa_{96}$ to $aa_{379}$ (Fig. 4A). The corresponding clones were termed *SDSIA-l* and *SDSIA-s*. They were introduced into the *pGEX-4T-2* plasmid containing the *Schistosoma japonicum* glutathione *S*-transferase (GST) gene and expressed with isopropyl β-D-galactopyranoside (IPTG) in the presence of 1 mM $ZnSO_4$ (Nair et al. 1991) for 4 or 6 h at 37 °C. The GST fusion proteins, termed rSIA_SUBDO-l (large form; $M_r$ 69 kDa) and rSIA_SUBDO-s (short form; $M_r$ 58 kDa), were purified by affinity chromatography on glutathione Sepharose 4B (Coligan et al. 2000). Then the fusion proteins were cleaved with thrombin (10 units/mg) to separate glutathione-*S*-transferase from the recombinant sponge silicase. Gel electrophoresis of the proteins was performed in polyacrylamide gels (10%) containing 0.1% $NaDodSO_4$ (PAGE) according to Laemmli (1970). Protein samples were subjected to gel electrophoresis in the presence of 2-mercaptoethanol and stained with Coomassie brilliant blue.

After cleavage, purification and subsequent PAGE, the purified recombinant proteins, the large form (rSIA_SUBDO-l; 43 kDa) and the short form (rSIA_SUBDO-s; 32 kDa) were obtained (Fig. 6).

## 6
## Enzymatic Activities of Recombinant Silicase

Due to (1) the high sequence similarity to carbonic anhydrase and (2) the fact that this gene is expressed after incubation at high silicon concentration, the

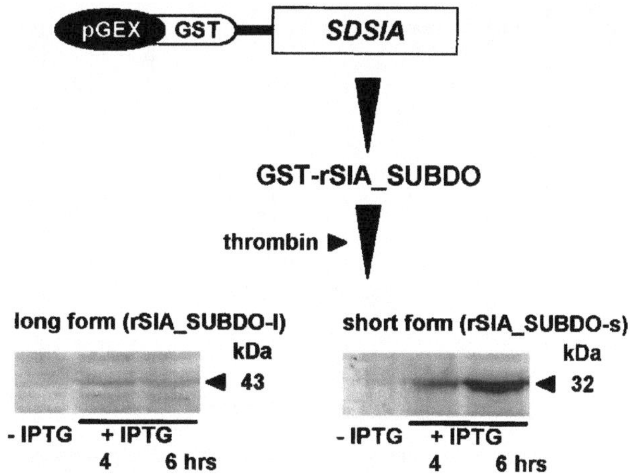


**Fig. 6.** Preparation of the recombinant silicase. The recombinant *S. domuncula* rSIA_SUBDO was prepared as GST fusion protein as described in the text. Both the long and the short *SDSIA* were cloned into a *pGEX-4T-2* plasmid containing the glutathione *S*-transferase (GST) gene. Either without IPTG (–IPTG) or after incubation with IPTG (+IPTG) for 4 or 6 h, the fusion protein was isolated, cleaved, purified and subsequently subjected to $NaDodSO_4$-PAGE (the gel was stained with Coomassie brilliant blue). The purified large form rSIA_SUBDO-l with a size of 43 kDa as well as the short form ($M_r$ 32 kDa) of the silicase were obtained

recombinant silicase was tested for potential activity on hydration of $CO_2$ and dissolution of amorphous silica (spicules). Here, only the activities found for the short form of the recombinant silicase (rSIA_SUBDO-s) are given.

## 6.1 Carbonic Anhydrase Activity

To test the carbonic anhydrase activity of rSIA_SUBDO-s, an established assay was applied (Armstrong et al. 1966). This assay is based on the hydrolysis of an ester, *p*-nitrophenyl acetate. 0.5 ml of a 30 mM solution of *p*-nitrophenyl acetate (Sigma) was mixed with 0.05 ml of a 0.3 mM Tris-HCl (pH 7.6) buffer. After a pre-incubation period (25 °C; 5 min), 50 µl of rSIA_SUBDO (silicase) were added. The increase in absorbance at 348 nm was measured. Over a period of time (5 min) the hydrolysis was determined. The enzyme activity is given in optical density (OD) units per min.

As shown in Fig. 7, the activity of the recombinant silicase depends on the concentration of the enzyme in the assay. Addition of 1 µg/assay (0.56 µl) resulted in a cleavage activity of 0.005 $OD_{348nm}$, a value that increased with higher protein concentrations up to 0.04 $OD_{348nm}$.

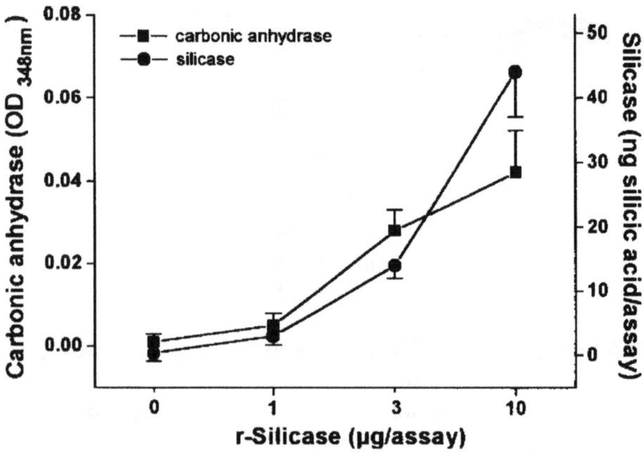


**Fig. 7.** Determination of the enzymatic activity of silicase in the carbonic anhydrase and in the silicase assay. Recombinant silicase was added to the reaction mixtures at concentrations between 1 and 10 µg/assay (0.56 µl). For the determination of the carbonic anhydrase activity (*filled squares*), the substrate *p*-nitrophenyl acetate was used and the released *p*-nitrophenol was measured at an absorbance of 348 nm. The activity of silicase (*filled circles*) was measured by using *S. domuncula* spicules. The free silicic acid, depolymerized from amorphous silica, was determined by "Silicon Test" colorimetric reaction

## 6.2 Silicase Activity

As a source/substrate for amorphous silica, spicules from *S. domuncula* were used. Spicules were obtained from tissue by incubation for 12 h in the presence of ethylenediaminetetraacetic acid (20 mM in PBS). After washing once with distilled water and twice with ethanol, the spicules were dried (56 °C). Then the spicules were ground in a mortar to a powder.

The silicase activity was determined as follows. In 2-ml Eppendorf tubes 100 µg of dry spicule powder was added to a 50 mM Tris-HCl buffer (pH 7.2), 10 mM of DL-dithiothreitol, 100 mM NaCl and 0.5 mM $ZnSO_4$; the final volume of the assay was 1 ml. Then 50 µl of the recombinant silicase was added and the reaction was allowed to proceed for 1 h (25 °C). To measure quantitatively the amount of dissolved silica, the nondissolved spicules were pelleted by centrifugation (14,000 × g; 15 min; 4 °C). The soluble released silicic acid was determined as described (Cha et al. 1999) using the "Silicon Test" colorimetric assay kit (Merck; 1.14794). Based on the absorbance values at 810 nm, the absolute amounts of silicic acid were calculated after construction of a calibration curve using a silicon standard (Merck 1.09947); the amounts of the reaction products are given in ng of silica released.

Activity determination revealed that the recombinant silicase catalyzes a strong depolymerization of amorphous silica. Quantitative analysis showed that at a concentration of 1 µg of recombinant silicase/assay a release free silicic acid of 3 ng/assay was measured. At higher protein concentrations, 3 or 10 μg/assay, the release of free silicic acid amounts to 16 or 43 ng/assay (Fig. 7).

## 7 Expression of Silicase in Response to Silicon

Based on the knowledge that in the presence of high silicon in the incubation medium, the primmorphs (1) form spicules and (2) react with an upregulation of the expressions of the *silicase* and the *isocitrate dehydrogenase* genes, as detected in the "differential display" procedure, it was reasonable to determine also the expression of the genes involved in the amorphous silica metabolism in *S. domuncula* by Northern blotting.

Previously, we found that at a higher silicon concentration in the incubation medium a change of the expression of the anabolic enzyme for amorphous silica synthesis, *silicatein* (Krasko et al. 2000), is seen. The cDNA for silicatein and its potential role in the synthesis of spicules have also been described (see Müller et al., this Vol.); the cDNA clone is termed *SDSILICA* (accession number AJ272013).

Primmorphs remained either nontreated with silicon or were incubated with 60 μM of silicon for 1–3 days. Then RNA was extracted. An amount of 5 μg of total RNA each was electrophoresed through a 1% formaldehyde/agarose gel and blotted onto Hybond-N$^+$ nylon membrane following the manufacturer's instructions (Amersham; Little Chalfont, Buckinghamshire, UK; Wiens et al. 1998). Hybridization was performed in parallel with 400–600-bp-large segments of the following probes; *SDSIA* (coding for silicase), *SDSILICA* (encoding silicatein; Krasko et al. 2000) or *SDIDH* (coding for isocitrate dehydrogenase, the α-subunit; see above). The probes were labeled with the PCR-DIG-Probe-Synthesis Kit according to the "Instruction Manual" (Roche). After washing, DIG-labeled nucleic acid was detected with anti-DIG Fab fragments (conjugated to alkaline phosphatase; dilution of 1:10,000) and visualized by chemiluminescence technique using CDP, the chemiluminescence substrate alkaline phosphatase, according to the instructions of the manufacturer (Roche).

Northern blots were performed (Fig. 8). It became evident that the genes encoding the two enzymes involved in the metabolism of amorphous silica in sponges, the *silicase* (catabolic enzyme) and the *silicatein* (anabolic enzyme) are strongly upregulated in response to higher silicon concentrations in the incubation assays. Furthermore, the *isocitrate dehydrogenase* gene is also upregulated, indicating that the formation of amorphous silica requires an increased metabolic rate of the cells.

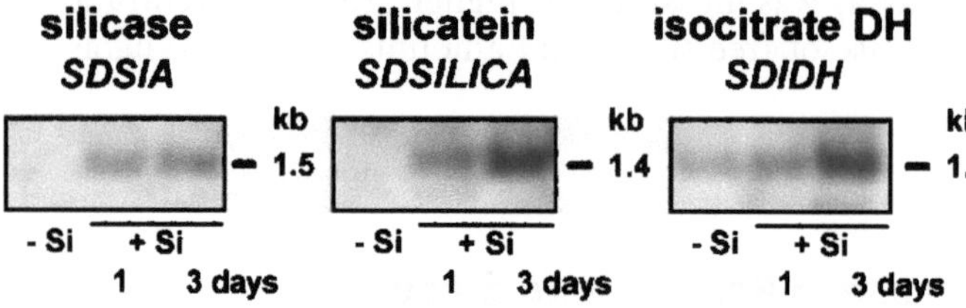


**Fig. 8.** Expression of the catabolic (*silicase*) as well as anabolic enzyme (*silicatein*) involved in amorphous silica metabolism in sponges by Northern blotting. RNA was extracted from primmorphs incubated in the absence of additional silicon (–Si) or the presence of 60 µM silicon (+Si) for 1–3 days. 5 µg total RNA was electrophoresed, blotted onto nylon membranes and finally hybridized with the following probes: *SDSIA* (silicase), *SDSILICA* (silicatein) or *SDIDH* (isocitrate dehydrogenase, the α-subunit). The sizes of the transcripts are given

# 8
# Proposed Mechanism of Action of Silicase

As shown here on the basis of sequence similarity, the silicase is a member of the family of carbonic anhydrases (carbonate hydrolase; EC 4.2.1.1). These enzymes catalyze the reversible hydration of carbon dioxide (Fig. 9, [1]). From studies on vertebrates it is known that carbon dioxide, as a metabolic end product of oxidative respiration, must be rapidly cleared from the body (see Taoka et al. 1994). The gas diffuses into the extracellular space where it is converted to $(HCO_3)^-$ and $H^+$ by the carbonic anhydrase. In mammalian cells the bicarbonate transport is mediated by an anion exchanger AE1 (Sterling et al. 2001).

Based on the finding summarized above, we could demonstrate that silicase also displays carbonic anhydrase activity, as measured by the colorimetric assay (Armstrong et al. 1966). In consequence, it is conceivable that silicase causes an alteration of the pH during the reaction of $CO_2$ to $(HCO_3)^-$. The consequence of this enzymic reaction might be an etching of the calcareous substratum. It is established that some sponge species can cause this reaction in a very efficient way. The species of the genus *Cliona* (Demospongiae; Tetractinomopha; Clionidae) are especially well known for their ability to dissolve calcium carbonate and excavate, burrow or bore into calcitic/aragonitic substrata. The enzyme (carbonic anhydrase) was localized on the outer surface of the etching cell filopodia and between cell processes (Pomponi 1979). It was hypothesized that the enzyme is secreted into the surrounding milieu (Rützler and Rieger 1973).

Some examples are given in Fig. 10. It is shown that the species *Cliona celata* drills holes into limestone, leaving open an approximately 2-mm-large space which is mainly occupied by the oscule (Fig. 10A). Meandering-like drilling traces can bee seen from *C. vermifera* (Fig. 10B); this species only rarely bores deeper than 3 mm into the stone, but covers large surfaces instead. The boring

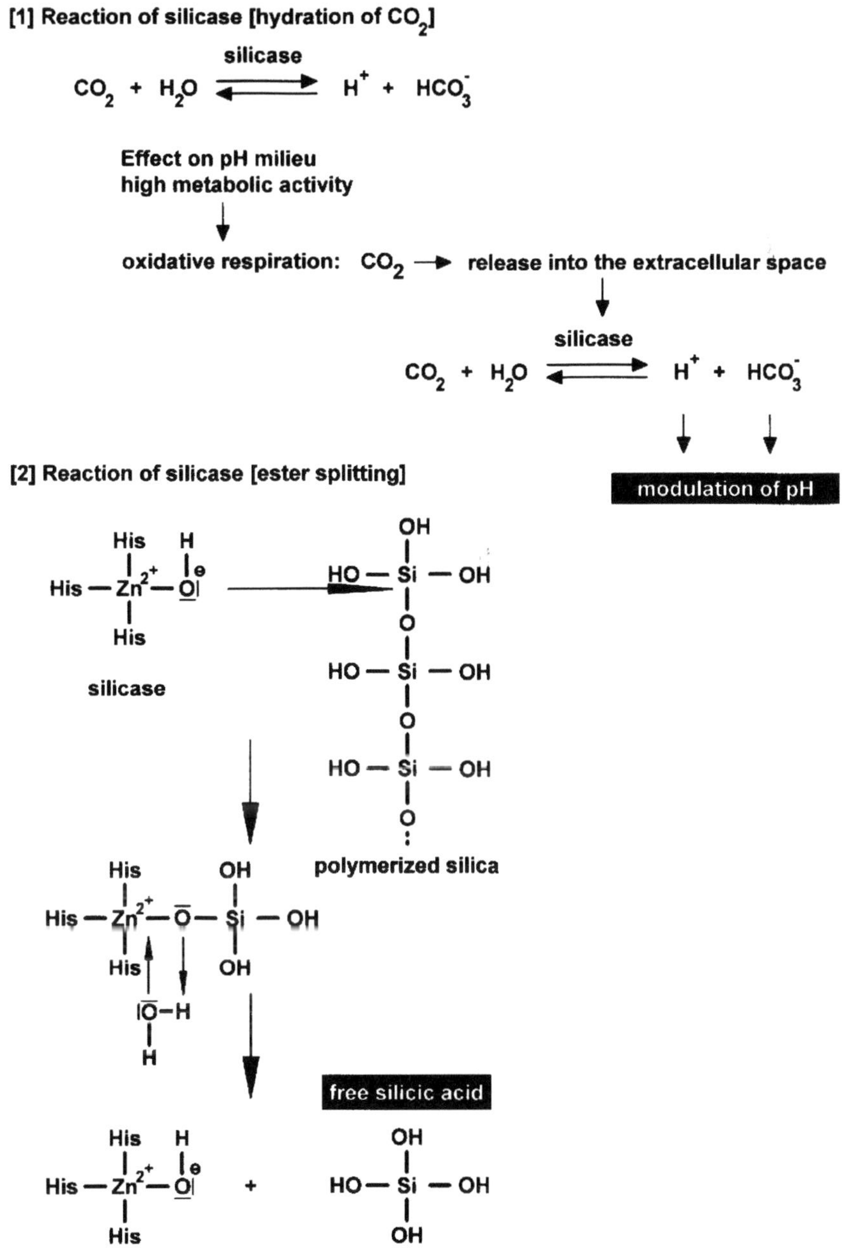


**Fig. 9.** Dual enzymatic reactions mediated by silicase (carbonic anhydrase-related enzyme) from *S. domuncula*. In [1], the conversion of $CO_2$ to $HCO_3^-$ is shown. It is proposed that high metabolic activity results in an increased production of $CO_2$ resulting in a modulation of the pH milieu. In [2], the reaction of the silicase is outlined. Silicase binds with its three histidine residues to one zinc ion. The zinc ion is a Lewis acid that interacts with water, a Lewis base. The zinc-bound hydroxide ion formed by splitting of the water molecule undertakes a nucleophilic attack at one of the silicon atoms linked by oxygen bond(s). This results in hydrolysis of the polymerized amorphous silica which remains, with one of the product halves, bound first to the enzyme. Through consumption of $H_2O$, the silicic acid product is released and the zinc-bound hydroxide is regenerated allowing the start of the heat catalytic cycle

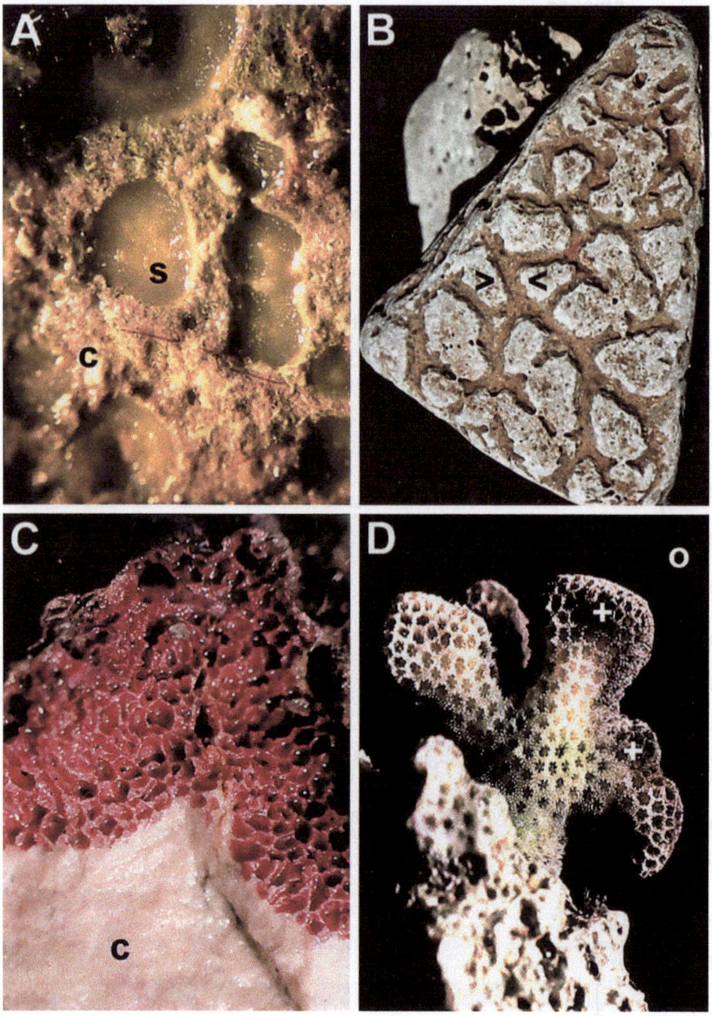


**Fig. 10A–D.** Boring, etching sponges/corals. Sponges: **A** The species *Cliona celata* (Demospongiae; Tetractinomopha; Clionidae) bores into calcite substrate (*c*), the part of the animal which is exposed to the surface is marked (*s*); magnification: ×6. **B Drilling traces** (> <) performed by *C. vermifera*; magnification: ×0.5. **C** Fracture through limestone (*c*) which had been excavated by the Clionidae *Thoosa istriaca* (Müller et al. 1979); magnification: ×0.3. Coral: **D** Within the contact zone of two allogeneic specimens from the scleractinian coral *Stylophora pistillata*, the polyps are destroyed and an increased activity of carbonic anhydrase is measured. By application of a colorimetric technique, the increase in the carbonic anhydrase activity can be visualized by a dark appearance (+) of the affected zone of calcareous skeleton of the animal. The two allogeneic animals grew for 6 weeks adjacent to each other; one specimen was taken and subjected to the color reaction [the site of the adjacent non-clone mate is indicated (*o*)]; magnification: ×0.5

activity of the sponge *Thoosa istriaca* (Müller et al. 1979) shows that a concerted invasion of the drilling activity of the sponge into the limestone occurs (Fig. 10C). In addition, it might be mentioned here that the carbonic anhydrase reaction is also extensively used in Cnidaria, e.g., in scleractinian corals (Müller et al. 1984), to cause cell death during histo-incompatibility reactions (Fig. 10D). It has been proven that in the coral *Stylophora pistillata* the polyps within the contact zone of two allogeneic tissues undergo necrotic/apoptotic alterations which parallels an increased carbonic anhydrase activity (Müller et al. 1984). During a period of 3–6 weeks, the allogeneic animals create a distance between themselves through dissolution of the calcareous material which forms the framework for the polyps.

In addition to the catalyzing function of hydration of $CO_2$, a second reaction is also mediated by the carbonic anhydrase, the hydrolysis of a series of esters, a reaction which is inhibited, e.g., by sulfonamides (see DeSantis 2000). This conversion involves the three histidine residues which are bound to the divalent zinc atom (Fig. 9, [2]). In the *S. domuncula* silicase the histidines are found in the deduced protein at $aa_{181}$, $aa_{183}$ and $aa_{206}$ (Fig. 4A). In water a zinc(II)-bound hydroxide ion is formed which functions as a Lewis acid. This undergoes nucleophilic attack to esters, here the "ester-like" bridging oxygen atom between two Si atoms (Fig. 9, [2]). In the next step the zinc-complex binds, within the Si polymer, to one silicon atom under cleavage of the oxygen linkage. Under the consumption of $H_2O$ the initial zinc(II)-bound hydroxide ion is formed again releasing the hydroxide-carrying Si and ultimately a free silicic acid (Fig. 9, [2]). This plausible mode of action, which has been proposed here for the function of silicase to depolymerize amorphous silica follows the reaction mechanism known for zinc-dependent enzymes involved in ester hydrolysis.

# 9 Conclusion

It is interesting to note that within the Porifera, some taxa exist which have a skeleton composed of both aragonite/calcite and silica. For example, the merliids (*Merlia*) have siliceous spicules and calcareous elements (Hartman 1983). This fact is important in the context of our results here. In general, the hard skeleton of all animals undergoes a pronounced metabolism, which includes of course anabolic processes, but also catabolic reactions. In the present study, we could demonstrate for the first time the existence of an enzyme which has the property to depolymerize the siliceous spicules, the major skeleton of Demospongiae and Hexactinellida.

The catabolic enzyme for amorphous silica described here, silicase, comprises very likely two functions; it has the ability (1) to dissolve calcareous material, in analogy to the carbonic anhydrase, and (2) to dissolve amorphous silica finally to silicic acid. Hence, the silicase found first in *S. domuncula* might be basically able to function both in the catabolism of their own siliceous spicules and – if present – also of endogenous calcareous material. The requirement of the silicase/carbonic anhydrase in the excavation of holes which may harbor some sponge specimens is established.

Besides the role of silicase in the maintenance of the skeleton of sponges, this enzyme may gain importance also in the field of biotechnology, e.g., as an enzyme which modifies silica matrices used in medicine and electronics at temperatures between ~10 and ~40 °C.

*Acknowledgements.* This work was supported by grants from the Bundesministerium für Bildung und Forschung (project: Center of Excellence *BIOTEC*marin) Germany, the European Commission (project: SILIBIOTEC) and the International Human Frontier Science Program (RG-333/96-M). Disclosure statement: "Abbau und Modifizierung von Silicaten und Siliconen durch Silicase und Verwendung des reversiblen Enzyms" (inventors: W.E.G Müller, A. Krasko and H.C. Schröder 2002) has been submitted for patent.

## References

Armstrong JM, Myers DV, Verpoorte JA, Edsall JT (1966) Purification and properties of human erythrocyte carbonic anhydrase. J Biol Chem 241:5137–5149

Ausubel FM, Brent R, Kingston RE, Moore DD, Smith JA, Seidmann JG, Struhl K (1995) Current protocols in molecular biology. Wiley, New York

Bavestrello G, Arillo A, Benatti U, Cerrano C, Cattaneo-Vietti R, Cortesogno L, Gaggero L, Giovine M, Tonetti M, Sarà M (1995) Quartz dissolution by the sponge *Chondrosia reniformis* (Porifera, Demospongiae). Nature 378:374–376

Brasier MD (1992) Nutrient-enriched waters and the early skeletal fossil records. J Geol Soc (Lond) 149:621–629

Brasier M, Green O, Shields G (1997) Ediacarian sponge spicule clusters from southwest Mongolia and the origins of the Cambrian fauna. Geology 25:303–306

Cha JN, Shimizu K, Zhou Y, Christianssen SC, Chmelka BF, Stucky GD, Morse DE (1999) Silicatein filaments and subunits from a marine sponge direct the polymerization of silica and silicones in vitro. Proc Natl Acad Sci USA 96:361–365

Coligan JE, Dunn BM, Ploegh HL, Speicher DW, Wingfield PT (2000) Current protocols in protein science. Wiley, Chichester

Cupp JR, McAlister-Henn L (1991) $NAD^+$-dependent isocitrate dehydrogenase. Cloning, nucleotide sequence, and disruption of the *IDH2* gene from *Saccharomyces cerevisiae*. J Biol Chem 266:22199–22205

Deitmer JM, Rose CR (1996) pH regulation and proton signaling by glial cells. Prog Neurobiol 48:73–103

DeSantis L (2000) Preclinical overview of Brinzolamide. Surv Ophthal 44 [Suppl 2]:S119–S129

Felsenstein J (1993) PHYLIP, ver. 3.5. Univ Washington, Seattle

Fujikawa-Adachi K, Nishimori I, Taguchi T, Yuri K, Onishi S (1999) cDNA sequence, mRNA expression, and chromosomal localization of human carbonic anhydrase-related protein, CA-RP XI. Biochim Biophys Acta 1431:518–524

Geigy Scientific Tables (1984) Physical chemistry - composition of blood, vol 3. Ciba-Geigy, Basel

Hartman WD (1983) Modern and ancient Sclerospongiae. In: Broadhead TW (ed) Sponges and spongiomorphs: notes for a short course. University of Tennessee (Department of Geological Sciences). Stud Geol 7:115–129

James NP, Gravestock DI (1990) Lower Cambrian shelf and shelf margin buildups, Flinders Ranges, South Australia. Sedimentology 37:455–480

Jones WC (1965) Photographic records of living oscular tubes of *Leucosolenia variabilis*. III. Irregular growth of the oscular tube. J Mar Biol Assoc UK 45:1–28

Kennish MJ (1994) Practical handbook of marine science. CRC Press, Boca Raton

Kim YO, Koh HJ, Kim SK, Jo SH, Huh JW, Jeong KS, Lee IJ, Song BJ, Huh TL (1999) Identification and functional characterization of a novel, tissue-specific $NAD^+$-dependent isocitrate dehydrogenase β subunit isoform. J Biol Chem 274:36866–36875

Knight CTG, Kinrade SD (2001) A primer on the aqueous chemistry of silicon. In: Datnoff LE, Snyder GH, Korndörfer GH (eds) Silicon in agriculture. Elsevier, Amsterdam, pp 57–84

Krasko A, Batel R, Schröder HC, Müller IM, Müller WEG (2000) Expression of silicatein and collagen genes in the marine sponge *Suberites domuncula* is controlled by silicate and myotrophin. Eur J Biochem 267:4878–4887

Krasko A, Schröder HC, Batel R, Grebenjuk VA, Steffen R, Müller IM, Müller WEG (2002) Iron induces proliferation and morphogenesis in primmorphs from the marine sponge *Suberites domuncula*. DNA Cell Biol 21:67–80

Kruse M, Leys SP, Müller IM, Müller WEG (1998) Phylogenetic position of the Hexactinellida within the phylum Porifera based on amino acid sequence of the protein kinase C from *Rhabdocalyptus dawsoni*. J Mol Evol 46:721–728

Laemmli UK (1970) Cleavage of structural proteins during the assembly of the head of bacteriophage T4. Nature 227:680–685

Mehl D, Müller I, Müller WEG (1998) Molecular biological and palaeontological evidence that Eumetazoa, including Porifera (sponges), are of monophyletic origin. In: Watanabe Y, Fusetani N (eds) Sponge science - multidisciplinary perspectives. Springer, Berlin Heidelberg New York, pp 133–156

Müller WEG (1997) Origin of metazoan adhesion molecules and adhesion receptors as deduced from their cDNA analyses from the marine sponge *Geodia cydonium*. Cell Tissue Res 289: 383–395

Müller WEG (2001) How was metazoan threshold crossed: the hypothetical Urmetazoa. Comp Biochem Physiol [A] 129:433–460

Müller WEG, Brümmer F (1998) Herstellung von Primmorphe aus dissoziierten Zellen von Schwämmen, Korallen und weiteren Invertebraten: Verfahren zur Kultivierung von Zellen von Schwämmen und weiteren Invertebraten zur Produktion und Detektion von bioaktiven Substanzen, zur Detektion von Umweltgiften und zur Kultivierung dieser Tiere in Aquarien und im Freiland. Patent application AZ 198 24 384 [30.05.1998]

Müller WEG, Zahn RK, Rijavec M, Britvic S, Kurelec B, Müller I (1979) Aggregation of sponge cells. The aggregation factor as a tool to establish species. Biochem System Ecol 7:49–55

Müller WEG, Zahn RK, Maidhof A (1982) *Spongilla gutenbergiana* n.sp., ein Süßwasserschwamm aus dem Mittel-Eozän von Messel. Senckenbergiana Lethaea 63:465–472

Müller WEG, Müller I, Zahn RK, Maidhof A (1984) Intraspecific recognition system in scleractinian corals: morphological and cytochemical description of the autolysis mechanism. J Histochem Cytochem 32:285–288

Müller WEG, Wiens M, Batel R, Steffen R, Borojevic R, Custodio MR (1999) Establishment of a primary cell culture from a sponge: primmorphs from *Suberites domuncula*. Mar Ecol Progr Ser 178:205–219

Müller WEG, Krasko A, Skorokhod A, Bünz C, Grebenjuk VA, Steffen R, Batel R, Müller IM, Schröder HC (2002a) Histocompatibility reaction in the sponge *Suberites domuncula* on tissue and cellular level: central role of the allograft inflammatory factor 1. Immunogenetics 54: 48–58

Müller WEG, Lorenz B, Krasko A, Schröder HC (2002b) Silicatein-vermittelte Synthese von amorphen Silikaten und Siloxanen und ihre Verwendung. Offenlegungsschrift (Deutsches Patentamt); C 12 P3/00

Nair SK, Calderone TL, Christianson DW, Fierke CA (1991) Altering the mouth of a hydrophobic pocket. Structure and kinetics of human carbonic anhydrase II mutants at residue Val-121. J Biol Chem 266:17320–17325

Nelson DM, Tréguer P, Brezinski MA, Leynaert A, Quéguiner B (1995) Production and dissolution of biogenic silica in the ocean: revised global estimates, comparison with regional data and relationship to biogenic sedimentation. Glob Biochem Cycles 9:359–372

Okamoto N, Fujikawa-Adachi K, Nishimori I, Taniuchi K, Onishi S (2001) cDNA sequence of human carbonic anhydrase-related protein CA-RP X and XI in human brain. Biochim Biophys Acta 1518:311–316

PC/GENE (1995) Data Banks CD-ROM; Release 14.0. IntelliGenetics, Mountain View, CA

Pomponi SA (1979) Ultrastructure and cytochemistry of the etching area of boring sponges. In: Lévi C, Boury-Esnault N (eds) Biologie des spongiaires, vol 291. Colloq Int CNRS, Paris, pp 319–323

Racki G (1999) Silica-secreting biota and mass extinction: survival patterns and processes. Palaeogeogr Palaeoclimatol Palaeoecol 154:107–132

Rützler K, Rieger G (1973) Sponge burrowing: fine structure of *Cliona lampa* penetrating calcareous substrata. Mar Biol 21:144–162

Schwab DW, Shore RE (1971) Mechanism of internal stratification of siliceous spicules. Nature 232:501–502

Sly WS, Hu PY (1995) Human carbonic anhydrases and carbonic anhydrase deficiencies. Annu Rev Biochem 64:375–401

Steiner M, Mehl D, Reitner J, Erdtmann BD (1993) Oldest entirely preserved sponges and other fossils from the Lowermost Cambrian and a new facies reconstruction of the Yangtze Platform (China). Berliner Geowiss Abh (E) 9:293–329

Sterling D, Reithmeier RAF, Casey JR (2001) A transport metabolon. Functional interaction of carbonic anhydrase II and chloride/bicarbonate exchangers. J Biol Chem 276:47886–47894

Sun MK, Alkon DL (2002) Carbonic anhydrase gating of attention: memory therapy and enhancement. Trends Pharmac Sci 23:83–89

Taoka S, Tu C, Kister KA, Silverman DN (1994) Comparison of intra- and intermolecular proton transfer in human carbonic anhydrase II. J Biol Chem 269:17988–17992

Uriz MJ, Turon X, Becerrro MA (2000) Silica deposition in demosponges: spiculogenesis in *Crambe crambe*. Cell Tissue Res 301:299–309

Vinogradov AP (1953) The elementary chemical composition of marine organisms. Sears Foundation, New Haven

Volkmer-Ribeiro C, Motta JFM (1995) Spongillite forming sponges from south-western Minas Gerais State and neighboring areas with an indication for habitat preservation. Biosciencias (Porto Alegre) 3:145–169

Weissenfels N (1989) Biologie und Mikroskopische Anatomie der Süβwasserschwämme (Spongillidae). Fischer, Stuttgart

Wiens M, Koziol C, Hassanein HMA, Batel R, Müller WEG (1998) Expression of the chaperones 14-3-3 and HSP70 induced by PCB 118 2,3′,4,4′,5-pentachlorobiphenyl in the marine sponge *Geodia cydonium*. Mar Ecol Prog Ser 165:247–257

# Studies of Biosilicas; Structural Aspects, Chemical Principles, Model Studies and the Future

Carole C. Perry, David Belton, and Kirill Shafran

## 1 Terminology

- Si: the chemical symbol for the element and the generic term used when the nature of the specific silicon compound is not known.
- $Si(OH)_4$: silicic acid, or more correctly orthosilicic acid, the fundamental building block used in the formation of silicas.
- $SiO_2 \cdot nH_2O$ or $SiO_{2-x}(OH)_{2x} \cdot 2H_2O$: amorphous, hydrated, polymerized material produced from orthosilicic acid.
- Oligomerization: the formation of dimers and small oligomers from orthosilicic acid by removal of water. e.g. $2Si(OH)_4 \rightarrow (HO)_3Si—O—Si(OH)_3 + H_2O$
- Polymerization: the mutual condensation of silicic acid to give molecularly coherent units of increasing size.
- Organosilicon compound: it must contain silicon covalently bonded to carbon within a distinct chemical species.

## 2 Introduction

Silica and silicates are extensively used in industry and medicine. The materials find use in paints, foods, medicines, adhesives, detergents, chromatography materials, catalysts and photonic materials (reviewed by Iler 1979). Silica may be produced at high temperature, via aqueous processing or by largely non-aqueous routes such as the low-temperature sol-gel process (reviewed in Brinker and Scherrer 1990; Hench and West 1990). Whatever the eventual use of the silica, it is its structure that determines its properties. By structure we mean order and organization on length scales from angstroms to the size of the final object, morphology, surface area, porosity and surface functionality. The essential building block is the $SiO_4$ tetrahedron although other structural units such as the $SiO_6$ octahedron are also used. These units can be put together

Department of Chemistry and Physics, The Nottingham Trent University, Clifton Lane, Nottingham NG11 8NS, UK

Progress in Molecular and Subcellular Biology, Vol. 33
W. E. G. Müller (Ed.)

in a wide range of patterns to yield both porous and non-porous crystalline materials including silica-based zeolite materials. As well as crystalline silica and silicates, an extremely diverse range of amorphous materials exist, such as disordered precipitates, gels, glasses and shaped objects (spheres, screws, hollow tubes etc.; Stöber et al. 1968; Yang et al. 1997; Miyaki et al. 1999), produced under a wide range of synthesis conditions. The amorphous materials, in contrast to their crystalline analogues, exhibit no long-range order and are built up from $SiO_4$ tetrahedra with variable Si—O—Si bond angles and Si—O bond distances. These materials are covalent inorganic polymers of general formula $[SiO_{n/2}(OH)_{4-n}]_m$ (Mann et al. 1983). This formula, where n = 0–4 and m is a large number, indicates the variation in the number of silanol functional groups within the condensed structure. There can also be a similar variation in the extent of hydration. The formation of 'regular' amorphous silica structures such as Stöber spheres requires great care during synthesis (Stöber et al. 1968) and many of the more novel silica structures (length scale for organization not defined) are produced under extreme conditions of acidity (silica screws; Yang et al. 1997) and basicity (silica-based zeolites and some MCM materials; Kresge et al. 1992). Figure 1 shows examples of siliceous

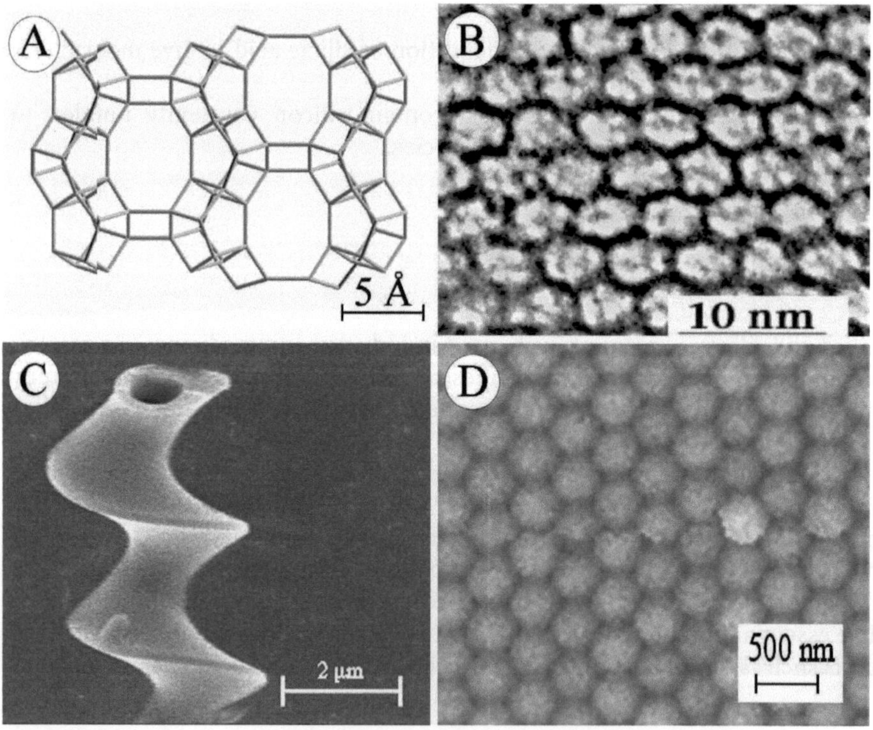


**Fig. 1A–D.** Computer graphics/scanning electron microscope images of **A** zeolite, **B** MCM mesoporous material (courtesy of Geoff Ozin), **C** silica spheres, **D** silica screw (courtesy of Geoff Ozin)

materials that are crystalline (silica-based zeolite) and amorphous on the angstrom scale, but exhibit higher order (MCM materials, silica spheres and screws).

# 3
# Structural Chemistry of Biosilicas

Biological organisms form much more silica than man (gigatons per annum as opposed to megatons produced by man). The silica is formed from an aqueous environment vastly under-saturated with respect to silicic acid at ca. atmospheric pressure and at temperatures ranging from ca. 4–40 °C in the presence of a vast range of cellular components including metal ions, carbohydrates, lipids and proteins. The organic phase, whether intracellular or extracellular, may play important roles in (1) the formation of primary particles by altering supersaturation requirements, (2) the formation of networks and morphological features and (3) in the stabilization and protection of the mineral structure from the environment. It is likely that the precise pathway followed during mineralization and the relative importance of the different stages will vary for each silicifying organism (Perry 1989). The biosilicas are not stoichiometric minerals and their density, hardness, solubility, viscosity and composition may vary considerably.

Biogenic silica is amorphous to X-rays and electrons at the angstrom level. No evidence has yet been found for short range order extending above 1 nm (ca. three Si—O bond lengths). High resolution electron microscopy images indicate that the silica structures are based upon a random network of $SiO_4$ tetrahedral units connected through covalently linked Si—O—Si bonds of variable bond angle rather than a microcrystalline/cluster structure composed of a random array of microcrystalline polyhedra (Mann and Perry 1986). Silicifying organisms also seem to actively exclude metal ions from the mineral phase, a factor that would promote the formation of the random network structure rather than the micro-crystalline cluster structure (Mann et al. 1986). Solid state $^{29}Si$ NMR studies of silica extracted from a range of organisms also point to a disordered material containing only tetrahedral structural units, but there are differences in the residual hydroxyl functionality (Table 1). Differences in the peak position for the $Si(=OSi)_4$ or $Q_4$ peak that relates to the mean Si—O—Si bond angle in the final material (Thomas et al. 1983; Table 1) are also observed, again suggesting variations in the molecular structure of the mineral phase.

$^{29}Si$ NMR experiments on siliceous materials are also able to provide information on the spatial location of silanol groups and can provide circumstantial evidence for the extent of interaction between the siliceous phase and any organic phase (Perry 1989). Of the siliceous materials listed above, the strongest interaction between inorganic and organic phases is proposed to exist for the diatom sample. The use of silylating agents with different surface chemistry requirements can provide information on the spatial distribution of

**Table 1.** $^{29}$Si Solid state NMR chemical shifts for biological silicas. Chemical shifts are quoted in parts per million (ppm) from tetramethylsilane (TMS). (Reprinted with permission from Perry 1989)

| | $Si(OSi\equiv)_4$ $Q^4$ | $Si(OSi\equiv)_3OH$ $Q^3$ | $Si(OSi\equiv)_2(OH)_2$ $Q^2$ |
|---|---|---|---|
| Synthetic gel | –109.3 | –99.8 | –90.6 |
| Grass hairs (*Phalaris canariensis*) | –109.8 | –101.2 | –92.6 |
| Limpet teeth (*Patella vulgata*) | –113.7 | –102.4 | - |
| Sponge fibres (*Halichondria panicea*) | –112.1 | –101.8 | –92.5 |
| Diatom frustule (*Navicula pelliculosa*) | –113.5 | –102.3 | –92.7 |

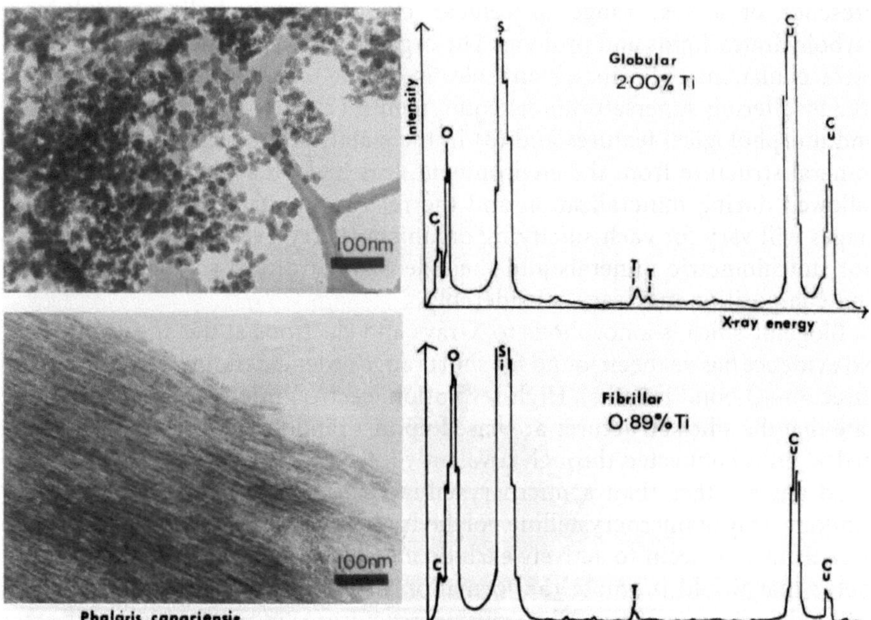


**Fig. 2.** Transmission electron micrographs and energy dispersive elemental analysis of the amount of a 'chemical stain' associated with different silica types present within a single hair cell from the lemma of the grass *Phalaris canariensis* L. (Reprinted with permission of the Royal Society; Perry et al. 1990)

silanol groups within a given sample (Mann et al. 1983) and studies on grass hair silica has shown that only 24% of the silanol groups are accessible to silylation with trimethylchlorosilane indicating that most of the silanol groups are found on the interior of the plant hairs in locations not accessible to the chosen reagent. Other studies using chemical 'stains' have provided evidence for the differences in surface functionality of different particulate silicas present within a single cell (Fig. 2; Perry et al. 1990).

Although the building blocks utilized in the fabrication of all biogenic silicas are essentially the same, a wide variety of structures showing evidence of organization on the microscopic and macroscopic scale exist. At the microscopic scale, gel, continuous and particulate structure types can be identified. The materials can be distinguished by measurement of their porosity (after removal of the organic matrix) and by structural analysis using transmission electron microscopy. For the particulate materials, the surface area may be of the order of several hundreds of $m^2 g^{-1}$, whereas for the gel-like and 'continuous' structures, the surface area measurements will be much lower and will arise from the external surface of the material only and/or from specific features such as small mesoporous (ca. 20-nm diameter) voids present in the material.

In true 'gel-like' phases such as costal strips from the marine protozoan, *Stephanoeca diplocostata* Ellis, the silica phase is a continuous network throughout the material (Mann and Williams 1982) and at no time during dissolution are distinct particles visible.

In contrast, the silica deposits in many organisms including higher plants comprise 'particulate' structures where structural motifs are readily recognized. These structures are built up from primary particles/initial aggregates of different sizes, shapes and surface chemistries (Perry et al. 1984; Mann et al. 1986). Examples of structural motifs include sheet-like, globular and fibrillar (non-spherical particles) motifs, their naming depending on the pattern of aggregation between the particles (Perry et al. 1984). Figure 2 shows examples of two of the different structural types. The range of particle sizes observed in a particular structural motif is limited and the great stability of the structural motifs under the electron beam (in contrast to materials prepared in the laboratory from aqueous solution) suggests surface stabilization (Perry et al. 1990). A noticeable feature of particulate silicas formed in the biological environment is the narrow particle size distribution for specific structural motifs and the lack of necking between particles, synonymous with controlled nucleation, particle growth and aggregation (Perry and Keeling-Tucker 2000, 2002).

An intermediate structural type termed 'continuous' silica is found in some sponges and diatoms. For these materials, fibrillar and hexagonal columnar arrangements built up from ca. 5-nm aggregates have been observed only during development of some diatom frustules (Li and Volcani 1985a, b). When silicification is complete in these organisms, the particles are no longer distinguishable. A similar situation is observed for silica from sponge spicules.

The beauty of siliceous structures has fascinated scientists for a long time. The complex, oftentimes geometric morphology of these structures still forms the basis for the taxonomy of a number of groups of lower plants and animals, e.g. Radiolaria and sponges (see Fig. 3). The architectural design of these intricate structures at the microscopic level can be mirrored by the ultrafine morphology on the nanoscopic scale, for example within each individual pore of a diatom frustule (Perry 1989).

**Fig. 3A,B.** Electron microscope images of **A** a sponge and **B** a diatom

It has been suggested that in order to obtain silica particles with narrow size and shape distributions such as those observed for silicas produced by living organisms, two criteria must be satisfied (Bogush and Zukovski 1990). Firstly, a constant number of growing particles must be formed early in the reaction, and secondly, particles must grow through a self-sharpening mechanism, with smaller particles growing more quickly than the larger ones. This suggests that kinetics are very important in the control of silica deposition (Sada et al. 1990). If we can understand the silica deposition process in biological organisms where silica is formed in an aqueous, 'green' environment, we may be able to take the principles used (control mechanisms) to generate tailor-made silicas for industrial and medical usage.

## 4 Organic Matrix-Controlled Silica Production in Biological Organisms

The finding that proteinaceous matrices can act as templates for the nucleation, growth and growth regulation of crystalline biominerals has prompted a search for similar regulatory molecules in silicifying organisms including diatoms, sponges and higher plants. The chemical nature of the biomolecules extracted from the different types of organism show some similarities, but there are also many differences in composition and structure.

Studies in the early 1970s on the organic casing of a range of mature diatoms (Hecky et al. 1973) showed that diatom cell wall proteins were consistently enriched in serine and threonine, comprising 16–34 mol% together with 16–23 mol% of aspartic/glutamic acid and 11–21 mol% glycine. It was proposed that the serine–threonine-enriched protein would present a layer of hydroxyls onto which orthosilicic acid molecules could condense.

Recent studies performed on one particular diatom, *Cylindrotheca fusiformis*, have identified three types of proteinaceous materials, frustulins (Kröger et al. 1994), HF-extractable proteins or HEPs (Kröger et al. 1997) and silaffins (Kröger et al. 1999) that differ in chemical composition, location and proposed function in the organism. The frustulins are proposed to play a structural role in the organic casing of diatoms (Kröger et al. 1999) and the HEPs are thought to be involved in the regulation of morphogenesis in subsets of the girdle bands that are structural components of the developing diatom frustule (Kröger et al. 1997). The silaffins have an interesting amino acid composition rich in basic amino acids with Lys–Lys and Arg–Arg clusters interspersed with the hydroxy-amino acids serine and tyrosine at regular intervals (Kröger et al. 1999). The silaffins also contain amino acids that have been subject to post-translational modification. The ε-*N*,*N*-dimethyl- and oligo-*N*-methyl-propylamine-lysine side chain functionalities both increase the positive charge on the molecules in the subneutral pH range (Kröger et al. 1999; Kröger et al. 2001). A role has been proposed for the low molecular weight silaffins that involves promotion of silicic acid polycondensation and precipitation of biogenic silica under neutral or slightly acidic pH conditions.

Sponges and higher plants have also been investigated for proteinaceous matrices capable of assisting in the silica formation process in these organisms (Harrison 1996; Krasko et al. 1997, 2000; Perry and Keeling-Tucker 1998; Shimizu et al. 1998). Spicules from the sponge *Tethya aurantia*, have a central protein filament from which silicatein α, β and γ have been isolated (Shimizu et al. 1998). The more abundant α subunit has been identified as a novel member of the cathepsin L subfamily of papain-like cysteine proteases. Structural variants of the silicatein protein produced using site-directed mutagenesis have shown that specific serine and histidine side chains are required for optimal activity in the catalysis of the formation of silica and hydrophobic regions on the surface of the silicatein protein molecule are necessary for inter-molecule interaction in the generation of protein filaments around which the silica is found (Zhou et al. 1999).

Higher plants including *Equisetum telmateia* (great horsetail), *E. arvense* (common horsetail) and hairs from the lemma of the grass *Phalaris canariensis*, have also yielded an organic matrix intimately associated with the silica phase (Harrison 1996; Perry and Keeling-Tucker 1998, 2002). Three extracts have been obtained with distinctive chemical composition. The most readily released is enriched serine/threonine (25 mol%), glycine (20 mol%) and acidic residues (25 mol%). The second extract has lower levels of glycine and hydroxyl-containing residues, with the glycine being largely replaced by proline and a ca. two-fold increase in the levels of lysine. A third, insoluble material is also obtained which contains even higher levels of lysine (up to 26 mol%), high levels of proline and high levels of aliphatic amino acids. This latter material may be involved in the regulation of nucleation. The other two groups of materials that contain higher levels of hydroxylated amino acids (such as serine) are probably involved in the regulation of particle growth.

## 5 The Chemistry of Silica Formation

Although all biological silica forms in an aqueous environment, many of the experiments designed to understand how silica is formed in the presence of biosilica-derived proteins and other solution adjuncts are performed in a largely non-aqueous environment using alkoxysilanes as the silicon precursor. It is, therefore, important to understand the chemical reactions and species present in both aqueous and non-aqueous environments so that the function of any bioextract can be correctly understood.

### 5.1 Silica Chemistry in Aqueous Solution

The simplest soluble form of silica is the monomer orthosilicic acid, which is silicon tetrahedrally co-ordinated to four hydroxyl groups and has the formula $Si(OH)_4$. It is only weakly acidic with a $pK_a$ of 9.8 (Iler 1979) and is found almost universally at concentrations of a few ppm. In water, at 25 °C, it is stable for long periods of time at levels below ca. 100 ppm. However, once the concentration exceeds the solubility of the amorphous solid phase, at 100–200 ppm, it undergoes autopolycondensation reactions. This process has been studied and reported by many authors (reviewed in Iler 1979; Tarutani 1989; Perry and Keeling-Tucker 2000), and can be divided into three distinct stages:

1. Polymerization of monomers to form stable nuclei of a critical size.
2. Growth of the nuclei to form spherical particles.
3. Aggregation of particles to form branched chains or structural motifs.

The term polymerization must be considered in its broadest sense, meaning the "... mutual condensation of silicic acid to give molecularly coherent units of increasing size ..." (Iler 1979), as the product does not retain the general formula of the precursors. An alternative term is oligomerization, which is particularly appropriate for the early stages of the reaction where discrete, identifiable units are formed.

There are various mechanisms proposed for the condensation reaction. The simplest involves the condensation of two molecules of silicic acid with the release of water, a process that is described in more 'chemical' terms below. This process results in no net change in pH for the reaction system. An alternative mechanism invokes a bimolecular collision between an ionized and an unionized silicic acid molecule (for example see Harrison and Loton 1995; Perry and Keeling-Tucker 2000) This process, although favoured in terms of the nucleophilic attack of a negatively charged oxygen atom on the electropositive silicon centre generates hydroxyl ions, thereby increasing the pH of the reaction system. At ca. neutral pH where silica is formed in biological organisms (or slightly to the acid side of neutral as proposed for silicification in diatoms; Vrieling et al. 1999) the proportion of ionized silicic acid molecules is likely to be very small (Table 2). As the oligomers grow in size, the $pK_a$ of the silanol groups decreases and the proportion of ionized species available for further condensation increases (Table 2).

This latter fact has serious implications for inter-particle interactions as any negative charge on the surface of each particle will need to be overcome in order for a flocculated and/or aggregated material to form. As a rule, silicic acid molecules condense in such a way as to produce a sufficient number of siloxane (Si—O—Si) bonds with oligomers cyclizing during the early stages of the process to give rings containing predominantly 3–6 silicon atoms linked by siloxane bonds. Almost all silicas except those formed under conditions of extreme acidity contain high proportions of cyclic species.

Once the cyclic species dominate, monomers and dimers and other small oligomers react preferentially with these (Tarutani 1989), as the oligomers will have a higher density of ionized silanol groups. A process known as Ostwald ripening also occurs where smaller, more soluble particles dissolve and release silicic acid that re-deposits onto the larger particles. At ca. neutral pH the particles bear a negative charge on the surface and grow as isolated sol particles until the levels of soluble silica reach the solubility of amorphous silica. If salts are present or under acidic conditions, however, the surface negative charge that causes repulsion between individual particles is reduced or eliminated and the particles aggregate to form dense three-dimensional networks.

**Table 2.** The effect of pH on the percentage of ionized silanol groups for silicic acid and colloidal particles of silica

| pH | Silanol groups as ionized species (~%) | |
|---|---|---|
| | $Si(OH)_4$ $pK_a = 9.8$ | Colloidal particles $pK_a = 6.8$ |
| 5 | 0.002 | 1.8 |
| 6 | 0.02 | 6.3 |
| 7 | 0.18 | 61 |
| 8 | 1.8 | 94 |

There are many studies of the kinetics of silica formation and the orders of reaction reported range from first to fifth (see references in Iler 1979; review by Perry and Keeling-Tucker 2000). Despite the differences in reported orders of reaction for silica deposition, it is universally acknowledged that orthosilicic acid concentration, temperature, pH and the presence of additional chemical species all affect the rate of oligomerization. The first two factors always cause a concomitant increase in the rate. The effect of pH is somewhat different in that the rate is either retarded or the reaction prevented altogether at high or low pH levels, owing to the formation of high concentrations of silicate ions or un-ionized silicic acid molecules, respectively. Below about pH 2, the rate again increases as acidity increases, possibly due to the formation of cationic species such as $(OH)_3SiOH_2^+$ (Frolov et al. 1983). The pH at which the maximum rate of condensation occurs is another area where there is some disagreement, with published values falling in the range pH 6.5–9 (Iler 1979; Perry and Keeling-Tucker 2000).

If we now consider the mechanism of silica formation by considering the electronegativity of the silicate species and constituent atoms we can have a better appreciation for the structural forms of silica that are formed under different pH regimes (Jolivet 2000). The process of silica formation requires the formation of oxo bridges between silicon atoms via a process known as oxolation. The likely mechanism for this reaction is a two-step $SN_2$ associative mechanism that may be catalyzed by both acids and bases. The mechanism for the reaction under neutral conditions is given in Fig. 4. Note the transition state involves a 5-co-ordinate intermediate, with proton transfer from the the -ol bridge to a terminal OH ligand leading to the loss of a water ligand. Above and below the isoelectronic point of silica at pH 3, the rate of condensation will be increased by basic or acid catalysis. Basic catalysis is due to hydroxyl ions from the medium reacting on the positively charged centres of the reactants, or deprotonation of the hydroxo ligands on the silicon cation. If the protons are abstracted by hydroxyl ions, the negative charge on the nucleophilic oxygen is increased and the first step of the oxolation process occurs more readily. Acid catalysis affects the negatively charged centres of the reactants via protonation of the hydroxo ligands leading to better leaving groups. The reactivity and structure type of the polymers formed can be assessed by considering the electronegativity and partial charges present on the species involved (Table 3).

In an alkaline medium, catalysis affects the first stage of the oxolation reaction. As the charge on all silicon atoms in the growing chain is higher than that for monomers, then attack will take place on the chain with preference for the sites with highest partial charge, i.e. the middle of the chain, leading to highly condensed three-dimensional structures. The low value of the electronegativity of the anion of silicic acid will make it the preferred reactant above pH 7 (Table 2). In an acidic medium, the second stage of the mechanism is facilitated. The OH groups involved are predominantly those at the ends of chains that bear the highest negative partial charge. The result of this is that chains are formed containing many unreacted silanol groups. The effects of acid and

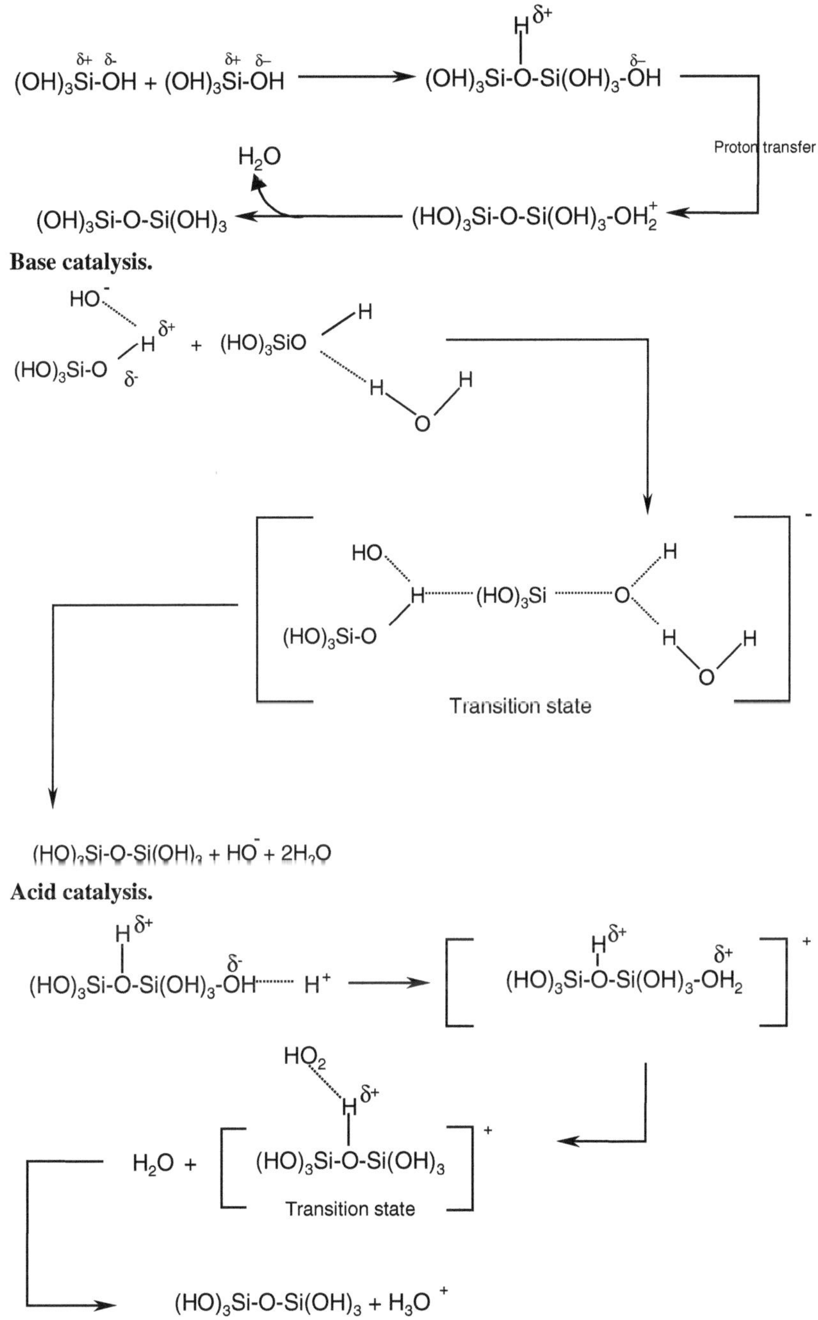


**Fig. 4.** Schematic of the oxolation reaction and the effects of acid and basic catalysis on the condensation of silicic acid molecules

**Table 3.** The electronegativity and partial charge on selected silicon species ($\chi$ = electronegativity, $\delta$ = partial charge on the particular atom)

| Site | $\chi$ | $\delta$(Si) | $\delta$(OH) |
|---|---|---|---|
| $Si(OH)_4$ | 2.58 | +0.47 | −0.12 |
| $Si(OH)_3O^-$ | 2.37 | +0.35 | −0.30 |
| $(OH)_3SiOH_2^+$ | 2.75 | +0.56 | +0.03 |
| $Si(OH)_3O\text{-}Si{\equiv}$ | 2.64 | +0.50 | −0.06 |
| $Si(OH)_2O_2\text{-}(Si{\equiv})_2$ | 2.71 | +0.54 | 0.0 |
| $Si(OH)O\text{-}(Si{\equiv})_3$ | 2.78 | +0.58 | +0.06 |

The average electronegativity of a chain with general chemical composition $[SiO_2(OH)_2]$ = 2.63.

base catalysis are also relevant when aggregation/flocculation phenomena are involved in the generation of the three-dimensional structures observed in the biological environment.

### 5.1.1
### *Effects of pH and $M^+$ Ion Identity on Speciation in Aqueous Solution*

In general, silica prepared under acidic conditions is built up from extended networks and silica formed under basic conditions is built up from highly branched networks. pH clearly affects the relative rates of condensation, particle growth and aggregation. For systems where reaction is rapid, it is impossible to identify the specific molecular species involved in the condensation process. Under certain conditions, however, the rates of the condensation processes are significantly reduced and it is possible to obtain information on the silicic acid-containing species that are present in solution. In particular, studies at high pH have contributed greatly to our understanding of how zeolites and mesoporous MCM materials are formed. Techniques that have been used in speciation studies include Raman spectroscopy, trimethylsilylation, mass spectrometry, reaction with molybdic acid and $^{29}$Si NMR spectroscopy. The chemical and spectroscopic techniques have shown that concentrated (e.g. 1.5 molar) solutions contain a variety of small silicate ions in dynamic equilibrium. The use of $^{29}$Si enrichment and $^{29}$Si NMR spectroscopy (including spin perturbation techniques such as $^{29}$Si–$^{29}$Si decoupling and two-dimensional shift-correlated $^{29}$Si homonuclear magnetic resonance experiments) have enabled the positive identification of 18 silicate anion structures (Harris and Knight 1983; Knight et al. 1986; Knight 1988) including those multisite silicate species depicted in Fig. 5. Other species, including unsubstituted cyclic and regular cage anions (these are species that yield only one signal in

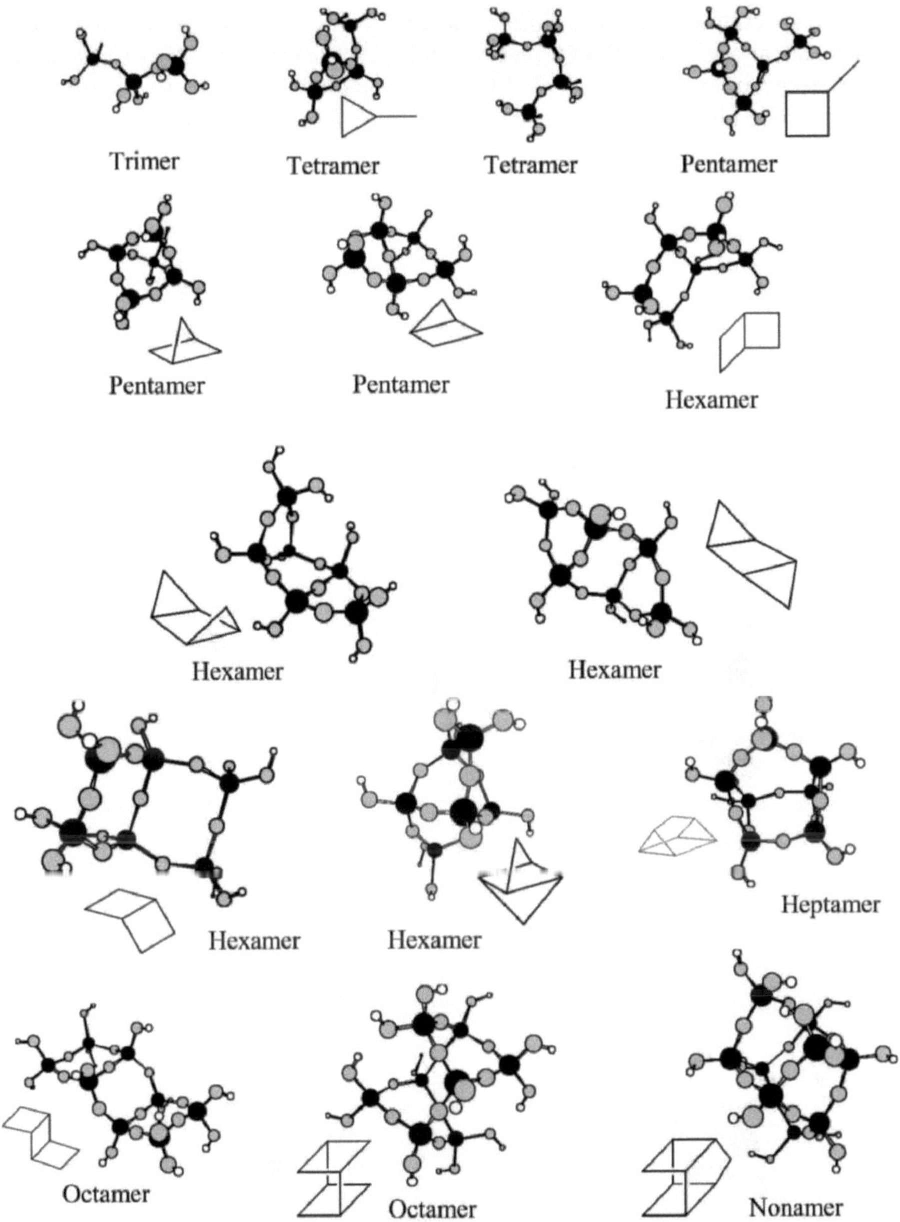


**Fig. 5.** Structures of multisite silicates suggested to occur in concentrated aqueous silicate solutions at high alkalinity showing energy-minimized structures (MM2 followed by application of the AM1 programme) and idealized shape. The energy minimization was performed with all silanols protonated although the extent of deprotonation vary according to the pH of the system. (Derived from Knight 1988)

the $^{29}Si$ spectrum), are more difficult to unequivocally identify due to extensive spectral overlap of the $Q^2$ (cyclic) and $Q^3$ (cage) regions. Use of 'one spin filter' techniques (Knight et al. 1989) has enabled these single silicon site species to be identified (Fig. 6), as spectral information from chemical species

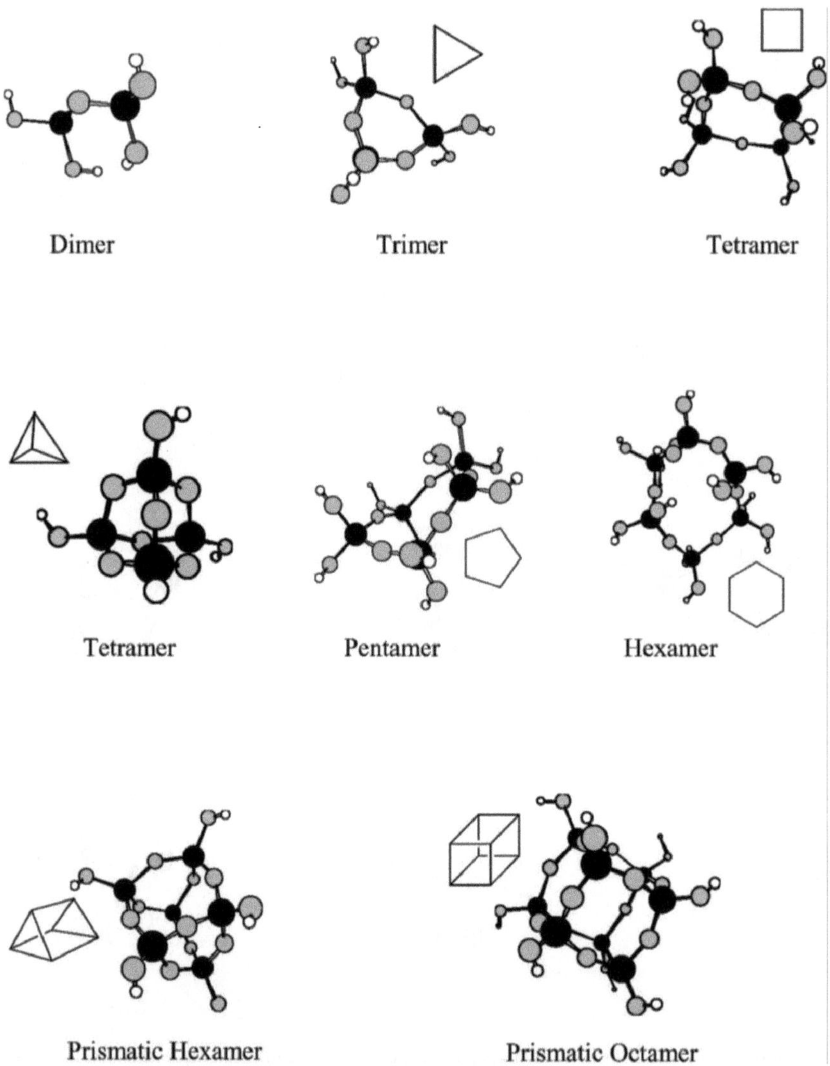


**Fig. 6.** Structures of single site silicates (excluding the monomer) showing energy minimized structures (MM2 followed by application of the AM1 programme) and idealized shape suggested following investigations by Knight et al. (1989). As in Fig. 5, the energy minimizations were performed with all silanols protonated. In reality, the pH of the solution will dictate the pattern of deprotonation observed. Also included are five- and six-membered rings that are thought to exist in the ca. neutral solutions relevant to biological silica precipitation

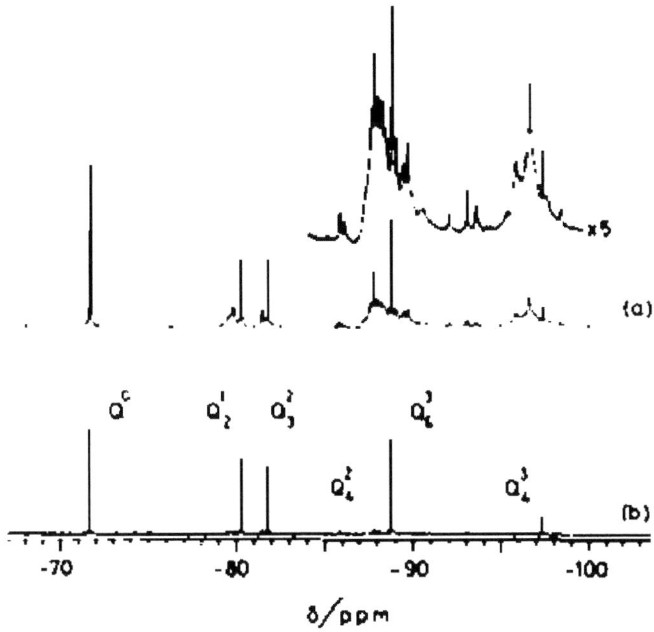


**Fig. 7a,b.** The 99.3 MHz $^{29}Si$ NMR spectrum of a nominally 1.5 M, 1 : 1 potassium silicate solution prepared with isotopically enriched $^{29}SiO_2$ at 22 °C. **a** Conventional single-pulse Fourier transform spectrum, **b** a 'one spin' filtered $^{29}Si$ NMR spectrum of the same solution as in **a** with the 'filter' set to exclude the signals of multisilicate silicate anions. The chemical shift scale is referenced externally to the signal of tetramethylsilane in [$^2H_6$]acetone (50 : 5••/v). (Reprinted with permission from Knight et al. 1989)

containing more than one silicon site are removed from the spectrum (Fig. 7). Assignment of the signals using the conventional Q terminology is given on the spectrum itself and has allowed for the confirmation of the presence of $Q^0$ species from the monomer, $Q^1$ species from the dimer, $Q^2{}_3$ from the cyclic trimer, $Q^2{}_4$ from the cyclic tetramer, $Q^3{}_6$ from the double three-membered ring or prismatic hexamer and a species, possibly identified as the tetrahedral tetramer (Knight et al. 1989). Such NMR studies have also been used to identify the effect of the counterion on speciation (see below).

The presence of cations can modify the rates of the various condensation reactions (Iler 1979; Harrison and Loton 1995). In experiments on the formation of silica from catecholate complexes, the size of the hydrated unicharged metal ions, present as counterions, was found to directly correlate with the rate of condensation processes, the smaller the cation and the larger the hydration sphere for a unicharged metal ion, the faster the reaction (i.e. condensation faster in the presence of $Li^+$ than $Na^+$ than $K^+$. The effect of cations derived from ammonium ions was less clear with ions such as ammonium and triethylammonium being able to reduce the solubility of any siliceous material formed in their presence. The behaviour of cations in aqueous media derives from their ability to modify the structure of the liquid. When no ions are present, the structure of water is largely a disordered three-dimensional network of hydrogen bonds that form statistical connections allowing fast proton and hydroxyl transfers (Stillinger 1980; Symons 1981). The presence of cations (and anions) in water induces local ordering of the water molecules owing to dipolar interactions continuing through the second and higher solvation spheres (reviewed in Jolivet 2000). Small ions such as $Li^+$, $Na^+$, $Mg^{2+}$, $F^-$ and $OH^-$ are known as 'type I structuring ions', as they are able to locally alter the primitive structure of water through breakage of hydrogen bonds and reorganisation of numerous solvent layers. Ions such as $K^+$, $Cs^+$, $NH_4^+$, $NO_3^-$ and $ClO_4^-$ are called 'destructuring or breaking ions', as they are able to break hydrogen bonds locally, but have little effect over longer distances. Ions such as the quaternary ammonium cations $[NR_4]^+$ ($R = CH_3, C_2H_5$ etc.) behave in a different manner and are known as 'type II structuring ions'. They are not able to break the network of hydrogen bonds in the liquid phase, but are incorporated within the liquid phase thereby altering the local water architecture dramatically. Thus, the differences in behaviour of catecholate and other aqueous silicon containing solutions in the presence of a range of cations can be attributed to the differences in hydration characteristics of the cations themselves.

The solution attributes of the different cations also have an effect on speciation and hence, on the precise reaction pathways to the materials produced, whether these be amorphous or crystalline. One major difference is in the effect of 'traditional' metal ions, such as sodium, found in the biological environment as opposed to tetraalkylammonium ions, which are often used in chemical synthesis. In the presence of ions such as $[NR_4^+]$ (R = methyl, ethyl, propyl etc.) cages made of double cycles of 3–5 $SiO_4$ tetrahedra are stabilised and can be observed in solution (Engelhardt et al. 1982; Knight et al. 1986) with $[N(CH_3)_4]^+$ particularly favouring the formation of the cubic octamer $[Si_8O_{20}]^{8-}$. In the presence of sodium ions, the transverse bonds are readily broken to yield simple $[Si_4O_{12}]^{4-}$ cycles (Jolivet 2000) probably because of the strong affinity of the solvated $Na^+$ ion for the basic oxygen of the silicate. More recent studies with 'less alkaline' solutions (pH ca. 12 as opposed to 14) have provided evidence for additional cage structures including two octamers linked by an Si—O—Si bridge (Kinrade et al. 2002a).

Further examples of the effect of solution components on speciation are given in the section on non-aqueous synthesis, where methyl vs. ethyl groups on the alkoxide used as the starting material affect the proportions of three- and four-membered ring species present during the formation of silica.

## 5.2 Silica Chemistry in Non-Aqueous Solution (Brinker and Scherrer 1990; Hench and West 1990)

Another common starting point to the formation of silica (and that often used in model studies of biosilicification) are the silicon alkoxides where the central silicon atom is attached via ether linkages to organic moieties including methyl, ethyl, phenyl and which may also include silicon directly bonded to carbon atoms of other functional groups. The alkoxy groups are generally hydrolysed when the compounds are placed in an aqueous environment, particularly under conditions of acid or base catalysis, but organic groups directly bonded to the silicon atom are not hydrolyzed. The molecular reactions that occur are given below:

$$\equiv\text{Si—OR} + H_2O \rightarrow \equiv\text{Si—OH} + \text{ROH} \qquad \text{Hydrolysis}$$

$$\equiv\text{Si—OH} + \equiv\text{Si—OR} \rightarrow \equiv\text{Si—O—Si}\equiv + \text{ROH} \qquad \text{Condensation}$$

$$\equiv\text{Si—OH} + \equiv\text{Si—OH} \rightarrow \equiv\text{Si—O—Si}\equiv + H_2O$$

Acid and base catalysis that is used to modify the rates of the hydrolysis and condensation reactions affects the speciation pattern of the oligomers that develop, thereby affecting the structure and physical properties of the material that forms. Factors that are known to affect hydrolysis and condensation include; (1) the chemical identity of the precursor, catalyst and solvent, (2) the relative amount of water present in the system, (3) the temperature, (4) the pH, (5) stirring and (6) the absolute amounts of any of the components in the precursor mixtures and the order in which they are added one to the other. Gel structures (acid hydrolysis leading to largely linear structures) and particulate materials (base catalysis leading to highly branched structures) can be formed. Solution $^{29}$Si NMR spectroscopy and computational studies have proved extremely useful in the identification of the molecular species present during the early stages of reaction. Many chemical species, including cyclic and linear ones can be identified and the effect of synthesis conditions on exact chemical shifts for particular species can be determined. For an example, see Table 4 where the effects of acidity and water content on the speciation of solutions of tetramethoxysilane are shown.

As for condensation in an aqueous medium, precursor identity has an effect on the formation of certain of the larger oligomeric units, thereby influencing the structure of the materials formed from such precursors. For example, reaction of tetramethoxysilane leads to a structure built up from a preponderence of three-membered rings, whereas if tetraethoxysilane is used there is a preponderance of four-membered ring structures. These differences in the building blocks available will have consequences for the nature of the materials formed from such precursors including effects on surface chemistry and porosity.

**Table 4.** $^{29}Si$ NMR assignments for Tetramethoxysilane:methanol:water:acid solutions at a range of solvent and acidity conditions. Collated from Brunet et al. (1991) and Brunet and Cabane (1993). Chemical shifts are quoted in parts per million (ppm) from tetramethylsilane set as 0 ppm

| Species | (OH,OMe) | TMOS:MeOH:$H_2O$ (pH) | | |
|---|---|---|---|---|
| | | 1:3.7:2.8 pH 3.0 | 1:3.7:8 pH 3.0 | 1:3.7:<2 pH 1.6 |
| $Si(OMe)_4$ | $Q^0(0,4)$[a] | −78.25 | | −78.29, −78.39[b] |
| $Si(OMe)_3OH$ | $Q^0(1,3)$ | −76.15 | | −76.07 |
| $Si(OMe)_2(OH)_2$ | $Q^0(2,2)$ | −74.60 | −74.70 | −74.46 |
| $Si(OMe)(OH)_3$ | $Q^0(3,1)$ | −73.45 | −73.60 | −73.22 |
| $Si(OH)_4$ | $Q^0(4,0)$ | −72.40 | −72.70 | −72.20 |
| $Si(OMe)_3OSi\equiv$ | $Q^1(0,3)$ | −85.85 | −84.40 | −85.80, −86.02[b] |
| $Si(OMe)_2(OH)OSi\equiv$ | $Q^1(1,2)$ | −84.15 | −83.60 | −84.04, −84.08[b] |
| $Si(OMe)(OH)_2\ OSi\equiv$ | $Q^1(2,1)$ | −82.80 | −83.05 | −82.65, −85.60[b] |
| $Si(OH)_3\ OSi\equiv$ | $Q^1(3,0)$ | −81.80 | −82.10 | −82.10 |
| Cyclic | $Q^2(2,0)$ | | −90.90 | |
| Cyclic | $Q^2(1,1)$ | −91.40 | −91.65 | −91.30, −92.08[b] |
| $Cyclic_{4c}$ | $Q^2(0,2)$ | | | −92.96, −93.20, −93.32 |
| Linear | $Q^2(2,0)$ | | | |
| Linear | $Q^2(1,1)$ | −92.40 | −92.75 | |
| Linear | $Q^2(0,2)$ | −93.70 | | |
| ??? | $Q^3(1,0)$ | | −98.60 | |
| ??? | $Q^3(0,1)$ | | −99.50, −100.65, −101.75 | |
| Linear | $Q^3(0,1)$ | | | −102.6, −102.33 |
| Cage | $Q^3(0,1)$ cf. tetramer | | | −101.38, −101.13 |
| Polyhedron | $Q^3(0,1)$ | | | −100.53, −100.72 |
| Other polyhedra | (More constrained) | | | −99.70 |

[a] $Q^0$, single silicon atom, $Q^1$ silicon atom attached through oxygen to one other silicon atom, $Q^2$ silicon atom attached to two other silicon atoms through oxygen atoms, $Q^3$ silicon atom attached to three other silicon atoms through oxygen atoms, $Q^4$ silicon atom attached to four silicon atoms through oxygen atoms.
[b] Observed under very low water ratios.

# 6 Solution Additives and Model Precipitation Reactions

The effects of additives on the oligomerization and aggregation of silica has been thoroughly reviewed by Iler (1979). Additives are able to affect all aspects of silica deposition from the kinetics through to the macroscopic properties including surface area, pore structure, particle size and aggregation. This makes the use of additives, from simple inorganic or organic molecules to complex organic polymers, a very powerful tool for the controlled production of silica. It should be noted that surfactants and small molecules such as ammonium-based cations are used in the production of mesoporous materials (for example Kresge et al. 1992) and silica-based zeolites (for example Lewis et al. 1996), respectively. Some of the most proficient systems at using this type of regulation are biological organisms. Their ability to tailor-make inorganic materials for a specific use is unsurpassed and has become the focus of much research effort in the search for ways to produce materials with modified properties for use by industry and academe alike.

## 6.1 Rationale for Use of Model Precipitation Reactions; Experimental Approaches

Biological organisms have demonstrated their ability to tailor-make biominerals that exhibit structural and physical properties that are distinct from materials produced in the laboratory. Biosilicas can exhibit controlled particle size distributions, hydration and aggregation characteristics. Control of these structural features tends to be attributed to the interaction of the mineral as it forms with an organic matrix, but the fact that the biomineralization environment is likely to contain small organic and/or inorganic chemical species cannot be ignored. The biological reaction environment is likely to be extremely complex and will contain the following; inorganic ions (up to 20 different ions) to maintain ionic strength if nothing else, organic molecules (possibly up to several hundred) and macromolecules, i.e. proteins, carbohydrates and lipids (possibly up to several thousand if their precursors are also included). It would be futile to try to mimic this chemical environment in its totality as the identity of the majority of species present are unknown (bearing in mind that the precise location for mineral formation is not even known with any certainty for most species currently studied). The idea behind a model system is to simplify, and even oversimplify the reaction environment so that the effects of individual chemical components can be identified/proposed. The disadvantage of this approach is that interactions between chemical species during the formation of the mineral phase are missed. Several different model systems have been utilized to study the effect of biosilica extracts on the formation of silica in an in vitro reaction environment.

One experimental approach utilizes a catecholato complex of silica $M_2[Si(C_6H_4O_2)_3]\cdot xH_2O$ (Perry and Lu 1992). At around physiological pH the complex partially decomposes to yield orthosilicic acid that immediately undergoes polycondensation reactions to relieve any resultant supersaturation. It should be noted that the oligomerizing system exhibits slow kinetics compared to the reactions of the alkoxy precursors, which allows the process to be studied more closely. The experimental method involves monitoring the orthosilicic acid and complex concentrations using a colorimetric molybdenum blue assay (Iler 1979; Perry and Lu 1992) and $^1H$ NMR spectroscopy respectively. Analysis of the loss of orthosilicic acid from solution using a range

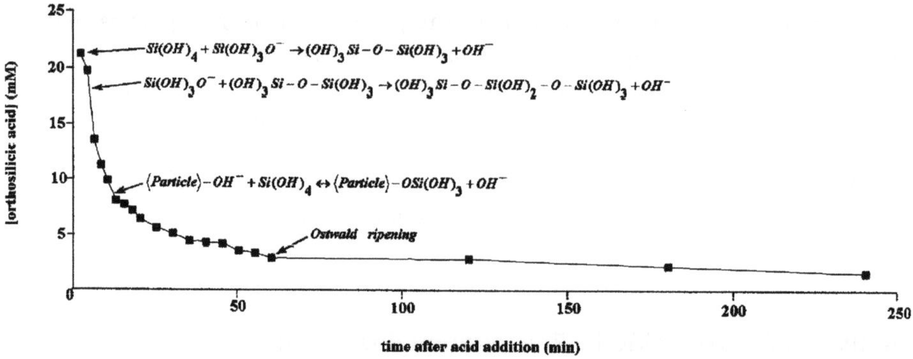


**Fig. 8.** The change in orthosilicic acid concentration with time for a 50-mM solution of $K_2(Si[C_6H_4O_2]_3)$ reduced to ca. pH 7 together with the time regions where dimers or trimers are formed, monomers are added to oligomers and Ostwald ripening occurs

of rate expressions (Harrison and Loton 1995) enables the oligomerization process to be considered in distinct stages. Information on the time periods when dominant molecular reactions are the formation of a dimer, the formation of a trimer, the addition/removal of a single silicic acid unit from larger oligomers and Ostwald ripening are all possible (see Fig. 8), as is the study of the development of particle sizes, aggregate characteristics and the porosity and surface area characteristics of the materials formed.

An alternative approach to the study of silica formation using aqueous chemistry is to utilize diluted and buffered Waterglass (27% $SiO_2$, 10% NaOH) as the source of silica and follow the loss of silicic acid from solution using a colorimetric molybdenum yellow assay together with studies of the materials formed using infrared spectroscopy and scanning electron microscopy (Coradin and Livage 2001). Other model reactions utilize the alkoxy compounds, TMOS, TEOS and alkyl silanes such as MTEOS in buffered solution at a range of pH values. The extent of reaction is generally assayed by the amount of centrifugable silica formed and the effect of the bioextract on morphology by the form of the precipitated materials (Shimizu et al. 1998; Cha et al. 1999; Kröger et al. 1999).

## 6.2
## Studies of the Effect of Biosilica Extracts on the in Vitro Formation of Silica

Proteinaceous extracts including silaffins from diatoms (Kröger et al. 1999), silicateins from sponges (Cha et al. 1999) and biosilica extracts from higher plants (Perry and Keeling-Tucker 1998, 2002) have been used in model studies of silicification in order to identify the role of these biomolecules.

Studies using silaffins (Kröger et al. 1999) have shown that these molecules accelerate the silica polycondensation process and aid in flocculation, the biopolymer extract being occluded in the silica formed. These molecules were able to cause a metastable solution of silicic acid to spontaneously aggregate with small, ca. 50-nm spherical particles being produced in the presence of a mixture of silaffins and larger ca. 500–700 nm particles being produced in the presence of a purified 4-kDa silaffin. The silaffins were found to be effective in promoting silicification at pH values as low as 4, due to the presence of highly positively charged lysine residues.

Studies using silicatein $\alpha$ extracted from sponge spicules on polysiloxane synthesis at neutral pH showed a ca. ten-fold increase in the amount of harvestable silica compared to experiments run in the presence of the denatured silicatein or proteins such as bovine serum albumin and trypsin (Cha et al. 1999). A catalytic role for the silicatein in the polycondensation of siloxanes was proposed in a manner analogous to that of serine- and cysteine-based hydrolytic proteases. The suggested mechanism was that of a general acid/base catalyst with a specific requirement for both serine and histidine amino acids at the active site (Zhou et al. 1999). A scaffolding or structure-directing activity was also reported, as the polymerized silica was able to form a layer following the contours of the underlying protein fibre, structures not observed on either cellulose or silk fibroin fibres (Cha et al. 1999).

Biosilica extracts from *Equisetum arvense* (Perry and Keeling-Tucker 1998, 2002) have been found to accelerate the rate of formation of small oligomers of silicic acid and to limit the distribution of particle sizes similar to that observed for the biological silica used to provide the biosilica extract (Perry and Keeling-Tucker 2002). Layered, crystalline materials were also produced that could have arisen from the epitaxial matching of initially formed oligomers on a $\beta$-sheet type material, the silica structure continuing to develop from the initial biopolymer-controlled nucleation event (Perry and Keeling-Tucker 1998, 2002).

## 6.3 Biomimetic Studies of Silica Precipitation

Using information obtained from silica precipitation studies performed using biosilica extracts as a starting point, several groups have been investigating the effect of amino acids, polyamines and synthesized fragments of the biosilica extracts as solution adjuncts in studies of silica formation from both aqueous (catecholates and diluted solutions of water glass; e.g. Harrison and Lu 1994; Coradin and Livage 2001; Coradin et al. 2002) and non-aqueous precursors (alkoxysilanes and their derivatives; Cha et al. 1999, 2000; Zhou et al. 1999). The results obtained show that, in accord with results obtained on simple carbohydrate materials (Harrison and Lu 1994), amino acids have little effect on the kinetics of silica formation (Cha et al. 2000; Coradin and Livage 2001; Coradin et al. 2002). Homopeptides of arginine (up to $Arg_{65}$) and lysine (up to $Lys_{1005}$; Coradin et al. 2002; Cha et al. 2000) affect the silica precipitation process at circumneutral pH with the effect being related to the length of the polypeptide chain. Electrostatic interactions have been proposed as the principal mode of interaction between negatively charged silicic acid species and positively charged polypeptide chains leading to gelation and/or precipitation according to the system investigated (Coradin et al. 2002). The obtained surface area, pore volume and average pore diameter were characteristic of dried flocculates. A polycationic peptide (derived from the *C. fusiformis* silaffin-1 protein (chemically synthesized with no post-translational modification of the lysine residues) has been used in the generation of a two-dimensional array of silica nanospheres within a hybrid polymer hologram (Brott et al. 2001).

# 7 Other Areas of Interest

## 7.1 Transport

It should be remembered that silicification occurs alongside other cellular processes and in order for the element to be used either in molecule-based cellular processes or in the production of silica it has to be transported from the external environment into the cell. Measurement of levels of silicic acid in the primitive plant, *Equisetum arvense*, has identified the presence of a supersaturated solution of silicic acid (1.6–2.5 mM) during the first few days of growth, suggesting that some means of maintaining the silicon in solution must be available to this plant (Weiss and Herzog 1978). Complexes of silicon with diols such as tropolone derivatives have been implicated although there is still the question as to whether the isolated silicon complex is present in the plant or whether it is formed from free thujaplicine and silica deposits or free silicic acid in the plant in the course of the extraction (Weiss and Herzog 1978). In diatoms, it is proposed that orthosilicic acid is the species transported (Del Amo and Brzezinski 1999). As the rate of growth of the frustule appears to be incredibly fast, many authors refer to the necessity of "active" (ATP/energy-dependent) transport with recent research on silicon transporter genes in diatoms suggesting that uptake is regulated by the number and location of such transporter molecules (Hildebrand et al. 1998). The transporter molecules are proposed to comprise a number of helical transmembrane domains with movement of silicic acid linked to the transport of sodium ions. The movement of silicic acid from the external to the internal cellular environment is alleged to be regulated by intracellular concentrations and by a demand for silicification (Hildebrand et al. 1997). However, the latter is not likely as injection of the messenger RNA transcribed from the cloned DNA of the proposed transporter into live frog eggs mediates the sodium-dependent flux of silicic acid across cell membranes into a system that has no requirement for silica formation!!! Nothing is known about the transport of silicic acid between cellular compartments within the diatom although it is thought that different transporter molecules could be used for this purpose (Hildebrand et al. 1998). It is also interesting to note that no definitive information on the cessation of growth other than by the shutdown of transport pathways has been proposed for any silicifying system.

Other researchers do not agree with the necessity for a specific molecular pathway of active transport of silicic acid (Wetherbee et al. 2000). The growth of a diatom on a timescale of minutes to hours can be successfully explained provided that diatoms are able to effectively sequester silicon by surface complexation and to transport it into the cytoplasm. This statement is based on estimates of diffusion limited growth:

$$J_{diff} = 4\pi aD/10^3 \cdot [Si]_{\infty mol\,s^{-1}}$$

where:

a = diatom radius (ca. 5-μm radius sphere with a silica shell thickness of 2 μm)
D = silicate ion diffusion coefficient ($D \sim 10^{-5}\,cm^2\,s^{-1}$)
$[Si]_\infty$ = bulk solution silicate concentration (ca. 10 μM)
$[Na^+]$ = effects minimal

$J_{diff}$ is ca. $5 \times 10^{-16}\,mol\,s^{-1}$.

The silica required to make the diatom shell is $1 \times 10^{-11}$ mol (based on the volume of silica required and silica's molar volume).

Time to accrue under optimal conditions is:

$$\begin{aligned} t_{growth} &= \text{moles required/diffusive flux} \\ &= 1 \times 10^{-11}/5 \times 10^{-16} \\ &= 2 \times 10^4\,s \text{ (at } 10\,\mu M) \\ &= 200\,s\ (1\,mM)!! \end{aligned}$$

N.B. estimates of soluble silicon concentrations inside diatoms are (450–750 nM; Sullivan 1976).

Diffusion limitation can explain silica production in diatoms on the timescales that are observed for frustule formation.

## 7.2 The Significance of Hypervalency in Biological Silicon Chemistry

The mechanisms proposed for the condensation of silicic acid species involve hypervalent silicon. Also, some of the literature describing the likely path for silicon uptake in higher plants suggests the involvement of hypervalent tropolone/catechol:silicon species that are partially stable at ca. neutral pH (Weiss and Herzog 1978; Perry and Lu 1992; Harrison and Loton 1995). In light of these observations, some experimental and theoretical studies have been undertaken to look at the generation of chemical species containing hypervalent silicon that might have some stability under reaction conditions approximating the natural environment under which silica is formed. Polyols containing a minimum of four hydroxyl groups with the middle two being in the threo configuration have been investigated as likely complexing agents for silicon (complexation results in the formation of seven-membered ring systems). $^{29}$Si NMR data obtained at high pH (non-biologically relevant conditions) have suggested the formation of short lived complexes on the NMR time scale that contain, 4-, 5- and 6-co-ordinate silicon species (Kinrade et al. 1999, 2001a) with 5-co-ordinate species being apparently favoured as the pH decreased. Similar experiments have also been conducted under neutral pH conditions at low silicon concentrations (1.4 mM) and in a vast excess of the polyol ligand (2500 times excess) that purportedly gave rise to NMR signals from 5-co-ordinate silicon species (Kinrade et al. 2001b).

Ab initio computational studies have been performed on some of the proposed hypervalent structures (and others besides) together with calculation of the expected $^{29}$Si NMR chemical shifts for the complexes and their relative stabilities (Sahai and Tossell 2002). These data provide an alternative interpretation of the $^{29}$Si NMR experimental data. Complexation involving smaller Si–O–C containing rings is proposed as well as oligomeric (trimers containing both 4- and 5- co-ordinate species) silicon–oxygen–polyol complexes rather than the relatively simple hypervalent silicon species co-ordinated to the polyol ligands with seven membered rings. Six co-ordinate species were also thought not to occur, although more 6-co-ordinate than 5-co-ordinate silicon–oxygen containing complexes are known! Further, calculation of the relative reaction energies of the different species indicated that the hyper co-ordinate species would only be favoured when the ionized form of silicic acid ($H_3SiO_4^-$) is the reactant. Under alkaline conditions such species can therefore form, albeit at extremely low concentrations. Under the mildly acidic conditions identified in the diatom silica deposition vesicle (Vreiling et al. 1999), it is extremely improbable that hypervalent silicon complexes are involved in the transport and utilization of dissolved silicon. Thermodynamic calculations do not favour the formation of silicon–polyalcohol complexes in aqueous solution under mild conditions, giving further support to the idea that soluble silicon uptake is via silicic acid rather than such complexes.

Is there evidence for the presence of hyper co-ordinate species in diatoms during frustule formation? One paper has been published where the authors using $^{29}$Si NMR spectroscopy on isotopically heavily enriched samples of *Navicula pelliculosa* report the observation of a transient signal at ca. −131.5 ppm during spectral accumulation (Kinrade et al. 2002b). This observation has created much interest in the peripheral scientific community; if it is true, it is the first direct evidence for the formation of an organosilicon complex during the life cycle of an organism. There is, however, much disagreement as to the interpretation of the data due to inconsistencies both in the experimental protocol described in the paper and the poor quality of the data presented. The court remains 'out' until better data are provided.

# 8
# Future Directions

Much remains to be understood concerning the formation of biosilicas in their natural environment. The approaches that will produce additional information are diverse and include theoretical chemical, biochemical and molecular engineering. A combination of the information gained from these different approaches will provide a much fuller picture of biosilica formation.

## 8.1
## Molecular Engineering in Diatoms

A recently discovered molecular engineering approach in diatoms may provide a new tool for analysing cellular biological processes including those involved in frustule formation (Kröger 2001) The work of Zaslavskaia et al. (2001) has produced a facultative heterotrophic diatom that grows in the dark by the introduction of a single gene (*glut*-1) encoding for the glucose transporter protein (glut-1) from human erythrocytes. Tagging the gene with a green fluorescent protein (termed eGFP) from *Aequorea victoria* enabled the location of the gene to be determined. It is proposed that transforming diatoms with genes encoding GFP-tagged cytoskeletal molecules should allow the dynamics of the diatom cytoskeleton to be monitored by fluorescence microscopy in vivo during cell wall formation. In addition, a combination of fluorescence and electron microscopy/energy dispersive elemental analysis would enable the identification of whether the formation of particular silica elements in the frustule coincides with the formation of specific cytoskeletal features (Kröger 2001).

## 8.2
## Isolation and Identification: Labelling

In our search to understand the chemistry of the biomineral produced by biological organisms, 'staining' methods have been used to locate the mineral and to provide additional information on the chemistry of the mineral phase. Chemical approaches have been used to identify the extent of hydroxylation of individual silica types by use of an organometallic staining agent (Perry et al.

1990; see Fig. 2). Bis(cyclopentadienyl)titanium dichloride specifically interacts with silanol groups on the surface of silica particles and has been used to obtain evidence for the variation in hydroxylation levels for three different silica structural types produced within single cell plant hairs. Siliceous materials have also been treated with silane-coupling agents such as 3-aminopropyl triethoxysilane prior to staining with fluorescein for the study of the interaction of mammalian cells with silica fibers, implicated in lung, skin and oesophageal cancer (Hodson et al. 1994). Experimental approaches have been used to locate silicon-containing species in biological organisms. These approaches have been used in living organisms, on embedded material and on the isolated mineral phase itself. Rhodamine 123 has been used with diatoms, but due to problems with autofluorescence (due to pigments in the diatom) it requires diatom cell washing and fixation prior to observation of the fluorescence produced. More recently, PDMPO (2-(4-pyridyl)-5-((4-(2-dimethylaminoethylaminocarbamoyl)methoxy)phenyl)oxazole, a fluorescent pH indicator has been used with organisms. The molecule has unique fluorescent properties when combined with polysilicic acid (bright green fluorescence) and on studies of diatoms it is specifically incorporated into newly synthesized frustules. The indicator is proposed as an ideal fluorescent tracer for Si (Shimizu et al. 2001).

## 8.3 Theoretical Studies

Extension of studies performed using the partial charge model (as briefly used to explain the reactivity of different silicon species in aqueous solution above) can currently provide structure-independent (for example, all silicon atoms and hydroxyl groups within a chosen structure are assumed to be identical and, thus, an average of the electronegativity and partial charge for the chosen functional groups or atoms is obtained) information on the charges and hence, reactivity and acid–base properties of silicic acids under chosen pH conditions. For example, calculations of the mean electronegativity, averaged partial charge distributions and estimated optimum pH for the formation of silica tetramers (linear, cyclic and tetrahedral) and their anions (Table 5) show that at a pH of ca. 7, doubly charged species will dominate the speciation pattern. Under conditions of high alkalinity, species with three or four negative charges can be readily formed. An extension of this approach is to use the more rigorous model (Henry 1997) that allows for the structure-dependent calculation of partial charges and, therefore, acid–base properties of specified atoms within a molecule. These are important parameters in defining the reactivity of individual groups and hence, the reaction pathways by which silica-based materials are formed and will be vital in extending our understanding of how biosilicas are formed de novo.

**Table 5.** Mean electronegativities of a series of protonated and deprotonated (many theoretical species) tetrameric silicic acid species. N.B. if $\delta(Si) < 0.3$, then the corresponding species are unable to condense further

| Species | $z^a$ | $\chi_t$ | $\delta(Si)$ | $\delta(OH)$ | $pH^b$ |
|---|---|---|---|---|---|
| Linear tetramer ($Q_1Q_2Q_2Q_1$; $\alpha^c = 0.75$) | 0 | 2.612 | 0.486 | −0.044 | 3.4 |
| | −1 | 2.549 | 0.451 | −0.072 | 5.2 |
| | −2 | 2.482 | 0.413 | −0.102 | 7.2 |
| | −3 | 2.408 | 0.372 | −0.136 | 9.3 |
| | −4 | 2.326 | 0.327 | −0.172 | 11.6 |
| Cyclic tetramer ($Q^2_4$; $\alpha = 1.00$) | 0 | 2.627 | 0.494 | −0.038 | 3.0 |
| | −1 | 2.557 | 0.456 | −0.069 | 5.0 |
| | −2 | 2.480 | 0.413 | −0.103 | 7.2 |
| | −3 | 2.395 | 0.365 | −0.141 | 9.6 |
| | −4 | 2.301 | 0.312 | −0.183 | 12.3 |
| Tetrahedral tetramer ($Q^3_4$; $\alpha = 1.25$) | 0 | 2.647 | 0.506 | −0.029 | 2.4 |
| | −1 | 2.568 | 0.461 | −0.064 | 4.7 |
| | −2 | 2.479 | 0.412 | −0.104 | 7.2 |
| | −3 | 2.379 | 0.356 | −0.148 | 10.1 |
| | −4 | 2.265 | 0.293 | −0.199 | 13.3 |

[a] z, The net charge of the species.
[b] Value of pH is based on the formula: $pH = 2.732 - 0.035\chi_t$ (Henry et al. 1992) and roughly estimates the optimum conditions for the protonated version of the particular silicate species to exist.
[c] $\alpha$, Degree of (poly)condensation, calculated as $\alpha = n(\text{-O-})/n(Si)$, where $n(\text{-O-})$ – the number of siloxane bridges, $n(Si)$ – the total number of silicon atoms in the species.

# 9
# Conclusions

Silica formation in the natural environment requires the uptake, transport, condensation, growth, precipitation and moulding of species involving silicic acid. Significant progress has been made in respect of transport and in respect of the role(s) played by biomacromolecules in the controlled formation of silica structures. However, much is not understood and there are many unresolved controversies concerning uptake, transport (the requirement for hypervalency or not) and the specific role of some of the extracted biomolecules such that there is plenty of scope for scientists wishing to the enter the field and to make a contribution to the understanding of the processes required to make and mould an amorphous mineral that is able to withstand pressure (sponges are found at depths of ca. 300 m), dissolution (all organisms found in water) and predation (water- and land-based species) by a range of animals. The information obtained in this search will, however, assist us in our understanding of the essentiality of silicon to life processes and in the generation of new materials with specific form and function for industrial application in the twenty-first century. Already, we have seen that the ability to create an ordered organic/inorganic array of silica nanospheres using a polycationic peptide derived from a diatom silaffin to promote localized condensation may be of practical use in the fabrication of photonic devices (Brott et al. 2001).

*Acknowledgements.* The authors would like to thank ESRC, AFRC, BBSRC and Crosfield Chemicals (now Ineos Chemicals) for their interest and funding of research programmes in silica chemistry.

## References

Bogush GH, Zukovski CF (1990) Studies of the kinetics of the precipitation of uniform silica particles through the hydrolysis and condensation of silicon alkoxides. J Colloid Interface Sci 142:1–18

Brinker CJ, Scherrer GW (1990) Sol-Gel Science-the physics and chemistry of sol-gel processing. Academic Press, London

Brott LL, Naik RR, Pikas DJ, Kirkpatrick SM, Tomlin DW, Whitlock PW, Clarson S, Stone MO (2001) Ultrafast holographic nanopatterning of biocatalytically formed silica. Nature 413: 291–293

Brunet F, Cabane B (1993) Population of oligomers in sol-gel condensation. J Non Cryst Solids 163:211–225

Brunet F, Cabane B, Dubois M, Perly B (1991) Sol-gel polymerization studied through $^{29}$Si NMR with polarization transfer. J Phys Chem 95:945–951

Brzezinski MA, Conley DJ (1994) Silicon deposition during the cell cycle of *Thalassiosira weissflogii* (bacillariophyceae) determined using dual rhodamine 123 and propidium iodide staining. J Phycol 30:45–55

Cha JN, Shimizu K, Zhou Y, Christiansen SC, Chmelka BF, Stucky GD, Morse DE (1999) Silicatein filaments and subunits from a marine sponge direct the polymerization of silica and silicones in vitro. Proc Natl Acad Sci USA 96:361–365

Cha JN, Stucky GD, Morse DE, Deming TJ (2000) Biomimetic synthesis of ordered silica structures mediated by block copolypeptides. Nature 403:289–292

Coradin T, Livage J (2001) Effect of some amino acids and peptides on silicic acid polymerisation. Colloids Surfaces 21:329–336

Coradin T, Duruphy O, Livage J (2002) Interactions of amino-containing peptides with sodium silicate and colloidal silica: a biomimetic approach to silicification. Langmuir 18:2331–2336

Del Amo Y, Brzezinski MA (1999) The chemical form of dissolved Si taken up by marine diatoms. J Phycol 35:1162–1170

Engelhardt G, Hoebbel D, Tarmale M, Samoson A, Lippmaa E (1982) Si-29 NMR investigations of the anion structure of crystalline tetramethylammonium-aluminosilicate and aluminosilicate solutions. Z Anorg Allg Chem 484:22–32

Frolov YG, Shabanova NA, Popov VV (1983) Effect of temperature and pH on the polycondensation of silicic acid in an aqueous medium. Kolloidn Zh 45:179–183

Harris RK, Knight CTG (1983) Si-29 nuclear magnetic resonance studies of aqueous silicate solutions; 2nd order patterns in potassium silicate solutions enriched with Si-29. J Chem Soc Faraday Trans 2:1539–1561

Harrison CC (now Perry) (1996) Evidence for intramineral macromolecules containing protein from plant silicas. J Phytochem 41:37–42

Harrison CC (now Perry), Lu Y (1994) In vivo and in vitro studies of polymer controlled silicification. BullInst Océanogr Monaco 14:151–158

Harrison CC (now Perry), Loton N (1995) Novel routes to designer silicas-studies of the decomposition of $(M^+)_2[Si(C_6H_4O_2)_3]\cdot xH_2O$, the importance of $M^+$ identity on the kinetics of

oligomerisation and the structural characteristics of the silicas produced. J Chem Soc Faraday Trans 91:4287–4297
Hecky RE, Mopper K, Kilham P, Degens ET (1973) The amino acid and sugar composition of diatom cell walls. Mar Biol 19:323–331
Hench LL, West JK (1990) The sol-gel process. Chem Rev 90:33–72
Henry M (1997) Application of the partial charge model to the aqueous chemistry for silica and silicates. In: Silvi B, D'Arco P (eds) Modelling of minerals and silicate materials. Kluwer, Dordrecht, pp 273–334
Henry M, Jolivet JP, Livage J (1992) Aqueous chemistry of metal cations – hydrolysis, condensation and complexation. Struct Bond 77:153–206
Hildebrand M, Volcani BE, Gassmann W, Schroeder JI (1997) A gene family of silicon transporters. Nature 385:688–689
Hildebrand M, Dahlin K, Volcani BE (1998) Characterization of a silicon transporter gene family in *Cylindrotheca fusiformis*: sequences, expression analysis, and identification of homologs in other diatoms. Mol Gen Genet 260:480–486
Hodson MJ, Smith RJ, van Bladeren A, Crafton T, O'Neill CH, (1994) Detecting plant silica fibres in animal tissue by confocal fluorescence microscopy. Ann Occup Hyg 38:149–160
Iler RK (1979) The chemistry of silica. Plenum Press, New York
Jolivet JP (2000) Metal oxide chemistry and synthesis from solution to solid state. Wiley, Chichester
Kinrade SD, Del Nin JW, Schach AS, Sloan TA, Wilson KL, Knight CTG (1999) Stable five- and six-coordinated silicate anions in aqueous solution. Science 285:1542–1545
Kinrade SD, Hamilton RJ, Schach AS, Knight CTG (2001a) Aqueous hypervalent silicon complexes with aliphatic sugar acids. J Chem Soc Dalton Trans 7:961–963
Kinrade SD, Schach AS, Hamilton RJ, Knight CTG (2001b) NMR evidence of pentaoxo organosilicon complexes in dilute neutral aqueous silicate solutions. Chem Commun 17:1564–1565
Kinrade SD, Gillson A-ME, Knight CTG (2002a) Silicon-29 NMR evidence of a transient hexavalent silicon complex in the diatom *Navicula pelliculosa*. J Chem Soc Dalton 3:307–309
Kinrade SD, Donovan JCH, Schach AS, Knight CTG (2002b) Two substituted cubic octameric silicate cages in aqueous solution. J Chem Soc Dalton 3:1250–1252
Knight CTG (1988) A two-dimensional silicon-29 nuclear magnetic resonance spectroscopic study of the structure of the silicate anions present in an aqueous potassium silicate solution. J Chem Soc Dalton Trans 6:1457–1460
Knight CTG, Kirkpatrick RJ, Oldfield E (1986) The unexpectedly slow approach to thermodynamic equilibrium of the silicate anions in aqueous tetramethylammonium silicate solutions. J Chem Soc Chem Commun 11:66–67
Knight CTG, Kirkpatrick RJ, Oldfield E (1989) Silicon-29 multiple quantum filtered N.M.R. spectroscopic evidence for the presence of only six single site silicate anions in a concentrated potassium silicate solution. J Chem Soc Chem Commun 14:919–921
Krasko A, Gamulin V, Seack J, Steffen R, Schröder HC, Müller WEG (1997) Cathepsin, a major protease of the marine sponge *Geodia cydonium*: purification of the enzyme and molecular cloning of cDNA. Mol Mar Biol Biotechnol 6:296–307
Krasko A, Batel R, Schröder HC, Müller WEG (2000) Expression of silicatein and collagen genes in the marine sponge *Suberites domuncula* is controlled by silicate and myotrophin. Eur J Biochem 267:4878–4887
Kresge CT, Leonowicz ME, Roth WJ, Vartuli JC, Beck JS (1992) Ordered mesoporous molecular sieves synthesized by a liquid-crystal template mechanism. Nature 359:710–712
Kröger N (2001) The sweetness of diatom molecular engineering. J Phycol 37:657–658
Kröger N, Bergsdorf C, Sumper M (1994) A new calcium binding glycoprotein family constitutes a major diatom cell wall component. EMBO J 13:4676–4683
Kröger N, Bergsdorf C, Sumper M (1996) Frustulins: domain conservation in a protein family associated with diatom cell walls. Eur J Biochem 239:259–264

Kröger N, Lehmann G, Rachel R, Sumper M (1997) Characterization of a 200-kDa diatom protein that is specifically associated with a silica-based substructure of the cell wall. Eur J Biochem 250:99–105

Kröger N, Deutzmann R, Sumper M (1999) Polycationic peptides from diatom biosilica that direct silica nanosphere formation. Science 286:1129–1132

Kröger N, Deutzmann R, Sumper M (2001) Silica-precipitating peptides from diatoms, the chemical structure of silaffin-1a from *Cylindotheca fusiformis.* J Biol Chem 276:26066–26070

Lewis DW, Wilcock DJ, Catlow CRA, Thomas JM, Hutchings GJ (1996) De novo design of structure-directing agents for the synthesis of microporous solids. Nature 382:604–606

Li CW, Volcani BE (1985a) Studies on the biochemistry and fine structure of silica shell formation in diatoms VIII, morphogenesis of the cell wall in a centric diatom *Ditylum brightwelli.* Protoplasma 124:10–29

Li CW, Volcani BE (1985b) Studies on the biochemistry and fine structure of silica shell formation in diatoms X, morphogenesis of the labiate processes in centric diatoms. Protoplasma 124:147–156

Mann S, Williams RJP (1982) High resolution electron microscopy studies of the silica lorica in the choanoflagellate *Stephanoeca diplocostata* Ellis. Proc Roy Soc Lond B 216:137–146

Mann S, Perry CC (1986) Structural aspects of biogenic silica. In: Evered D, O'Connor M (eds) Silicon biochemistry. CIBA Foundation Symposium 121. Wiley, Chichester, pp 40–58

Mann S, Perry CC, Williams RJP, Fyfe CA, Gobbi GC, Kennedy GJ (1983) The characterisation of the nature of silica in biological systems. J Chem Soc Chem Commun 4:168–170

Mann S, Perry CC, Webb J, Luke B, Williams RJP (1986) Structure, morphology, composition and organization of biogenic minerals in limpet teeth. Proc R Soc Lond B 227:179–190

Miyaji F, Davis SA, Charmant JPH, Mann S (1999) Organic-crystal templating of hollow silica fibres. Chem Mater 11:3021–3024

Mizutani T, Nagase H, Fujiwara N, Ogoshi H (1998) Silicic acid polymerisation catalysed by amines and polyamines. Bull Chem Soc Jpn 71:2017–2022

Perry CC (1989) Chemical studies of biogenic silica. In: Mann S, Webb J, Williams RJP (eds) Biomineralisation, chemical and biological perspectives. VCH, Weinheim, pp 233–256

Perry CC, Lu Y (1992) Preparation of silica from silicon complexes – role of cellulose in polymerisation and aggregation control. J Chem Soc Faraday Trans 88:2915–2921

Perry CC, Keeling-Tucker T (1998) Crystalline silica prepared at room temperature from aqueous solution in the presence of intrasilica bioextracts. J Chem Soc Chem Commun 23:2587–2588

Perry CC, Keeling-Tucker T (2000) Biosilicification: the role of the organic matrix in structure control. J Biol Inorg Chem 5:537–550

Perry CC, Keeling-Tucker T (2003) Model studies of colloidal silica precipitation using biosilica extracts from *Equisetum telmateia.* Colloid and Polymer Science. DOI 10.1007/S00396-002-0816-7

Perry CC, Mann S, Williams RJP (1984) Structural and analytical studies of silicified macrohairs from the lemma of the grass *Phalaris canariensis* L. Proc R Soc Lond B 222:427–438

Perry CC, Moss EJ, Williams RJP (1990) A staining agent for biological silica. Proc R Soc Lond B 241:47–50

Sada E, Kumazawo H, Koresawa E (1990) reaction-kinetics and size control in the formation of monosized silica spheres by controlled hydrolysis of tetraethylorthosilicate in ethanol. Chem Eng J Biochem Eng J 44:133–139

Sahai N, Tossell JA (2002) $^{29}$Si NMR shifts and relative stabilities calculated for hypercoordinated silicon-polyalcohol complexes: role in sol-gel and biogenic silica synthesis. Inorg Chem 41: 748–756

Shimizu K, Cha J, Stucky GD, Morse DE (1998) Silicatein $\alpha$: cathepsin L-like protein in sponge biosilica. Proc Natl Acad Sci USA 95:6234–6238

Shimizu K, del Amo Y, Brzezinski MA, Stucky GD, Morse DE, (2001) A novel fluorescent silica tracer for biological silicification studies. Chem Biol 8:1051–1060

Simpson TL, Volcani BE (eds) (1981) Silicon and siliceous structures in biological systems. Springer, Berlin Heidelberg New York

Stillinger FH (1980) Water revisited. Science 209:451–457
Stöber W, Fink A, Bohn E (1968) Controlled growth of monodisperse silica spheres in the micron size range. Colloid Interface Sci 26:62–69
Sullivan CW (1976) Diatom mineralization of silicic acid (1) silicic acid transport characteristics in *Navicula pelliculosa.* J Phycol 12:390–396
Symons MCR (1981) Water structure and reactivity. Acc Chem Res 14:179–187
Tarutani T (1989) Polymerization of silicic acid: a review. Anal Sci 5:245–252
Thomas JM, Klinowski J, Ramdas S, Hunter BK, Tenakoon DT (1983) The evaluation of non-equivalent tetrahedral sites from Si-29 NMR chemical shifts in zeolites and related alumino-silicates. Chem Phys Lett 102:158–162
Vrieling EG, Gieskes WWC, Beelen TPM (1999) Silica deposition in diatoms: control by the pH inside the silica deposition vesicle. J Phycol 35:548–559
Weiss A, Herzog A (1978) Isolation and characterisation of a silicon-organic complex from plants. In: Bendz G, Lindqvist I (eds) Biochemistry of silicon and related problems. Plenum Press, New York, pp 109–127
Wetherbee R, Crawford S, Mulvaney P (2000) The nanostructure and development of diatom biosilica. In: Bauerlein E (ed) Biomineralization; from biological to biotechnology and medical applications. Wiley, VCH, Weinheim, pp 189–206
Yang H, Coombs N, Ozin GA (1997) Morphogenesis of shape and surface patterns in mesoporous silica. Nature 386:692–695
Zaslavskaia LA, Lippmeier JC, Shih C, Ehrhardt D, Grossman AR, Apt KE (2001) Trophic conversion of an obligate photoautotrophic organism through metabolic engineering. Science 292: 2073–2075
Zhou Y, Shimizu K, Cha JN, Stucky GD, Morse DE (1999) Efficient catalysis of polysiloxane synthesis by silicatein $\alpha$ requires specific hydroxy and imidazole functionalities. Angew Chem Int Ed 38:780–782

# Silicon Biomineralisation: Towards Mimicking Biogenic Silica Formation in Diatoms

Engel G. Vrieling, Sandra Hazelaar, Winfried W.C. Gieskes,[1] Qianyao Sun, Theo P.M. Beelen, and Rutger A. van Santen[2]

## 1 Introduction

Silica is by far the most abundant compound in the earth's crust. It serves various functions in nature: organisms ranging from lower eukaryotes to plants use it as one of the biominerals that provide support, protection, strength, or tools in various functions. Biogenic silica is the least studied of the biominerals. The best-studied ones (Hunter 1996) are composed of calcium (apatite, enamel, calcite); these are present in the bone tissue and teeth of mammals and in the shells of shellfish. In contrast to calcium minerals, relatively little is known about the formation of biogenic silica. The most likely function of biogenic silica in unicellular organisms seems to be giving strength to the living cell; good examples from the aquatic environment are siliceous (exo)skeletons or cell casings of the microalgal groups of diatoms (Fig. 1) and silicoflagellates, and of Radiolaria (Simpson and Volcani 1981; Pickett-Heaps et al. 1990; Round et al. 1990).

Silica and silica-based materials (e.g. zeolites and MCM-type materials) are also widely used in a variety of industrial and technological applications. A limited number of micro- and mesoporous silicas serve as fillers in rubber materials (e.g. in tyres), as catalyst supports, as drying agents, as abrasives (e.g. toothpastes), or as filters in separation technologies (Iler 1979; Brinker and Scherer 1990; Bergna 1994). A prerequisite for each application is that a silica with specific properties of mechanical strength, pore volume, pore-size distribution, or specific surface area is required. The current demand for "the best" silicas in these respects is increasing and a search for new, improved or alternative artificial silica materials has been going on for a few years. Biogenic silica sources also have received more attention lately (Morse 1999; Vrieling et al. 1999b). It is expected that if nature can be mimicked, a wide range of innovative and new tailor-made silicas can be produced approaching the variety in architecture seen in nature, e.g. in the algal class of diatoms (Vrieling et al.

[1] Department of Marine Biology, Centre for Ecological and Evolutionary Studies, University of Groningen, Biological Centre, P.O. Box 14, 9750 AA Haren, The Netherlands
[2] Laboratory of Inorganic Chemistry and Catalysis, Schuit Institute of Catalysis, Eindhoven University of Technology, P.O. Box 513, 5600 MB Eindhoven, The Netherlands

Progress in Molecular and Subcellular Biology, Vol. 33
W. E. G. Müller (Ed.)

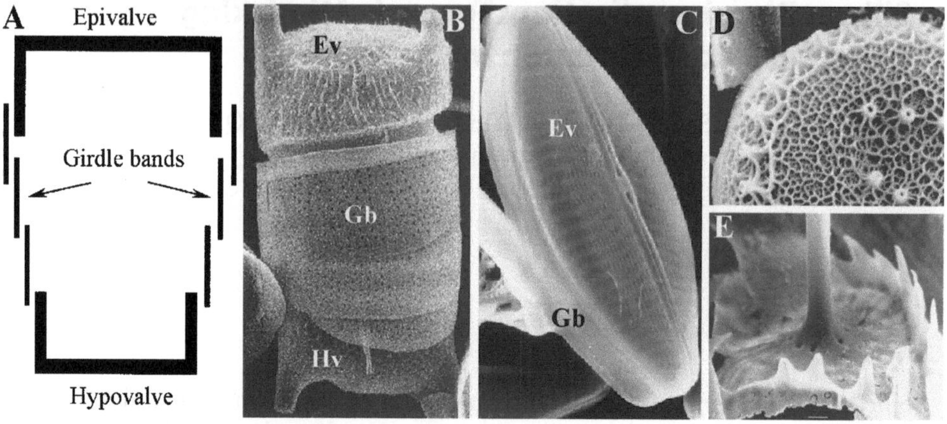


**Fig. 1A–E.** Overview of diatom exoskeletons. **A** A schematic drawing of the frustule components. **B** The centric species *Odontella* spp. **C** The pennate species *Navicula salinarum*. **D, E** Higher magnifications of the species *Thalassiosira weissflogii* and *Ditylum brightwellii*, revealing the variety of silica structures and pores. *Ev* Epivalve, *Hv* hypovalve, and *Gb* girdle band. **A–C** were reproduced from Vrieling et al. (1999b), with permission of Elsevier Science

1999b). In fact, biomimicking appears to be the best choice when tailor-made new materials are to be designed (Hunter 1996; Ozin et al. 1996; Yang et al. 1996; Zaremba and Stucky 1996; Almqvist et al. 1999).

The most familiar source of biogenic silica for industrial use is diatomaceous earth or "Kieselguhr", which in fact is a geological deposit of minute exoskeletons that are the remnants of diatoms. These unicellular photosynthesizing microalgae of the class Bacillariophyceae are the dominant component of the plant biomass in marine and freshwater environments. They are best known for their silica cell wall, with morphologies that are characteristic for each one of over 10,000 species (Round et al. 1990). Although diatomaceous earth has been studied quite well (Iler 1979), the application of this type of silica is limited nowadays, because of the development of superior man-made silicas. Indeed, the deposits of commercially exploited diatomaceous earth are geologically very old (a few hundred million years), so processes of aging, sintering, and dissolution have proceeded for a very long time. Consequently, during the detailed structures that are present in freshly prepared diatomaceous, silicas have been lost completely.

Moreover, this source of diatomaceous silica has become contaminated over such a long period with catalytic elements such as Al, Fe, Ti, or Ca, so it has become inappropriate as catalyst support. Especially aluminium is known to bind to silica remnants of diatoms (van Bennekom et al. 1989, 1991) and on basaltic ocean floors enhanced aluminium contamination inhibits the dissolution of exoskeletons in seawater (van Capellen and Qiu 1997a, b). Finally, diatomaceous earth contains a collection of exoskeletons from many different species that each has its own physico-chemical characteristics (Perry 1989;

Gordon and Drum 1994; Vrieling et al. 2000, 2003; Crawford et al. 2001), because of the distinct species-specific morphological features of the exoskeleton (Round et al. 1990).

This morphological feature in architecture of the diatoms is much greater than the variety of required features that are present in artificial silicas. Diatom silica formation is now recognized as an attractive model that may eventually result in successful biomimicking of the natural silicification processes (Mann and Ozin 1996; Ozin et al. 1996; Yang et al. 1996; Morse 1999; Parkinson and Gordon 1999; Vrieling et al. 1999b). Given the rapid pace of development in this field of science, it seems appropriate at this time to give an overview of research efforts in approaches and the analysis of the way the formation of silica proceeds in diatoms, a process that is expected to inspire new approaches in silica biomimicking.

## 2 Biochemical and Physico-Chemical Characteristics of Diatomaceous Silica

### 2.1 Organic Composition

Although the role of organic molecules in silica transformations has been demonstrated (Holt and Bowcott 1954; Kröger et al. 1999, 2000; Cha et al. 2000; Vrieling et al. 2002), we will shortly discuss the occurrence of organic matter in the frustules of diatoms; a more detailed overview on the molecular biology of diatom silicon biomineralization is presented by Ma (this Vol.).

Based on the organized distribution of pores and channels in MCM-type silicas (Biz and Ocelli 1998) and the fact that the nano-structural characteristics of diatoms are quite comparable (Vrieling et al. 2000), it is tempting to suggest that diatoms use structure-directing molecules (templates) to form the pores. As demonstrated by others, the cell wall of diatoms consists of different organic compounds such as carbohydrates, peptides and polyamines (Hecky et al. 1973; Swift and Wheeler 1992; Kröger et al. 1994, 1997, 1999, 2000). In other organisms proteinaceous matter is also associated with the silica; in plants (e.g., Harrison 1996) and inside the needles of marine sponges (Cha et al. 1999). It should be noted, however, that not all of these compounds play an essential role. Some peptides are only part of the casing that covers the silica (van de Poll et al. 1999; Kröger and Wetherbee 2000) and can be removed completely by oxidation. However, following the removal of all organic matter from the surface of the silica from diatoms, a substantial portion of proteinaceous matter remains encapsulated in the silica. This is demonstrated after dissolution of the silica and analysis of peptides that were recovered. Even with $^{13}C$-NMR the typical -O–C–O– bonds can be detected without dissolution (see also

Fig. 4C). From the recovered molecules the exact function has not been completely determined yet, but it is shown that silica does interact with some of these proteins by quickly inducing precipitation (Kröger et al. 1999, 2000). For diatoms computational studies provide additional data for the energetically favorable synthesis of silica (Lobel et al. 1996). Since proteins or polypeptides are able to interact with silica, it is obvious to assume that such compounds are of interest to be studied in biomimicking diatom silica formation.

## 2.2
## Chemical Aspects

### *2.2.1*
### *Principles of Silica Synthesis*

For most applications, silica is prepared from water glass, a concentrated solution of silica in water at high pH. Monomeric silica is present as a mixture of ions of silicic acid:

$$Si(OH)_4 \leftrightarrow Si(OH)_3O^- + H^+ \leftrightarrow SiO_4^{\;4-} + 4H^+ \tag{1}$$

Due to condensation/hydrolysis reactions, dimers and oligomers are also present:

$$\equiv Si-O^- + HO-Si\equiv \leftrightarrow \equiv Si-O-Si\equiv + OH^- \tag{2}$$

with three horizontal bars at Si representing bonds with -OH, $-O^-$ and/or -O–Si groups.

Because also -OH groups of dimers may be ionized, reactions with monomers or other dimers result in the formation of trimers, quadrimers, etc. These oligomers promote ring formation by the easy Si–O–Si bond bending, while the three- to six-membered rings are precursors for three-dimensional structures as the prismatic hexamer or the cubic octamers. As may be expected from the condensation/hydrolysis equilibrium (Eq. 2), decreasing the pH of the silica solution will favor condensation reactions, resulting in bigger oligomers or polymers; interestingly, in diatoms it appears that the silica deposition vesicle becomes acidic in the course of valve formation (Vrieling et al. 1999a).

In combination with the flexible Si–O–Si bonds and the resulting tendency for ring formation and cross-linking, three-dimensional polymers and ultimately, spherical particles are expected, the surface covered with ≡Si–OH and $\equiv Si-O^-$ groups. Diatomaceous silica happens to be assembled from silica spheres, which seem to coagulate to form the silica cell wall (Schmid and Schulze 1979; Crawford et al. 2001). The condensation reaction requires the combination of ≡Si–OH with $\equiv Si-O^-$, whereas the continuation of growth of these silica particles depends on many factors influencing the condensation/solvolysis equilibrium, especially the pH and the presence of ions (Fig. 2). This results in a negligible growth of particles larger than 2–4 nm at low pH values and in the presence of salts, but if the ions have been removed and at higher pH values, stable sols of 10–100-nm sized particles may be obtained (Iler 1979). However, already before the particles have obtained their optimal size, another process has started. The condensation reaction not only combines silica monomers, oligomers or polymers, but also can link colliding particles together, with the ≡Si–O$^-$ group at the surface of one particle and the ≡Si–OH group at the other. This weak connection between the nanometer-sized particles is reinforced by the so-called aging reactions. Because of the difference in surface energy at the convex surface of the particle and the concave surface in the "necks" between particles, silica will dissolve relatively easily from rather flat surfaces of the particles and recondensate preferentially in the necks, strengthening the interparticle bond. This effect (know as Ostwald ripening) is also responsible for the dissolution of small particles in favor of larger particles. Since the distribution of reactive groups on the surface is stochastic, directions of particle–particle bonds are rather arbitrary and

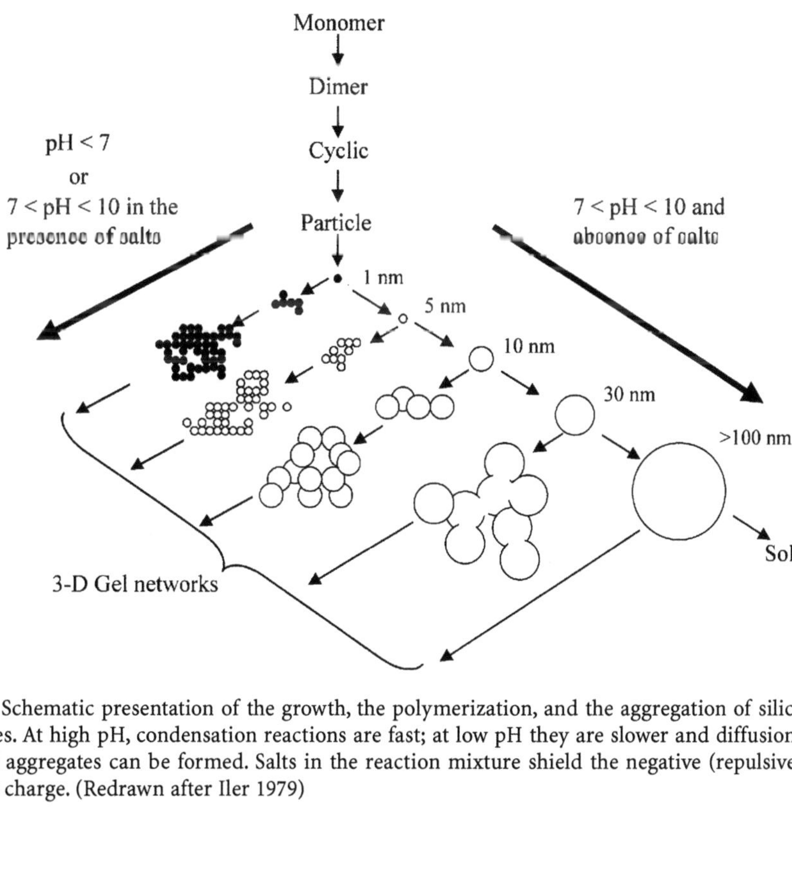


**Fig. 2.** Schematic presentation of the growth, the polymerization, and the aggregation of silica particles. At high pH, condensation reactions are fast; at low pH they are slower and diffusion-limited aggregates can be formed. Salts in the reaction mixture shield the negative (repulsive) surface charge. (Redrawn after Iler 1979)

therefore particle–particle interactions result in tenuous aggregates with an amorphous structure. Interactions are not restricted to particle–particle or particle–aggregate bonds, but also interactions between aggregates or clusters of particles are possible, leading to highly ramified and extended aggregates (Fig. 2).

The structure of the aggregates depends on the relative rates of the formation of interparticle bonds or chemical reactivity and the diffusion of particles and aggregates. If the chemical reactivity is relatively high ("every" collision results in the formation of a bond), the formation of aggregates is controlled by the diffusion of the clusters and we are dealing with a diffusion-limited cluster–cluster mechanism (DLCC). For diatoms the same principle has been proposed by Gordon and Drum (1994). In contrast, if the chemical reactivity is rather low (only after many collisions a bond is formed) the formation is controlled by the chemical reactivity and in that case we are dealing with a reaction-limited cluster–cluster mechanism (RLCC). If the tips and branches of the growing aggregates touch each other and link the aggregates, an interconnected network of aggregates will result and a gel is formed. This gel point will be recognized by a sudden increase in the viscosity. Beyond the gel point the gel will stiffen gradually by the expansion of both the contacts between the aggregates and the continued growth of the aggregates in the empty spaces in the gel. Because water occupies (even in concentrated silicas) most of the space of the gel; drying of the gel will reduce the volume considerably. Moreover, due to capillary effects the weak aggregate structures will collapse, leaving no more than micropores at best. Only if reinforcement of the gel structure by aging has been applied before drying, the collapse is only partial, and mesoporous silica will be the result.

The most important parameter during silica preparation is the pH, not only because changes in $[OH^-]$ will shift the equilibria in Eqs. (1) and (2), but especially because $OH^-$ is a very efficient catalyst for both the condensation and the solvolysis reaction. Therefore, kinetics is much faster at high pH. Since silica preparation is a multi-step process, silicas produced at different pH are often totally different and in combination with appropriate temperatures and silica concentrations, many different porous silicas can be prepared (Iler 1979; Bergna 1994). A second important parameter is the presence of electrolytes, organic or solvent molecules. Now the stochastic character of the condensation reaction may be restricted, and with big organic molecules or (bio)polymers even the formation of aggregates or the gelation process itself can vary considerably. Because of the ability to control the morphology of silica, these organic molecules behave as "structure-directing devices" and are called templates. This important item is further discussed in Section 4.1.

### 2.2.2
### *Silica Synthesis in Diatoms*

Although the growing diatom is a very complicated "reactor" to synthesize porous silica, the aforementioned principles are still valid. The reaction rate and equilibria are driven by the availability of reactive silica monomers (and oligomers) and possibly affectors (ions, template molecules) to allow the formation of the larger silica particles, gels and finally the "aged" frustule. Diffusion limitation in diatoms only occurs when the concentration of silica precursors in the silica deposition vesicle (SDV) is the limiting factor (Gordon and Drum 1994). In diatom cells the internal concentration of reactive silica can be very high. As recalculated from a large series of data presented by Conley et al. (1989), different species contain less than 1 to well over 20 mM cellular silica concentrations (Vrieling et al. 1999b). Occasionally, cytoplasmic concentrations of 400–700 nM per cell were found (Robinson and Sullivan 1987). The amount of 20 mM cellular silica is equivalent to 1 wt% of dissolved $SiO_2$, which is too low to allow the fast silica aggregation that is observed in diatom valve formation – therefore, there is a problem. It is likely that in the diatom cell local concentration gradients are present. This could mean that in the proximity of the SDV the concentration of silica precursors is built up prior to being transported into the expanding SDV during the process of valve formation. It should be noted, however, that the presence of such local concentrated sources of silica is under debate. Although it has been proposed that so-called silica transport vesicles (STVs) are involved (Pickett-Heaps et al. 1990) – STVs are supposed to fuse with the edges of the expanding SDV – their existence in the diatom cells has not been demonstrated beyond doubt. On the other hand, when local concentrations are higher than micromolar levels, silica should polymerize (Iler 1979), unless the silica precursors are somehow stabilized by binding to or association with organic constituents before or during translocation over the SDV membrane. In that case, polymerization reactions inside the cytoplasm will not occur. Unfortunately, the exact location and the composition of the silica precursors in diatom cells are not really known: knowledge is only derived from indirect evidence (Pickett-Heaps et al. 1990). Consequently, only approximations of the real silica concentrations can be used for *in situ* silica synthesis experiments.

The second affector of silica polymerization is temperature. In the industrial synthesis of silicas and silica-based materials, elevated temperatures are often required to produce the desired structure. Especially in the formation of crystalline products (the zeolites), temperature is important to start the nucleation reaction (van Bekkum et al. 1991; de Moor et al. 1997, 1998; Biz and Ocelli 1998). In many other processes, higher temperatures ($\Delta T$ >80 °C) are only applied to accelerate processing times, without changing equilibria or product properties considerably. In diatoms the role of temperature is negligible, since diatoms are restricted to moderate temperature boundaries of the aquatic environment (about −2 to 30 °C). Of course, temperature can be used very well to manipulate the reaction rate in in situ silica synthesis, e.g., to design conditions to study silica transformations over time (de Moor et al. 1997, 1998; Sun et al. 2002; Vrieling et al. 2002). For time-resolved small angle X-ray (SAXS) analysis of (bio)polymer-silica particle formations, the concentration of silica can be as low as 1% (w/v) $SiO_2$, resulting in polymerization at room temperature followed by a very slow aging of the silica. In order to speed-up the process of this aging, the temperature can be elevated to follow silica transformations over short periods of a few hours.

Furthermore, the pH of the reaction mixture is important, because it influences the equilibrium described in Eq. (2) (Iler 1979). The SDV membranes in diatoms have a membrane potential, as has been demonstrated with molecular probes. In view of the ability to translocate the fluorescent dye Rhodamine 123, this membrane potential must be quite strong (Li et al. 1989; Brzezinski and Conley 1994; Vrieling et al. 1999a). Using the pH sensitive probe 3-(2,4-dinitroanilino)-3′-amino-*N*-methylpropylamine (DAMP) in combination with immunocytochemical localization, it has been shown that the SDV acidifies in the course of valve formation (Vrieling et al. 1999a), in line with the role of pH in silica aggregation (Iler 1979). In experiments to synthesize silica in the presence of small proteins recovered from diatom exoskeletons, an acidic environment (pH 4.0) was required for optimal silica precipitation (Kröger et al. 1999). Acidification should force the equilibrium of the polymerization reaction (Eq. 2) to the right, resulting in a thermodynamically driven mass aggregation of the silica (see also Fig. 2), although at much lower reaction rates, because $OH^-$ is an effective catalyst for the condensation/solvolysis reactions. When silica precursors are constantly imported, however, this local elevated silica concentration allows a very fast polymerization. In the production of silica from water glass as the silica resource, often the final step to precipitate remaining nonpolymerized silica is acidification. If we wish to mimic diatom silica polymerization, the pH of the reaction mixture must be considered to be important and based on the data now available, we assume that it should be acidic with values chosen in the narrow range between pH 4 and 6.

The final factor that affects silica polymerization is the role of organic and inorganic additives or templates. The importance of inorganic compounds in diatom silicon biomineralization is less well understood than that of organic molecules. For salts, however, it appeared that for each species the salinity of the culture medium or natural environment affects the silicification grade in terms of the total amount of biogenic silica per cell (Tuchman et al. 1984; Conley et al. 1989; Romero 1994; Vrieling et al. 1999c). It is proposed that at decreased salt concentrations, aggregation of silica particles is much faster (Iler 1979), and when these particles are smaller, this may lead to a more compact (dense) silica network (Fig. 2). When salts do have a direct role in diatom silicon biomineralization (Vrieling et al. 1999c) and if the volume of the siliceous parts of diatoms remains approximately equal, the increase in silicification grade per cell at lower salinities can be explained by densification. In fact, small angle X-ray scattering (SAXS) studies of diatomaceous silica revealed shifting slope transitions in the SAXS spectra towards smaller pores/particles at lower salinities, confirming this densification (Fig. 3). With densified and more heavily silicified cell walls, the diatom cell may compensate for osmotic pressure (Olsen and Paasche 1986; Romero 1994). Recently, it has been demonstrated that in diatom species the valves are generated by a network of silica spheres, the size of which significantly differs per species (Crawford et al. 2001). It would be worthwhile, using the method of atomic force microscopy (AFM) in combination with scanning electron microscopy (SEM), to study whether within species the sphere size depends on the salinity of the culture medium and, thus, on certain cations that may play a role. Cations (and especially Al) determine the dissolution rate of biogenic silica (van Bennekom et al. 1989, 1991; van Capellen and Qiu 1997a,b), but a precise role of cations in diatom silica biomineralization has not been established yet. In some experiments, it has been shown that introduction of cations gives various responses of diatoms with respect to silicification. Germanium oxide is chemically the best competitor for silicon, but it can only be taken up in very low quantities because of its toxicity (Lewin 1966). In studies with aluminium the results obtained contradict each other: from no effects to an increase in the biogenic silica content (Gensemer 1990; Vrieling et al. 1999c).

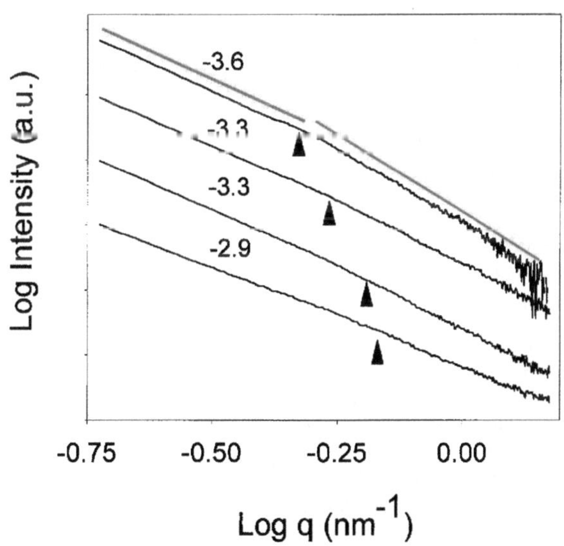


**Fig. 3.** By lowering the salinity the primary pores become smaller as indicated by the shift of the transition point (*arrowheads* indicate the upper limit of the Porod region) to the *right*, while at the same time the surface becomes rougher indicated by the decreasing slope above the Porod region. (Reproduced from Vrieling et al. 2000, with permission of the Phycological Society of America)

## 2.3 Nanoscale Structure

### 2.3.1 *Specific Surface Area*

Many studies have been performed to estimate the specific surface area (calculated as $m^2 \cdot g^{-1}$) of diatomaceous silica (for an overview see Vrieling et al. 1999b). Even within one species (Table 1), a wide range has been determined. Although we may expect that each species has its own specific surface area, the data presented so far are not always reliable. This is partly due to the way samples have been pre-treated to remove organic matter and partly because different instruments have been used (Vrieling et al. 1999b). The specific area of diatomaceous silica tends to be lower when very strong acids (concentrated $H_2SO_4$, nitric acid) and oxidants (saturated permanganate or oxalic acid) are

**Table 1.** Overview of typical specific surface areas and pore diameters determined for different biogenic silica and (bio)polymer-silica particles produced with biomimicking approaches. Note that for diatomaceous silica quite a few estimates have been made (reviewed in Vrieling et al. 1999b), of which a few representative genera are presented here

| | Surface area ($m^2 g^{-1}$) | Pore diameter (Å) |
|---|---|---|
| Diatoms | | |
| Benthic natural populations | 17–165[a] | No data |
| *Cyclotella cryptica* | 26–28[b] | No data |
| *Navicula pelliculosa* | 100[c] | 100 |
| *Nitzschia sigma* | 9–25[a] | 10–100 |
| *Rhizoselenia* species | 50–60 | No data |
| *Thalassiosira* species | 15–258[a] | <100 |
| Sponge | | |
| *Euplectella*[c] | 1 | No data |
| Plant hairs | | |
| *Phalaris canariensis*[c] | 241 | 35 |
| (Bio)polymer-silicas | | |
| Synthetic silica gels | 100–1200 | 10–1000 |
| PEG-600-silica (low PEG/$SiO_2$) | 258 | 200 |
| PEG-600-silica (high PEG/$SiO_2$) | 636 | 40 |
| PEG-2000-silica (low PEG/$SiO_2$) | 687 | <20 |
| PEG-2000-silica (high PEG/$SiO_2$) | 1030 | <20 |
| PEG-20,000-silica (low PEG/$SiO_2$) | 720 | <20 and 35 |
| PEG-20,000-silica (high PEG/$SiO_2$) | 994 | <20 |
| Block-copolypeptide-silica[d] | 436 | No data |

[a] From Vrieling et al. (1999b).
[b] From Csögör et al. (1999).
[c] From Perry (1989).
[d] From Cha et al. (2000).

used (Vrieling et al. 1999b). In contrast, with less harsh chemicals ($H_2O_2$, ozon) or procedures (low temperature ashing) higher values could be obtained (Vrieling et al. 1999b). Typically, the specific surface area of diatomaceous silica is low, corresponding to the fact that micropores (<1 nm) are not present in comparison to many other microporous synthetically derived silicas.

### 2.3.2
### *Coordination of Molecules of Biogenic Silica*

A powerful method to follow silica transformations in the course of silica or zeolite synthesis and to study the coordinations of molecule bonds is NMR (Wijnen et al. 1993; Burkett and Davis 1994). For dry silica powders, for example diatomaceous silica, the coordination of the silicon and carbon atoms can be revealed readily (Perry 1989; Fig. 4). Using $^{29}$Si-NMR on dried cultured biomass of the diatom *Ditylum brightwellii* revealed two dominant peaks, corresponding to the silicon coordination of Si(OSi =)OH and Si(OSi =) that are denoted as $Q_3$ and $Q_4$, respectively (Fig. 4A). Using $^{29}$Si-CP-NMR, however, the coordination of silicon at the surface of the silica particles can be studied in more detail, indicating that a relatively large portion of silicon is present as $Q_3$, while also Si(OSi =)$(OH)_2$ (denoted as $Q_2$) is observed (Fig. 4B). The presence of relatively large amounts of $Q_3$ and $Q_2$ is of interest for industry, because not only are $Q_3$ and $Q_2$ indications for the size of the surface area, they are also potential reactive sites to bind other materials; for instance diatomaceous silica can be used as a carrier to build zeolites on the surface (Anderson et al. 2000). In comparison, in MCM-type silicas catalytic molecules can be covalently bound using silsequiloxane compounds (Krijnen et al. 1999). With respect to the coordination of the silicon atoms in diatoms, we noticed that between untreated and acid-cleaned (with all cellular organic matter removed) diatom frustules no differences appear (Fig. 4A, B). These $^{29}$Si-(CP)-NMR observations on *D. brightwellii* are in line with those made for the species *Navicula pelliculosa* (Perry 1989), showing that diatomaceous silica is mainly $Q_4$-oriented with the potential reactive sites ($Q_3$ and $Q_2$) at the boundaries of the particles (Table 2).

When carbon is studied using $^{13}$C-NMR, there is a clear difference between the raw and acid-cleaned biomass (Fig. 4C). Overall, the most obvious carbon bonds with –C, –O, and –H are properly detected in the raw diatom sample. When all cellular matter has been removed, the remaining carbon must originate from encapsulated organic matter in the biogenic silica; the identified carbon coordinations most probably coincide with the polyamines and peptides (silaffins) that have been recovered from diatomaceous silica (Kröger et al. 1999, 2000; see also Ma, this Vol.).

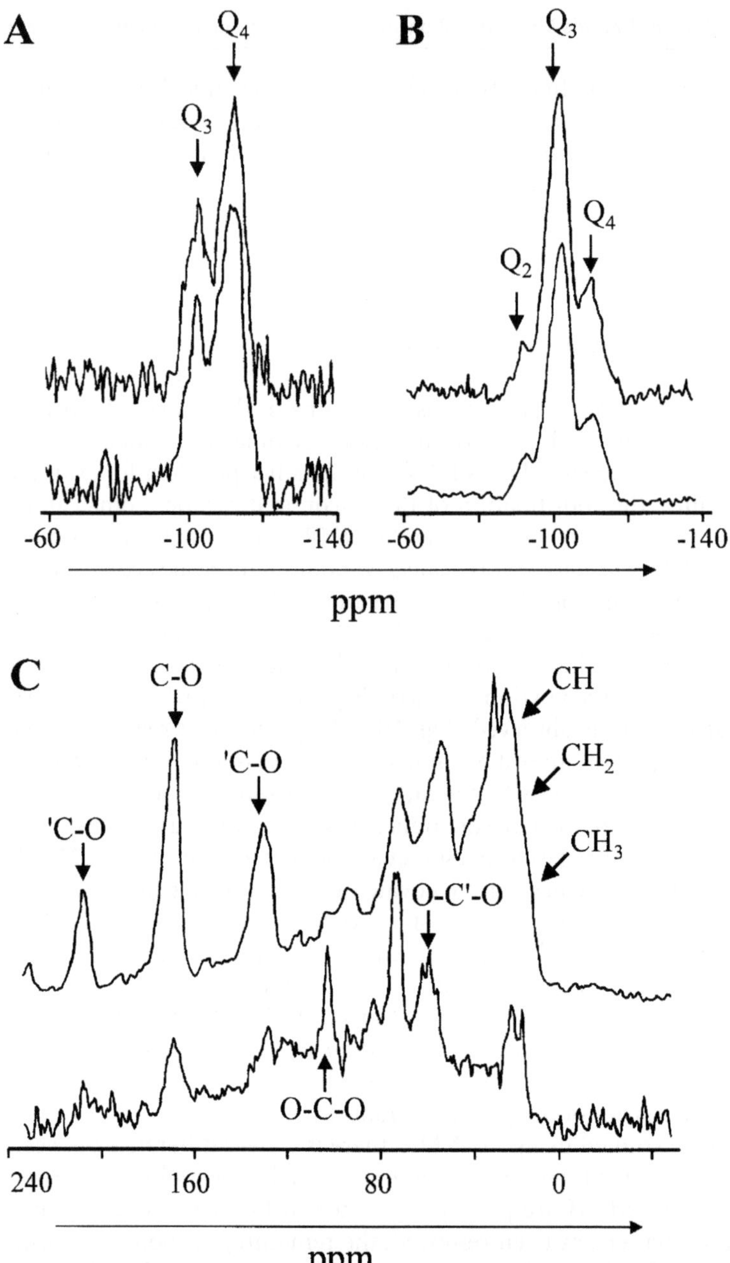


**Fig. 4A–C.** NMR spectra of dried, nontreated upper spectra and acid-cleaned lower spectra biomass of the centric diatom *Ditylum brightwellii*; following acid treatment, only the siliceous exoskeleton parts remain. **A** $^{29}$Si-NMR showing the typical coordinations $Q_3$ and $Q_4$. **B** $^{29}$Si CP-NMR show that a relative large portion of the surface has coordinations of $Q_2$ and $Q_3$ which are potential reactive sites for binding catalysts (Krijnen et al. 1999) or zeolitization (Anderson et al. 2000). The difference between nontreated and acid-treated biomass was marginal for both $^{29}$Si-NMR and $^{29}$Si CP-NMR. **C** Using $^{13}$C-NMR, however, the difference between the treated and non-treated biomass is obvious. In nontreated material, peaks for numerous C-bonds are obtained, whereas in the acid-treated material only a few are observed. Typical are the O–C–O and O–C′–O bonds, representing organic matter that is encapsulated inside the silica, most probably from peptides or polyamines recovered after dissolution. (Kröger et al. 1999, 2000)

**Table 2.** Overview of chemical shifts of $^{29}Si$ bonds (in ppm) observed by NMR for different biogenic silica sources and synthetically produced (bio)polymer-silica particles

| | Si(OSi=)(OH)$_2$ ($Q_2$) | Si(OSi=)OH ($Q_3$) | (Si(OSi=) ($Q_4$) |
|---|---|---|---|
| Diatoms | | | |
| *Ditylum brightwellii* (untreated) | −91.8 | −101.6 | −111.4 |
| *D. brightwellii* (acid cleaned) | −92.4 | −101.6 | −110.8 |
| *Navicula pelliculosa** | −92.7 | −102.3 | −113.5 |
| Limpet teeth | | | |
| Patella vulgata[a] | n.d. | −102.4 | −113.7 |
| Sponge | | | |
| *Halichondria panicca*[a] | −92.5 | −101.8 | −112.1 |
| Plants | | | |
| *Phalaris canariensis* (untreated)[a] | −92.6 | −101.2 | −109.8 |
| *P. canariensis* (acid cleaned)[a] | −92.7 | −101.1 | −109.8 |
| (Bio)polymer-silicas | | | |
| Synthetic silica gel[a] | −90.6 | −99.8 | −109.3 |
| Silicatein-silica[b] | −94.0 | −102.0 | −110.0 |

[a] From Perry (1989).
[b] From Cha et al. (1999).

### 2.3.3
### *X-Ray Diffraction and Wide-Angle X-Ray Scattering*

Crystallinity of solid silica can be determined by diffraction methods and reveal important information of the Bragg reflections of the crystal lattices in zeolites and crystalline silicas (van Bekkum et al. 1991). X-ray diffraction (XRD) was interpreted in such a way that diatomaceous silica is amorphous (Kamatani 1971; Perry 1989), but it was suggested, or at least expected, that below the resolution of traditional XRD ($2\theta < 10°$, corresponding to >10 nm in resolution) crystalline phases still would be present (Pickett-Heaps et al. 1990). Recently, we have shown with the powerful approach of wide angle X-ray scattering (WAXS; Fig. 5A) that also at the atomic level no Bragg reflections are present in diatomaceous silica (Vrieling et al. 2000, 2003). Therefore, we can state that on the atomic and colloidal scale diatomaceous silica is amorphous. Consequently, in mimicking diatomaceous silica, we do not have to face difficulties of the formation of crystals, which is essential in zeolite synthesis (Jansen et al. 1991; van Bekkum et al. 1991; Zones et al. 1998).

### 2.3.4
### *Small-Angle X-Ray Scattering*

Diffraction methods can also be applied in noncrystalline systems if the atom or electron density is not homogeneous. However, now the intensity of the scat-

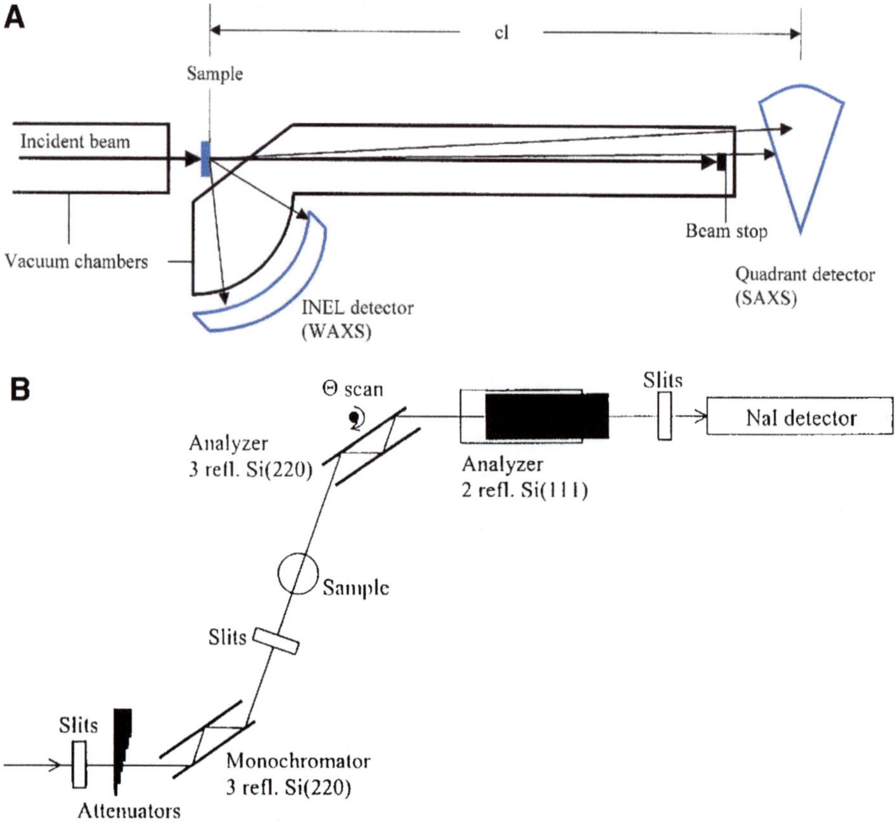


**Fig. 5.** **A** Schematic presentation of the setup for combined small- and wide-angle X-ray scattering analysis. The resolution of SAXS depends on the length between the sample and the detector (*cl*), affecting the scattering angle 2Θ. With cl = 3.3 m the resolution of SAXS is $0.07 < q < 2.5\,nm^{-1}$ (corresponding to $2.5 < d < 82$ nm). **B** Bonse-Hart setup (for USAXS) at the high-brilliance beam line ID2/BL4 at the European Synchrotron Radiation Facility in Grenoble (France). By rotation of the first analyzer crystal, the USAXS spectrum is scanned between $0.001 < q < 0.15\,nm^{-1}$ (corresponding to $45 < d < 6200$ nm). Due to a substantial overlap between the upper limit of SAXS and the lower limit of USAXS, spectra can be merged (see also Fig. 11)

tered radiation is not only exclusively very strong or zero, resulting in sharp peaks and zero intensity between the peaks respectively as in crystals, but the intensity versus scattering angle (θ) plots often show shallow "hills and valleys". The scattering depends on the distribution of mass (X-rays are scattered by electrons, by neutrons, and by atomic nuclei) and the corresponding contributions of the positive and negative interference of scattered radiation. Thus, scattered intensity is only observed if positive interference is not cancelled by negative interference. This only happens when density gradients are present and a homogeneous liquid or solid will give no small-angle scattering. In contrast, surfaces of particles will always be important in the contribution of scattered intensity, because here density gradients are optimal. Therefore, the size and morphology of particles or pores will be reflected in the intensity pattern of scattered radiation. However, if we apply Bragg's law (first-order reflections): $\lambda = 2d.\sin\theta$ and use X-rays with $\lambda = \pm 0.1$ nm, then for colloidal particles with size $d > 5$ nm we have to expect scattering angles less than 0.01 rad or 0.60°. To measure these small angles conventional XRD cannot be applied, and special cameras (for example the Kratky camera; Glatter and Kratky 1982) have been developed for laboratory use. A second difficulty is the relatively weak intensity of the scattered radiation. If sharp peaks are present as in XRD, the conventional X-ray sources are appropriate, but their intensity is much too low for the "hill-like" spectra observed in SAXS. For this purpose, dedicated SAXS setups (Fig. 5A) at large synchrotrons have been built, with the high-brilliance beam of the synchrotron as the X-ray source. The sample detector distance or camera length (denoted cl in Fig. 5A) is important, because with appropriate camera lengths optimal d-ranges can be covered. For example, with cl = 1 or 10 m, d-ranges of 1–20 or 10–200 nm might be covered, respectively. If, in combination with a one- or two-dimensional position-sensitive detector, use can be made of the high-brilliance source of a synchrotron, acquisition times of less than seconds are possible, creating possibilities for high-quality dynamic experiments.

Deducing sizes and morphologies of colloidal systems from SAXS is quite difficult. The general expressions for the intensity are too complicated to be solved, but for many systems one may write in a simplified way (Teixera 1986):

$$I(q) = n \cdot P(q) \cdot S(q) \tag{3}$$

where q is the wavelength-corrected scattering angle [$q = (4\pi/\lambda).\sin\theta$, in which $\lambda$ = the wavelength and $2\theta$ = the scattering angle], n is dependent on the number of scatterers, P is the form factor, describing the relation between the geometry of the particle and the scattering and S is the structure factor, containing information on the correlation of the particles. For example, for a sphere (radius R), we can write:

$$P(q) = V^2(\rho - \rho_o)^2 \cdot \left[3(\sin(qR) - qR.\cos(qR))/(qR)^3\right]^2 \tag{4}$$

where $\rho$ and $\rho_o$ are the electron densities inside and outside the particle, respectively, and V is the volume of the scattering sphere. Because $P(q) \propto (\rho - \rho_o)^2$, a silica sphere (with radius R) in air will show exactly the same SAXS pattern as a pore (with radius R) within a solid silica. This Babinette principle – where pores and particles with the same morphology show the same SAXS (Glatter and Kratky 1982) – makes SAXS a very suitable technique to investigate porous materials. However, because of this Babinette principle, sometimes it is very difficult to decide if the observed scattering has to be attributed to scattering by pores or by particles. In this case, contrast variation might be helpful: if pores are filled with a liquid of equal density, the scattering will be decreased. For example, some halogenated hydrocarbons (e.g. dibromomethane) mimic the scattering of X-rays by $SiO_2$ (Hua et al. 1994; Vrieling et al. 2000).

In real experiments with silica, we are never dealing with one isolated sphere, but with a colloidal dispersion of a collection of silica spheres. If the polydispersity is not too large (for example, with a standard deviation in the order of 0.1), the peaks in Fig. 6 are replaced by a periodic pattern of small bumps, indicative for the characteristic peaks of the spheres. However, two other important properties can also be recognized. The slope of the straight line over the tops is –4 (black line in Fig. 6), a hallmark for the smooth surface size of a sphere. Also perfect cubes, ellipsoids and other three-dimensional objects with a smooth surface show a slope of –4 in the high q-region of the scattering pattern: the so-called Porod-region (Glatter and Kratky 1982). If the surface is rough (fractal), the slope is between –3 and –4 and $D_s = 6 - (-\text{slope})$, the fractal surface dimension $2 < D_s < 3$ (Martin and Hurd 1987). If $D_s = 2$ (corresponding to slope = –4), we are dealing with a nonfractal smooth surface with 2 as the appropriate Euclidian dimension of a surface. If the dimension of the particles is 1 (needles) or 2 (discs), the slope of the Porod-plot will be –1 and –2, respectively. These slopes can be interpreted as the signature of the primary particles in a colloidal system (Glatter and Kratky 1982; Pedersen 1997).

The second important feature is the transition from the Porod-region to the low-q part of the scattering pattern (arrow heads in Fig. 8A). The corresponding length ($d = 2\pi/q$) at this transition point is a good approximation of the size of the sphere. If the particles form a homogeneous distribution without preferred distances between the particles, the low-q SAXS region shows a nearly horizontal and featureless curve. However, during the preparation of silica, first the primary particles are formed, followed by a combination of these particles into aggregates. The formation of diffusion- and reaction-limited aggregates (respectively, DLCC and RLCC) were discussed earlier (Sect. 2.3). These aggregates do have different gradients in mass and are fractal with appropriate fractal dimensions. Because scattering density depends on mass gradients (i.e. the surface of a sphere!) a straight curve in the log(I)–log(q) plot turns up, instead of the featureless pattern of the homogeneous distribution. The (negative) slope $= -D_m$, with $D_m$ = fractal dimension of the aggregate and $1 < D_m < 3$. As confirmed by computer simulations, fractal dimensions below 1.9 have to be attributed to DLCC, with $D_m$ from 2.0–2.2 to RLCC (Teixera 1986; Martin and Hurd 1987). If aging (redistribution of silica from the tiny tips or small particles to the necks and crevices in the aggregates) is important, the fractal dimensions are slightly higher. Again, the transition at low q-values of the straight line correspond to the size $\xi$ of the aggregate ($\xi = 2\pi/q$). If the growing aggregates touch each other forming a gel, on the length scale $L_g$ ($L_g > \xi$), this gel will be rather homogeneous without mass gradients, and therefore I(q) with $q < \xi$ will show a horizontal and featureless curve. For the same reason, the sol-gel transition will not be visible in a dynamic SAXS investigation (Fig. 7). Because silica aggregates are often much bigger than 100 nm, only

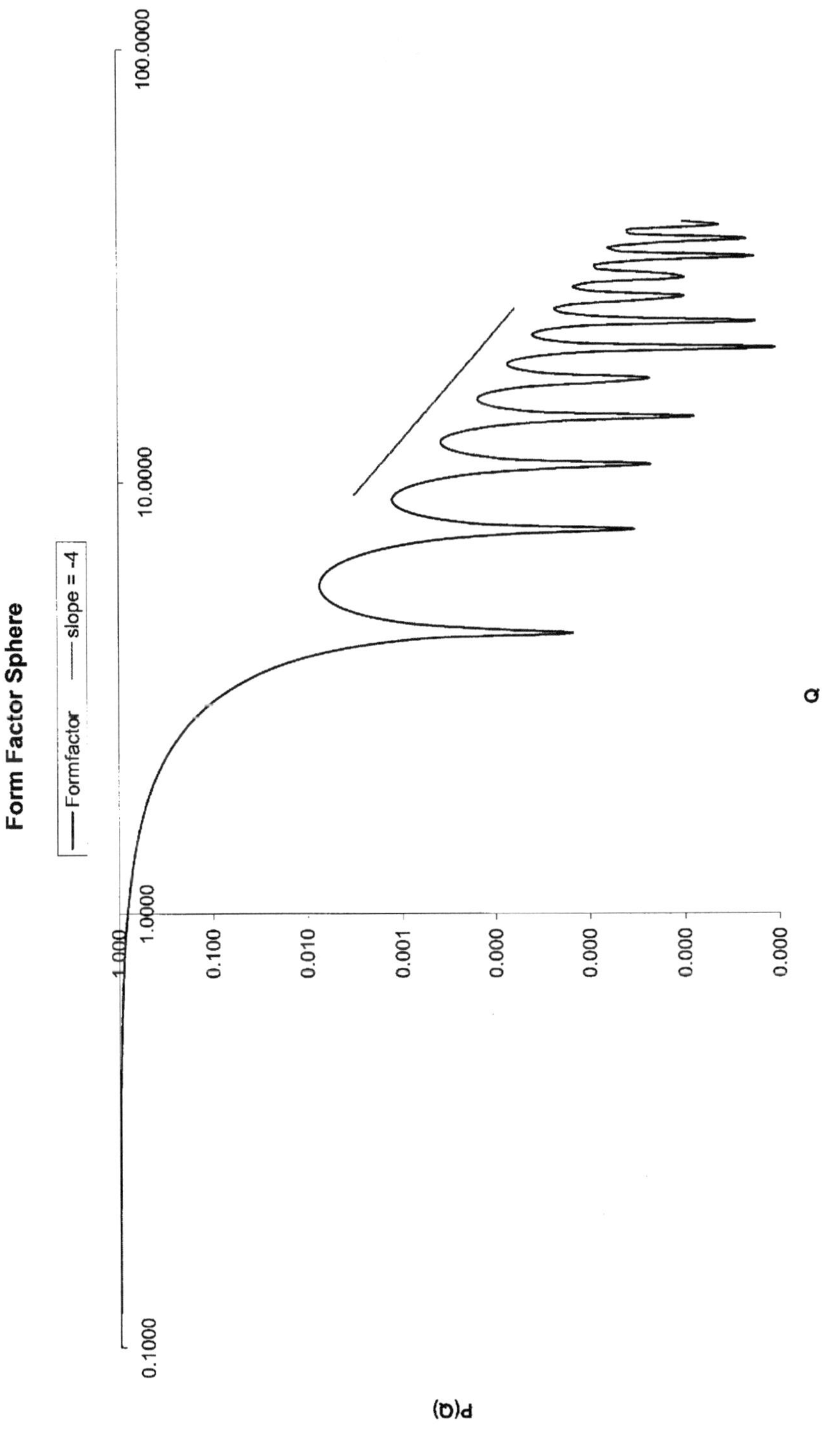


**Fig. 6.** Calculated SAXS spectrum (form factor) of a perfect sphere, with q = (4π/λ) sinθ, and the intensity I in arbitrary units

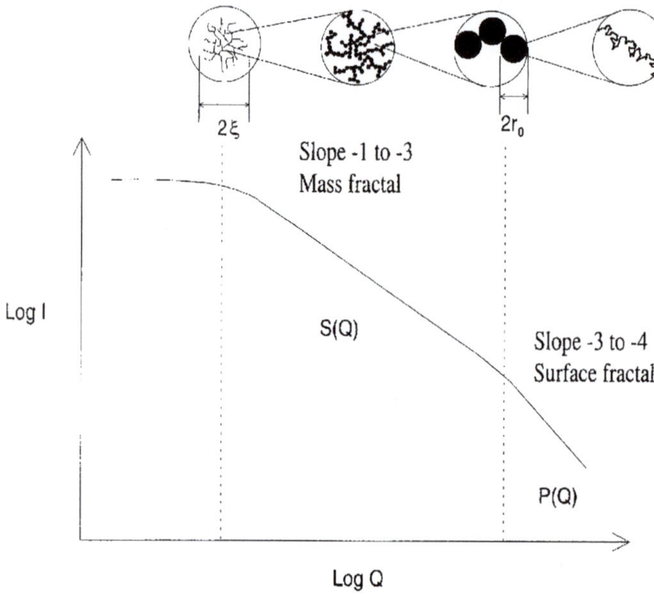


**Fig. 7.** *Above* Schematic representation of an aggregate of size ξ (with mass fractal properties), which is built of primary particles of size $r_o$ having surface fractal properties. *Below* The matching small-angle X-ray scattering pattern. For nonfractal systems the slopes will be –4 for both *P*(*Q*) and *S*(*Q*)

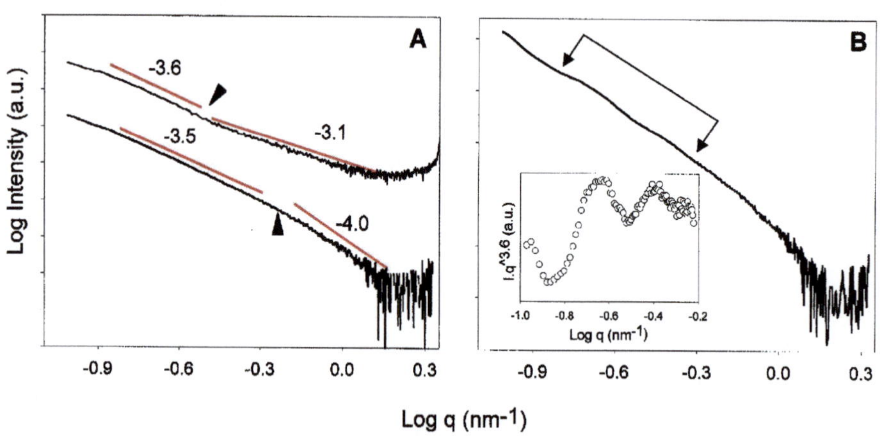


**Fig. 8A,B.** Representative small-angle X-ray scattering spectra for silica of the diatoms *Asterionalla japonica*, *Coscinodiscus granii*, and *Thalassiosira* spp. **A** Clearly identified are two regions with different slopes as indicated for *A. japonica* (*lower spectrum*) and *C. granii* (*upper spectrum*), although the transition between the regions (*arrowheads*) are at different q-values for each species. **B** In some silica, oscillation patterns appear as shown for *Thalassiosira* spp. (indicated by the *connected arrows*); these oscillations represent regions with preferential pore distributions and can be examined more carefully when the descending nature of the scattering is cancelled by plotting q vs. $I \cdot q^{-\text{slope}}$ (*inset* in **B**). (Reproduced from Vrieling et al. 2000, with permission of the Phycological Society of America)

the onset of the growth of the aggregates may be studied by small-angle scattering methods, because in general the upper limit for SAXS is less then 100 nm. To study features of length scales up to several microns, ultra small SAXS (USAXS) may be applied, e.g. using the Bonse-Hart setup (Fig. 5B) at the European Synchrotron Radiation Facility, in Grenoble, France (Diat et al. 1997). In contrast to the position-sensitive detector in SAXS, the scattering pattern has to be scanned in USAXS and therewith only every 10–15 min a complete USAXS spectrum can be obtained. Most important is that a total length scale of 1–6000 nm can be obtained if SAXS with both short and long camera length is combined with USAXS, then merging the scattering spectra (see further in Sect. 3.2, Fig. 11).

Using SAXS for the analysis of diatomaceous silica, it appeared that the silica specimen of every diatom species has its own scattering spectrum. The observed sample-specific characteristics also apply for silica samples of mixed populations. Typical examples of scattering patterns are shown for the species *Asterionella japonica*, *Coscinodiscuss granii*, and a *Thalassiosira* species (Fig. 8A); the independent straight regions (red lines) and typical transition points (arrows) are indicated. In the spectrum of the *Thalassiosira* species a region ($-1.0 < \log q < -0.2\,\mathrm{nm}^{-1}$) with weak oscillations is present (indicated by the arrows). When the effect of the descending scattering spectrum is cancelled out (by plotting log q against $I \cdot q^{-\mathrm{slope}}$), these oscillations can be examined more precisely (Fig. 8B). Since scattering from diatomaceous silica is caused by the pores (Vrieling et al. 2000) and with the help of the Babinette principle, from each transition point the radius of gyration ($R_g$) of the pores, analyzed in the size domain of SAXS, could be determined with $d = 2\pi \cdot q^{-1}$. Pores sizes that have been determined by SAXS have also been found in the exoskeletons using electron microscopy (Vrieling et al. 2000; Fig. 8C, D). Despite the distinct scattering patterns of each species, of which the underlying and prominent straight slope of the form factor [P(q)] is very strong, there seems to be a general pattern in all silica specimens that were studied (Fig. 9A). In the SAXS domain, two q-regions are present that each represent a quite uniform fractral dimension ($D_m$). It was, therefore, concluded that nano-porosity in diatomaceous silica is basically uniform (Vrieling et al. 2000). In agreement with SAXS is the observation that the silica of each diatom species also has it own scattering pattern in the USAXS domain (Fig. 9B). Here, the spectra also, revealed one or two straight regions (red lines), but in some cases dominant and distinct scattering peaks are clearly superimposed on these straight lines (arrows in Fig. 9B). For *Thalassiosira punctigera* (upper spectrum in Fig. 9B), the most dominant peak is $q = 0.016\,\mathrm{nm}^{-1}$ ($R_g = 396\,\mathrm{nm}$), whereas in *Nitzschia sigma* (lower spectrum in Fig. 9B) two dominant peaks are at $q = 0.021$ and $0.041\,\mathrm{nm}^{-1}$ (corresponding to $R_g = 300$ and $150\,\mathrm{nm}$, respectively). The determined $R_g$ for the pores of the species examined coincided with uniform pore distributions at a preferential distances as determined by EM (Vrieling et al. 2003).

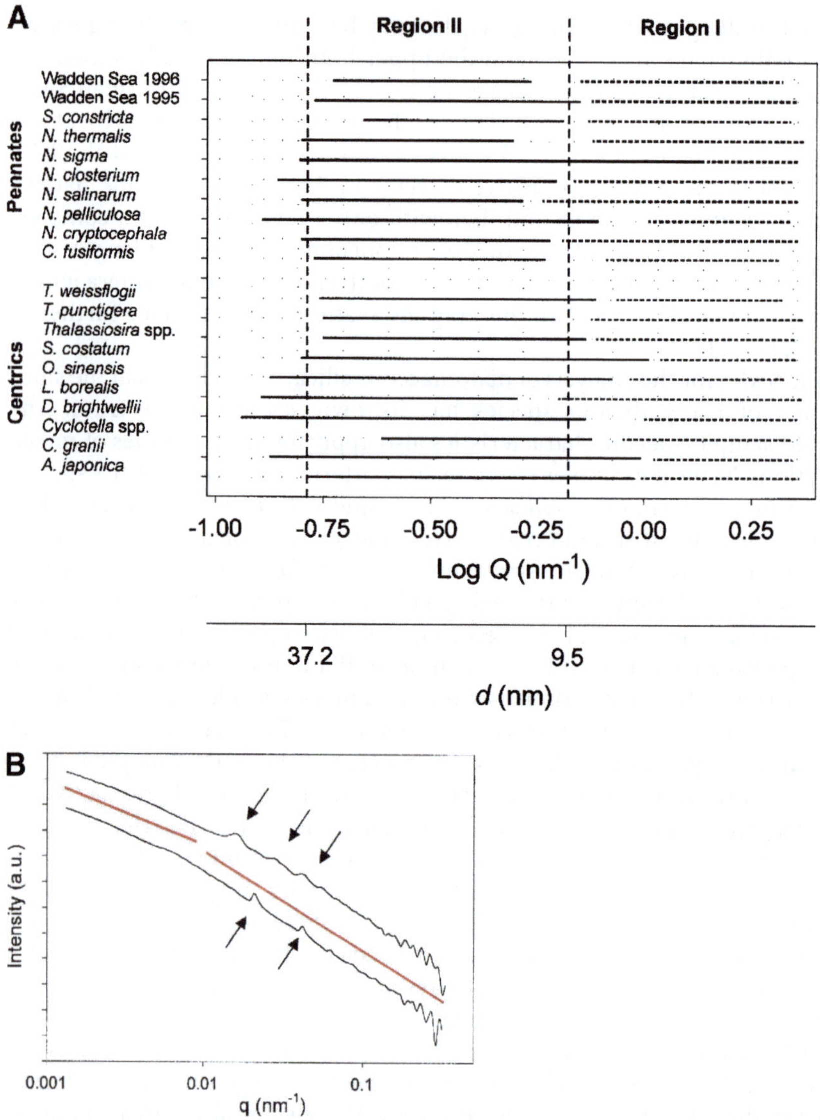


**Fig. 9.** **A** Common nanoscale features determined for a series of silicas retrieved from different diatom species and natural populations. Two distinct regions can be observed; the Porod region (slope about –4.0) indicating spherical pores (*Region I*: d < 9.5 nm) and *Region II* (9.5 < d < 37.2 nm) with a fractal dimension between 3 and 3.8 (reproduced from Vrieling et al. 2000, with permission of the Phycological Society of America). **B** USAXS spectra of the centric species *Thalassiosira punctigera* (*upper spectrum*) and the pennate species *Nitzschia sigma* (*lower spectrum*). Clearly visible are peaks (*arrows*) representing the scattering of dominant pores in the silica

# 3 In Situ Silica Synthesis

## 3.1 The Application of Templates

Silica preparations as described in Section 3.1 have two serious shortcomings if we want to make use of products for industrial purposes: the pores are too small for many applications and, moreover, the pores are not monodispersed. The second difficulty is solved in zeolites and molecular sieves: using a choice of organic molecules as templates, many crystalline structures can be obtained. Although there is often a close relation between size and morphology of the templates and the pores, the templates are more than just space-filling molecules. There are many indications that there are specific interactions between templates and silica that result in primary silica-template particles, which are the building blocks for the zeolites. For example, during the preparation of the zeolite silicalite, the size of the primary particles corresponds to the presence of four template molecules (de Moor et al. 1999), strongly suggesting that four template molecules together act as a "super template" to be surrounded by silica for the formation of the primary particles. Also, the fact that the same molecule under different conditions may be a template for different zeolites suggests that the template-silica interaction can be highly specific. Probably, the stochastic reactions during the preparation of silica are directed by the template, providing very specific primary particles. Templates can thus safely be considered as structure-directing devices. The experimental conditions during the preparation of zeolites are often quite different from silicas: high temperatures and high pH are almost always necessary, probably for the conversion of primary particles into crystals.

Zeolites or molecular sieves are silicate or alumino-silicate crystals with an excellent pore monodispersity, but their pore size is too small for many applications. For example, to prepare fine chemicals on an industrial scale large-pore catalysts are required to accommodate the organic molecules without diffusion limitation. Therefore, the preparation of MCM-type materials was considered a huge step forwards by the industry (Biz and Ocelli 1998). In these systems, detergent-like molecules such as quaternary ammonium compounds with $C_{16}$ tails form micelles, and silica is condensed around them. Using tubular micelles, MCM-41 is prepared, a silica with very long parallel tubes with nanometer-sized diameter. Although the silica between the tubes is amorphous, the tubes themselves form a hexagonal crystalline system (Biz and Ocelli 1998; Sayari and Hamoudi 2001). Because the pores of MCM-41 are still small compared with pores in diatomaceous silica, the presence of templates with structures similar to micelles during the preparation gives a strong indication about the nature of templates necessary for the formation of diatomaceous silica. Just as in MCM-41, in diatoms there are also no proteins or protein-carbohydrate molecules big enough to be a "single-molecule template" for the 100–1000-nm-scaled pores in diatoms. In contrast, when small single molecules are involved in the synthesis of amorphous silicas, they become totally encapsulated (Gill and Ballesteros 2000). Consequently, we assume again that only micelles or vesicles built from specific proteins can provide the required structure-directing systems (Lasic et al. 2001; Vasilescu et al. 2001).

This hypothesis is confirmed by experiments with both polymers and proteins. Already, many papers have been published concerning the "phase separation" effects of polyethyleneglycol (PEG) and similar compounds (Ågren et al. 2000; Wang et al. 2000; Cho et al. 2001; Kipkemboi et al. 2001; Lee et al. 2001; Richer and Mercier 2001; Tolbert et al. 2001; Sun et al. 2002) or collected in a special issue on this topic (Eckert and Ward 2001). Takahashi et al. (2000) even distinguished four different gels by increasing the PEG concentration in a low-pH TEOS/water solution. Without silica relatively compact units of PEG in water are formed, but with silica not only silica--PEG interactions are shown, but at high PEG concentrations also phase separation was observed: one phase rich in solvent and the other phase rich in PEG and PEG–silica. The driving force in this process is the interaction between hydrophobic polymer and the hydrophobic part of the silica condensation products: $(-O-Si-)_n$. These observations were confirmed by us using USAXS (Sun et al. 2002; Vrieling et al. 2002b). Both with PEG and PEI (polyethylene-imine) during the silica preparation, the "in situ" results corresponded with the formation of very big "particles" and in some cases occurrence of separation of mesophases, with a clear distinction between the PEG- and the PEI-system. With two model proteins myoglobin (Mb: 17 kDa) and horseradish peroxidase (HRP: 40 kDa), USAXS analysis also showed that particles of a size >100 nm were formed (Vrieling et al. 2002), with a protein-rich mesophase that is many times the size of one Mb- or HRP- protein molecule (the calculated dimensions of both Mb and HRP are $<100\,nm^3$).

## 3.2 Synthesis of PEG-Templated Silicas

As a typical example of the application of polymers as templates using phase separation, a few preparations of porous silica using PEG are discussed here. To study the influence of the PEG/silica ratio, experiments were performed both at $C–C–O/SiO_2 = 0.5$ mol/mol and $C–C–O/SiO_2 = 2.0$ mol/mol, with C–C–O as the monomeric unit of PEG. Because hydrophobic interactions are also crucial in the formation of PEG–silica intermediates, the hydrophobicity of "template-vesicles" was varied by applying different polymer weights, notably: PEG-600, PEG-2000 and PEG-20,000 provided increasing hydrophobicity.

The hydrophilic PEG-600 showed at low $PEG/SiO_2$ ratios a fractal growth and rough surfaces after calcination (Fig. 10A) and a specific surface area of $258\,m^2\cdot g^{-1}$ (Table 1). The aggregates were only slightly different from pure silica,

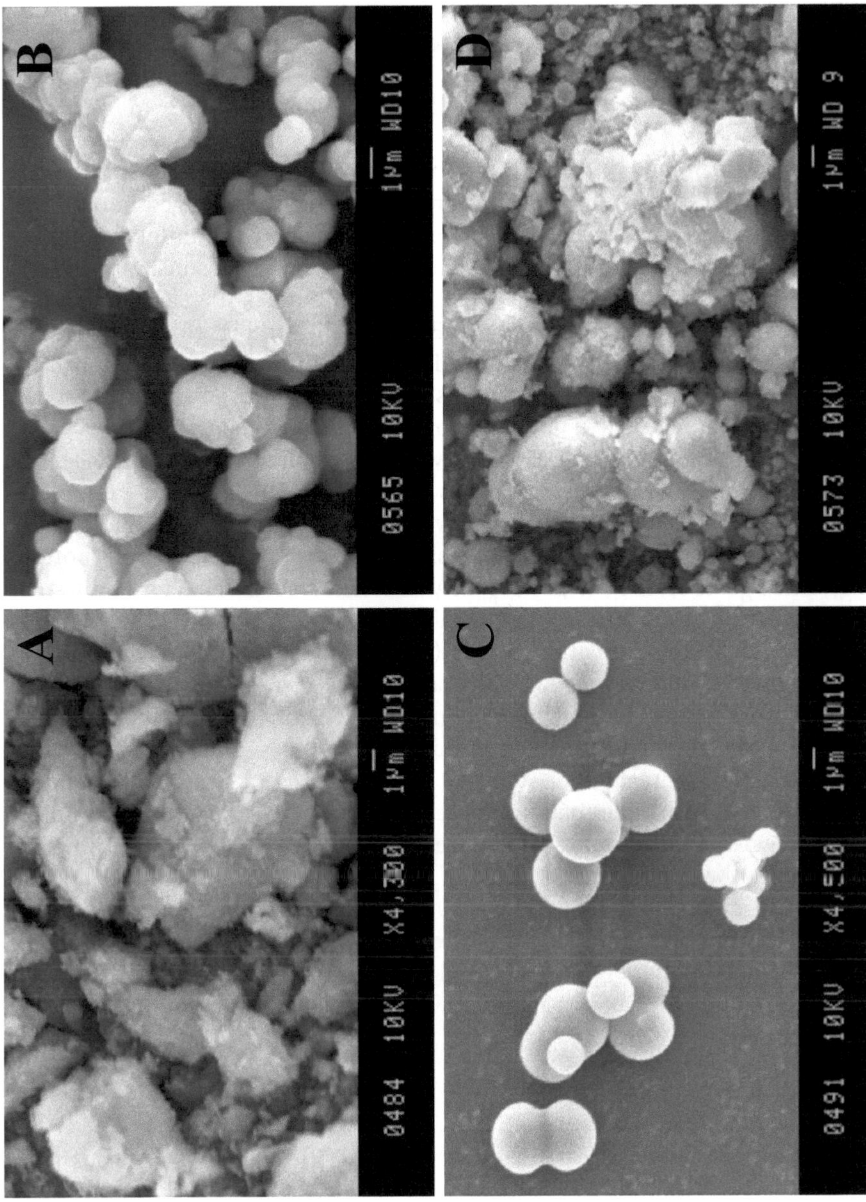


**Fig. 10A–D.** Morphologies of silica synthesized from water glass in the presence of different molecule sizes of polyethleneglycol (PEG). PEG-600-silica at a ratio of 0.5 (**A**) revealed a very fractal silica, which becomes smoother for PEG-600-silica at a ratio of 2.0 (**B**). Much larger and smoother particles are formed with PEG-2000-silica (**C**), while in PEG-20,000-silica both large and small particles appear with fractal properties (**D**)

but the rate of growth was increased considerably. If a surplus of PEG-600 was applied (thus high PEG/$SiO_2$-ratios), only the "elementary particles" were fractal, but the surface of the resulting spheres was changed and was almost smooth (Fig. 10B) and the specific surface area increased to $636\,m^2 \cdot g^{-1}$ (Table 1). With the more hydrophobic PEG-2000; only very smooth silica spheres were obtained, both at high and low PEG/silica ratios (Fig. 10C). In addition, the elementary particles were nonfractal, so fractal growth has changed into a PEG-directed growth. However, PEG-2000 seems to be near an optimum concerning the interaction with silica. With PEG-20,000 and a low PEG/silica ratio, fractal elementary particles were again observed (although both size and fractal dimension did increase compared to PEG-600/silica in Fig. 10A) and also the spheres were roughened again (Fig. 10D). As shown in Table 1, BET surfaces and pore size distributions (micropores versus mesopores) also corresponded with these different growth and pore formation mechanisms.

Time-resolved SAXS and USAXS was also applied on these PEG-silica systems to follow the transformations and growth of silica; an example is shown for PEG-600-silica (ratio 2.0) with merged data of SAXS and USAXS (Fig. 11). From the onset of the synthesis to 1 h at room temperature (bottom line), silica gel is formed with obviously different fractal behavior at different length scales over the whole SAXS/USAXS region (3–6200 nm). When aging is started at 90 °C, the transition points (arrows) shift towards lower q-values (thus higher d), indicating that silica aggregates grow. Over time scattering appears more intense in the USAXS region ($q < 0.2\,nm^{-1}$), whereas the smaller particles have been consumed to build the growing aggregates. At the final stage, particles/aggregates of different size are present: the small building blocks ($R_g < 6.5\,nm$), the medium large aggregates ($R_g = 400\,nm$), and finally the aggregates and networks which are much bigger with $R_g > 6200\,nm$. The latter cannot be determined because of their size if beyond the upper limit of the USAXS setup. Because a PEG-600 polymer has approximately only 15 -C-C-O- monomer units, a sphere with a diameter of 5 nm may contain 60–80 PEG-600 polymers. As discussed in Section 4.1, the growing silica with its hydrophobic -Si-O-Si- backbone will attract PEG, and these silica-PEG

**Fig. 11.** Merged small- and ultrasmall X-ray scattering (SAXS/USAXS) snapshots during in situ synthesis of PEG-600-templated silica ([C-C-O]/[$SiO_2$] = 2). To avoid overlap of the patterns, the energy scales were shifted vertically, while in the *inset* (only the SAXS spectra) the energy scales are not shifted. In the *lower spectrum* (1 h at RT), primary particles (radius <5 nm) have formed the 115-nm-sized fractal aggregates (D = 2.2); these aggregates form homogeneous gel particles (R > 6000 nm) with a rather rough (fractal) surface ($D_{surface} = 6 - 3.6 = 2.4$). In the *middle spectrum* (1 h at RT followed by aging for 0.5 h at 90 °C), both the primary particles and the fractal aggregates have grown to 6 and 260 nm, respectively. At the kernel of the aggregates, aging effects can be observed resulting in a local densification and an increase in the fractal dimension to 2.7. In the *upper spectrum* (1 h at RT followed by 1 h aging at 90 °C), the aging of the fractal aggregates is completed (D = 2.6) and the aggregates have grown to 400 nm

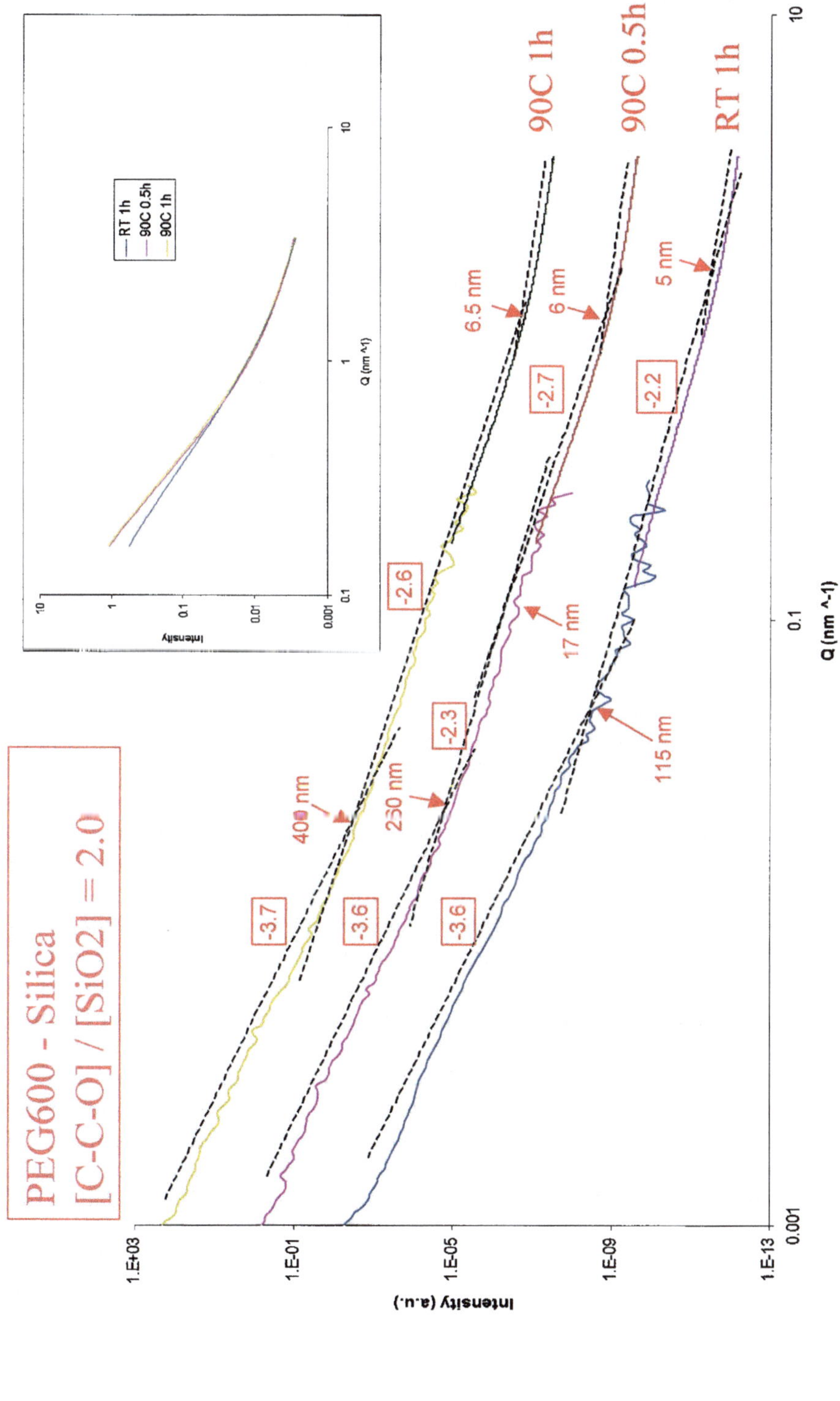
PEG600 - Silica
[C-C-O] / [SiO2] = 2.0
90C 1h
90C 0.5h
RT 1h
-3.7
-3.6
-3.6
-2.6
-2.3
-2.7
-2.2
400 nm
230 nm
115 nm
17 nm
6.5 nm
6 nm
5 nm
Intensity (a.u.)
1.E+03
1.E-01
1.E-05
1.E-09
1.E-13
0.001
0.1
10
Q (nm ^-1)
RT 1h
90C 0.5h
90C 1h
Intensity
Q (nm ^-1)

intermediates will transfer more PEG polymers from the water phase to the growing silica–PEG particle (Takahashi et al. 2000; Sun et al. 2002). At the same time, the added PEG polymers will change the stoichastic silica growth into a PEG-directed growth. The diffusion and interactions of these silica–PEG particles will result in fractal aggregates (Martin and Hurd 1987; Brinker and Scherer 1990) and the growing aggregates touch each other, forming a continuous network, the gel. However, in contrast to gels formed by nontemplated silica (Brinker and Scherer 1990; Wijnen et al. 1991) – because of phase separation – big gel particles precipitate (so-called flocculation; Iler 1979) with irregular growth at the surface, resulting in the observed fractal surface dimension of 2.4.

## 4
## A New Concept: Mesophases in Structure-Directing Processes in Diatom Silica Biomineralization

As discussed earlier, the formation of silica is directed by stochastic processes, the properties of a porous product determined by reaction conditions (such as pH, temperature, concentration), and the presence of (small) electrolytes and ions. Although many different micro- and mesoporous silicas can be prepared by varying the experimental conditions, pore sizes are limited to tens of nanometers, too small to mimic diatomaceous silica. As shown by molecular sieves and zeolites, templates may diminish the influence of the stochastic processes during the preparation; they can be considered to be structure-directing devices which create strong relations between the template morphology and the porous structure. However, recent experiments with silica and templates have shown that in order to form really big pores at the (sub)micrometer scale, even large (bio)polymers are much too small if only a single template molecule has to form this pore. This difficulty can be overcome if "superstructures" of organized templates are applied, forming micelles, vesicles or even micro-droplets. In the aqueous silica systems, the hydrophobic interactions between the templates define the thermodynamic conditions to create the superstructures. Because also silica oligomers and small polymers are very hydrophobic, template-silica can promote the formation of vesicles and micro-droplets, resulting in phase separation. Now the silica formation is directed by the growing superstructure of associated template molecules, resulting in (sub)micron-sized pores. Since the hydrophobic interactions are the driving force for this phase separation, both polymers and proteins with comparable hydrophobic properties can be applied. This knowledge is the starting point of a model for silica biomineralization (Fig. 12).

When small proteinaceous molecules, i.e. silaffins and/or polyamines (red curved structures), are present in the immature silica deposition vesicle together with silica precursors, fast silica polymerization is induced (Kröger

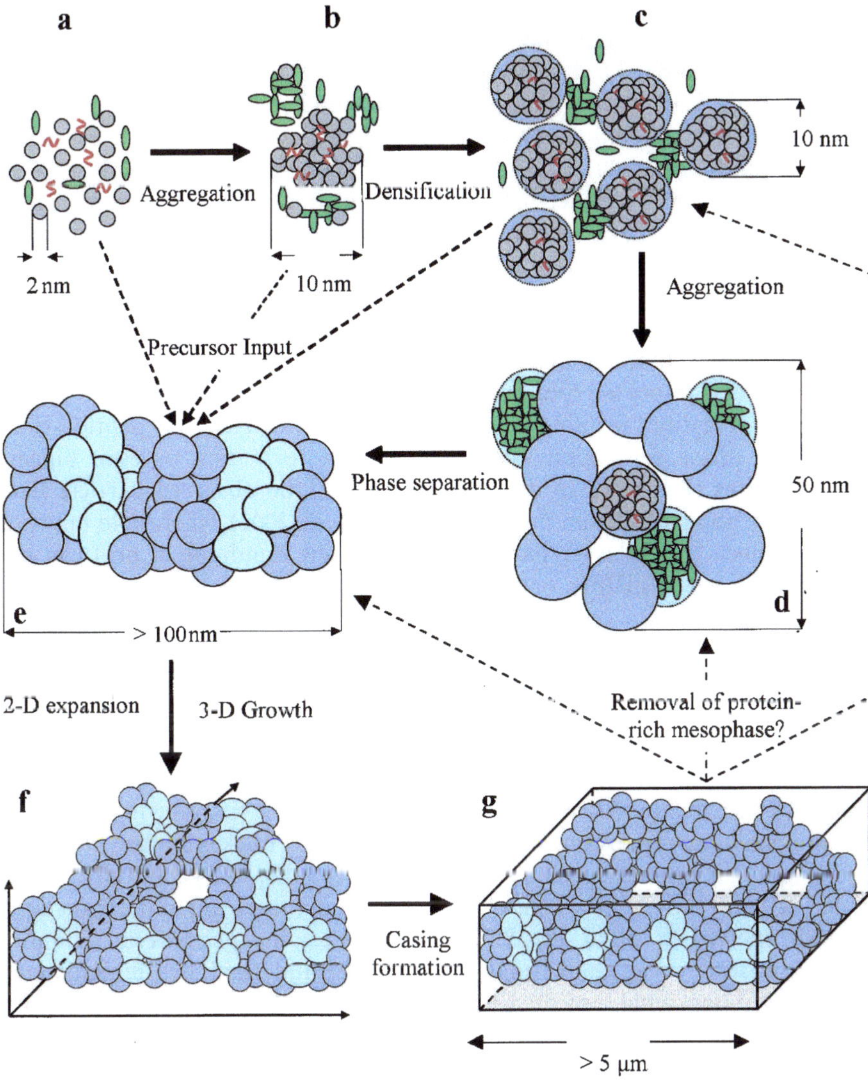


**Fig. 12a–g.** Schematic presentation of the described model for diatom silicon biomineralization in which mesophase separation is included. At the onset of valve formation (**a**) silica precursors and peptides are imported into the SDV, where precipitation of silica is induced by small organic molecules (e.g., silaffins and/or polyamines indicated by the *red curved structures*) to form silica sols shown as *grey spheres* (**b**). These sols further densify and grow to larger silica particles (*bluish spheres*), while larger peptides (*green ellipsoids*) start to interact with the silica (**c**). When aggregation continues (**d**), silica form larger particles (up to 50 nm). At this stage silica and protein aggregates are transferred to silica- and protein-rich mesophases (*greenish ovals*) by phase separation (**e**). This process proceeds until the SDV has reached it final two-dimensional and three-dimensional size (**f**) and the protective casing has been formed (**g**). Somehow the protein-rich mesophase (the "super templates") is removed prior to assembly of the casing. (Reproduced from Vrieling et al. 2002, with permission of Wiley-VCH)

et al. 1999, 2000). The intracompartmental pH of the SDV decreases (Vrieling et al. 1999a), resulting in the thermodynamically accelerated mass aggregation of the small silica particles assisted by a high input rate of reactive silica into the expanding SDV. During this process, the small polycationic silaffins and polyamines become completely encapsulated in the primary silica particles (Gill and Ballesteros 2000), while these particles coagulate (see Fig. 2; Iler 1979; Crawford et al. 2001) to silica sols (grey spheres in Fig. 12a, b) of several nm in size (i.e. <5 nm). These sols interact and densify/grow to larger structure of about 10 nm (blue circles in Fig. 12c). In order to allow formation of mesophases, as observed in the template-directed synthesis of amorphous silica (Biz and Ocelli 1998; Takahashi et al. 2000; Sun et al. 2002; Vrieling et al. 2002), we assume that other larger organic molecules (possibly peptides) are co-imported continuously (green ellipsoids in Fig. 12a–c). During ongoing aggregation, forming larger interacting silica-rich and peptide-rich aggregates, phase separation occurs during the formation of even bigger silica particles (Fig. 12d). In diatomaceous silica, the observed particle size differs from about 30–45 nm in size (Crawford et al. 2001), whereas in in vitro silica synthesis much larger particles (up to a few hundreds of nm) can be formed, even when silaffins or polyamines are used (Kröger et al. 1999; Cha et al. 2000).

In the process of phase separation, silica-rich (blueish spheres) and protein-rich (greenish ovals) mesophases are formed (Fig. 12e). Since this protein-rich mesophase is much larger than a single protein molecule, it allows structure-direction of the silica during further growth and densification of the silica mesophase to particles with sizes larger than 100 nm. Following continuous input of peptides and silica precursors to the edges of the growing SDV, the two-dimensional expansion and three-dimensional thickening are realized, resulting in the species-specific size of the silica walls. The genetically determined cellular activity (the role of microtubuli, cytoskeleton and organelles) of each species to mold the SDV during macromorphogenesis (Schmid 1986; Crawford et al. 2001) may well control the final two-dimensional and three-dimensional size of the SDV (Fig. 12f). When the silica parts are completed, but before a protective organic layer (the casing) covers them, the protein-rich mesophases may be removed (for re-use?) or may just degenerate (Fig. 12g). With respect to the latter, one of the recently obtained proteins with a high affinity to solid silica is a typical homologue of ubiquitin (Hazelaar et al., unpubl. data). This highly conserved peptide appears to be abundant in all eukaryotic cells, where it is involved in DNA repair, in regulation of differential gene expression, or to act as chaperone. The most important role of ubiquitine, however, is that in selective protein degradation (Pickart 2000). Our ubiquitine homologue has a high affinity for silica and preliminary immunocytochemical studies suggest a random scattered localization over the siliceous parts of the cell wall (Hazelaar et al., unpubl. data). Still its specific role in the silica biomineralization pathway remains to be determined.

The model generally describes diatom silica biomineralization. Nevertheless, AFM, NMR, Infrared spectroscopy and small-angle X-ray scattering methods have discerned, as we have seen, different silicas from each individual diatom species (Kamatani 1971; Perry 1989; Vrieling et al. 2000, 2003; Crawford et al. 2001). Therefore, we expect that each species also controls the micro-morphogenesis genetically, by releasing different types or species-specific sets of proteins and related molecules in the SDV. Kröger et al. (1999) show that by using mixtures of silaffins the silica particle size can be changed, which agrees with the finding that amongst species significantly different particle sizes can be found (Crawford et al. 2001).

# 5 Conclusions and Future Perspectives

Comparing nontemplated silica synthesis in water and the formation of the nano/micro-porous siliceous frustules of diatoms, it is clear that aggregation and gelation processes have to be essentially different. To create the large-porous diatomaceous silica, the stochastic growth of silica has to be changed into a template-directed growth. We argue that primarily the template molecules originate from proteins, which form larger clusters (the "super-templates"). Due to the hydrophobic interactions of both the proteins and the growing silica, separation of mesophases is the result. As shown by similar reactions with polymers, these phase separations are very sensitive with respect to small changes in hydrophobic properties of the templates. Although at the nanoscale the taxonomically different diatom species show several similar general features, at the (sub)micron scale species-specific proteins (or a set of different proteins) can facilitate the formation of the abundant and beautiful silica architectures.

To observe the silica growth in diatoms in vivo will be extremely difficult, but with well-chosen model templates the formation of porous silicas, which are very similar to the diatom frustules, may become possible. An important challenge here is to isolate and purify the silica deposition vesicles for a complete analysis of all organic and inorganic parameters that are involved in valve formation and silica deposition. Especially newly identified biomolecules (we assume proteins) that are larger than the silaffins and polyamines, which have been identified so far, require attention for their role in structure-direction. Preliminary studies on silica-binding capacities not only identified the ubiquitine homolog, but also several other peptides (Hazelaar et al. and van der Strate et al., unpubl. data). These peptides are under investigation, but reveal very tight binding to solid silica and are in the size range (25–80 kDa) that may enable phase-separation similar to the previously discussed PEG/silica systems and protein/silica systems (Sun et al. 2002; Vrieling et al. 2002).

Only when sufficient quantities of diatom proteins can be isolated or produced – depending on essential modifications, peptides can be produced synthetically or via gene expression methods – they can be applied as templates to test our hypothesis in in vitro silica syntheses. However, probably more useful information may be obtained with preparations with tailor-made proteins (or analogues) or polymers. In such experiments many features of the templates may be tested (hydrophobicity, ionic interactions, influence of chain length, etc.). Because of the extreme vulnerability of the reacting system, it is only possible to follow the silica transformations in in situ experiments. Due to the size range of the particles and pores that have to be formed and opacity of the samples, it only seems useful to apply USAXS or USANS. However, an alternative for high-resolution structural analysis of silica intermediates might be cryo-TEM (Jung et al. 1999; Lasic et al. 2001). In this method, the application of sophisticated cooling methods makes the freezing of the reaction mixture extremely fast, preventing any change, and most importantly, any damage of vulnerable silica intermediates as seen in conventional freezing and other treatments prior to physico-chemical characterization.

Using proteins or polymers as templates in synthesizing silicas can also have industrial applications. However, to isolate and purify commercial quantities of diatom proteins will most probably be too expensive and the application of a relatively cheap analogous material (e.g. polymers such as PEG) is a more promising alternative. Therefore, we suggest that the study of the natural processes of silicon biomineralization will inspire the design of appropriate chemical compounds for the production of tailor-made silicas, potentially mimicking those seen in nature.

*Acknowledgements.* We thank Dr. B.U. Komanschek (CCLRC, Daresbury, UK), Drs. W. Bras and T. Narayanan (ESRF, Grenoble, France), and Dr. P.-P.A.M. de Moor, C. Houssin, F. Gümüsburun (all at EUT, Eindhoven, NL) for assistance during various synchrotron experiments. EGV, SH, and QS were financially supported by the Netherlands Technology Foundation (STW; grants Gbi55.3883 and GFc4983) that is subsidized by the Netherlands Organization for the Advancement of Pure Research (NWO).

## References

Ågren P, Counter J, Laggner P (2000) A light and X-ray scattering study of the acid catalyzed silica synthesis in presence of polyethylene glycol. J Non-Cryst Solut 261:195–203

Almqvist A, Thomson NH, Smith BL, Stucky GD, Morse DE, Hansma PK (1999) Methods for fabricating and characterizing a new generation of biomimetic materials. Mater Sci Eng C 7:37–43

Anderson MW, Holmes SM, Hanif N, Cundy CS (2000) Hierarchial pore structures through diatom zeolitisation. Angew Chem Int Ed 39:2707–2710

Beelen TPM, Wijnen PWJG, Vonk CG, van Santen RA (1989) SAXS study of silica gel formation from aqueous silicate solutions. Catal Lett 3:209–216

Bergna H (1994) The colloid chemistry of silica. Adv Chem Ser, vol 234. American Chemical Society, Washington, DC

Biz S, Ocelli M (1998) Synthesis and characterization of meso-structured materials. Catal Rev Sci Eng 40:329–407

Brinker CJ, Scherer GW (1990) Sol-gel science. Academic Press, New York

Brzezinski MA, Conley DJ (1994) Silicon deposition during the cell cycle of *Thalassiosira weissflogii* (Bacillariophyceae) determined using dual rhodamine 123 and propidium iodide staining. J Phycol 30:45–55
Burkett SL, Davis ME (1994) Mechanism of structure direction in the synthesis of Si-ZSM-5: an investigation by intermolecular $^{1}H$-$^{29}Si$ CP MAS NMR. J Phys Chem B 98:4647–4653
Cha JN, Shimizu K, Zhou Y, Christiansen SC, Chmelka BF, Stucky GD, Morse DE (1999) Silicatein filaments and subunits from a marine sponge direct the polymerization of silica and silicones in vitro. Proc Natl Acad Sci USA 96:361–365
Cha JN, Stucky GD, Morse DE, Deming TJ (2000) Biomimetic synthesis of ordered structures mediated by block copolypeptides. Nature 403:289–292
Cho EB, Kwon KW, Char K (2001) Mesoporous organosilicas with PEO-containing triblock copolymers with different hydrophobic moieties. Chem Mater 13:3837–3839
Conley DJ, Kilham SS, Theriot E (1989) Differences in silica content between marine and freshwater diatoms. Limnol Oceanogr 34:205–213
Crawford SA, Higgins MJ, Mulvaney P, Wetherbee R (2001) Nanostructure of the diatom frustule as revealed by atomic force microscopy and scanning electron microscopy. J Phycol 37:543–554
Csögör Z, Melgar D, Schmidt K, Posten C (1999) Production and particles characterization of frustules of *Cyclotella cryptica* in comparison with siliceous earth. J Biotechnol 70:71–75
de Moor P-PEA, Beelen TPM, Komanschek BU, van Santen RA (1998) Nanometer scale precursors in the crystallization of Si-TPA-MFI. Micropor Mesopor Mater 21:263–269
de Moor P-PEA, Beelen TPM, van Santen RA (1997) SAXS/WAXS study on the formation of precursor and crystallization of silicalite. Micropor Mater 9:117–130
de Moor, P-PEA, Beelen TPM, van Santen RA (1999) In situ observation of nucleation and crystal growth in zeolite synthesis. A small-angle X-ray scattering investigation on Si-TPA-MFI. J Phys Chem B 103:1639–1650
Diat O, Bösecke P, Lambard J, de Moor P-PEA (1997) High-angular-resolution camera coupled with an undulator source at the European Synchrotron Radiation Facility High Brilliance Beamline. J Appl Crystallogr 30:862–886
Dokter WH, van Garderen HF, Beelen TPM, van Santen RA, Bras W (1995) Homogeneous versus heterogeneous zeolite nucleation. Angew Chem Int Ed 34:73–75
Eckert H, Ward M (2001) Controlling the length scale through "soft" chemistry: from organic-inorganic nanocomposites to functional materials. Chem Mater 13:3059–3809
Gensemer RW (1990) Role of aluminium and growth rate on changes on cell size and silica content of silica-limited populations of *Asterionella ralfsii* var. *americana* (Bacillariophyceae). J Phycol 26:250–258
Glatter O, Kratky O (1982) Small angle X-ray scattering. Academic Press, New York
Gordon R, Drum RW (1994) The chemical basis of diatom morphogenesis. Int Rev Cytol 150: 243–372
Harrison CC (1996) Evidence for intramineral macromolecules containing protein from plant silica. Phytochemistry 41:37–42
Hecky RE, Mopper K, Kilham P, Degens ET (1973) The amino acid and sugar composition of diatom cell-walls. Mar Biol 19:323–331
Holt PF, Bowcott JEL (1954) The interaction of proteins with silicic acid. Biochem J 57:471–476
Hua DW, D'Souza JV, Schmidt PW, Smith DM (1994) Pore structure analysis via small angle X-ray scattering and contrast matching. In: Rouquerol J, Rodriguez-Reinoso F, Sing KSW, Unger KK (eds) Characterization of porous solids III. Studies in surface science and catalysis. Elsevier, Amsterdam, pp 255–262
Hunter GK (1996) Interfacial aspects of biomineralization. Curr Opin Solid State Mater 1:30–35
Iler RK (1979) The chemistry of silica. Wiley, New York
Gill I, Ballesteros A (2000) Bioencapsulation within synthetic polymers, part 1. Sol-gel encapsulated biologicals. TIBTECH 18:282–296
Jansen JC, Stöcker M, Karge HG, Weitkamp J (1991) Advanced zeolite science and applications. Studies on surface sciences and catalysis, vol 85. Elsevier, Amsterdam

Jung M, Hubert DHW, Frederik PM, Fischer H, van Herk AM, German AL (1999) Vesicle-polymer hybrids. Rec Res Dev Macromol Res 4:137–156
Kamatani A (1971) Physical and chemical characteristics of biogenous silica. Mar Biol 8:89–95
Kipkemboi P, Fogden A, Alfredsson V, Flodstrom K (2001) Triblock copolymers as templates in mesoporous silica formation: structural dependence on polymer chain length and synthesis temperature. Langmuir 17:5398–5402
Krijnen S, Mojet BL, Abbenhuis HCL, van Hooff JHC, van Santen RA (1999) MCM-41 heterogenised titanium silsesquioxane epoxidation catalysts: a spectroscopic investigation of the adsorption characteristics. Phys Chem Chem Phys 1:361–365
Kröger N, Wetherbee R (2000) Pleuralins are involved in theca differentiation in the diatom *Cylindrotheca fusiformis*. Protist 151:262–273
Kröger N, Bergsdorf C, Sumper M (1994) A new calcium binding glycoprotein family constitutes a major diatom cell wall component. EMBO J 13:4676–4683
Kröger N, Lehmann G, Rachel R, Sumper M (1997) Characterization of a 200 kDa diatom protein that is specifically associated with a silica-based substructure of the cell wall. Eur J Biochem 250:99–105
Kröger N, Deutzmann R, Sumper M (1999) Polycationic peptides from diatom biosilica that direct silica nanosphere formation. Science 286:1129–1132
Kröger N, Deutzmann R, Bergsdorf C, Sumper M (2000) Species-specific polyamines from diatoms control silica morphology. Proc Natl Acad Sci USA 97:14133–14138
Lasic DD, Joannic R, Keller BC, Frederik PM, Auvray L (2001) Spontaneous vesiculation. Adv Coll Interf Sci 89/90:337–349
Lee CH, Park SS, Choe SJ, Park DH (2001) Synthesis of periodic mesoporous organosilica with remarkable morphologies. Micropor Mesopor Mater 46:257–264
Lewin JC (1966) Silicon metabolism in diatoms V. Germanium dioxide, a specific inhibitor of diatom growth. Phycologia 6:1–12
Li C-W, Chu S, Lee M (1989) Characterising the silica depositioning vesicle of diatoms. Protoplasma 151:158–163
Lobel KD, West JK, Hench LL (1996) Computational model for protein-mediated biomineralization of the diatom frustule. Mar Biol 126:353–360
Mann S, Ozin GA (1996) Synthesis of inorganic materials with complex form. Nature 382:313–318
Martin JA, Hurd AJ (1987) Scattering from fractals. J Appl Crystallogr 20:61–78
Morse DE (1999) Silicon biotechnology: harnessing biological silica production to construct new materials. TIBTECH 17:230–232
Olsen S, Paasche E (1986) Variable kinetics of silicon-limited growth in *Thalassiosira pseudonana* (Bacillariophyceae) in response to changed chemical composition of the growth medium. Br Phycol J 21:183–190
Ozin GA, Yang H, Sokolov I, Coombs N (1996) Shell mimetics. Adv Mater 9:662–667
Parkinson J, Gordon R (1999) Beyond micromachining: the potential of diatoms. TIBTECH 17: 190–196
Pedersen JS (1997) Analysis of small-angle scattering data from colloids and polymer solutions: modeling and least-squares fitting. Adv Coll Interface Sci 70:171–210
Perry CC (1989) Chemical studies of biogenic silica. In: Mann S, Webb J, Williams RJP (eds) Biomineralization: chemical and biochemical perspectives. VCH, New York, pp 223–256
Perry CC, Fraser MA, Hughes NP (1991) Macromolecular assemblages in controlled biomineralization. In: Sikes CS, Wheeler AP (eds) Surface reactive peptides and polymers: discovery and commercialization, vol 444. ACS Symposium Series Dallas TX USA, pp 316–339
Pickart CM (2000) Ubiquitin in chains. Trends Biol Sci 25:544–548
Picket-Heaps J, Schmid AM, Edgar LA (1990) The cell biology of diatom valve formation. Progr Phycol Res 7:1–168
Richer R, Mercier L (2001) Direct synthesis of functional mesoporous silica by neutral pH nonionic surfactant assembly: factors affecting framework structure and composition. Chem Mater 13:2999–3008

Robinson DH, Sullivan CW (1987) How do diatoms make silicon biominerals. Trends Biol Sci 12: 151–154

Romero O (1994) Morphological variation of *Skeletonema costatum* (Greville) Cleve (Bacillariophyceae) in Aysen Fiord, Chile. Rev Biol Mar 29:1–21

Round FE, Crawford RM, Mann DG (1990) Diatoms, the biology and morphology of the genera. Cambridge Univ Press, Cambridge

Sayari A, Hamoudi S (2001) Periodic mesoporous silica-based organic-inorganic nanocomposite materials. Chem Mater 13:3151–3168

Schmid AM (1986) Organization and function of cell structures in diatoms and their morphogenesis. In: Ricard M (ed) Proceedings of the 8th International Symposium on Living and fossil diatoms. Koeltz, Koenigstein, Germany, pp 271–292

Schmid AMM, Schulze D (1979) Wall morphogenesis in diatoms: deposition of silica by cytoplasmic vesicles. Protoplasma 100:267–288

Simpson TL, Volcani BE (1981) Silicon and siliceous structures in biological systems. Springer, Berlin Heidelberg New York

Sun Q, Beelen TPM, van Santen RA, Hazelaar S, Vrieling EG, Gieskes WWC (2002) Preparation of polyethyleneglycol-mediated silicas, resembling properties of diatomaceous silicas. J Phys Chem B 106:11539–11548

Swift DM, Wheeler AP (1992) Evidence of an organic matrix from diatom biosilica. J Phycol 28: 202–209

Takahashi R, Nakanishi K, Soga N (2000) Aggregation behavior of alkoxide-derived silica in sol-gel process in presence of polyethyleneoxide. J Sol-Gel Sci Technol 17:7–18

Teixera J (1986) Experimental methods for studying fractal aggregates. In: Stanley HE, Ostrowsky N (eds.) On growth and form: fractal and non-fractal patterns in physics. NATO-ASI Series, vol E100. Nijhoff, Dordrecht, pp 145–162

Thompson RB, Ginzburg VV, Matsen MW, Balazs AC (2001) Predicting the mesophases of copolymer-nanoparticle composites. Science 292:2469–2472

Tolbert SH, Landry CC, Stucky GD, Chmelka BF, Norby P, Hanson JC, Monnier A (2001) Phase transitions in mesostructured silica/surfactant composites: surfactant packing and the role of charge density matching. Chem Mater 13:2247–2256

Tuchman ML, Theriot EL, Stoermer E (1984) Effects of low level salinity concentrations on the growth of *Cyclotella meneghiniana* Kütz. (Bacillariophyta). Arch Protistenkd 128:319–326

van Bekkum H, Flanigen EM, Jansen JC (1991) Introduction to zeolite science and practice. Studies in surface science and catalysis, vol 58. Elsevier, Amsterdam

van Bennekom AJ, Jansen JHF, van der Gaast SJ, van Iperen JM, Pieters J (1989) Aluminium-rich opal: an intermediate in the preservation of biogenic silica in the Zaire (Congo) deep-sea fan. Deep Sea Res 36:173–190

van Bennekom AJ, Buma AGJ, Nolting RF (1991) Dissolved aluminium in the Weddel-Scotia Confluence and the effect of Al on dissolution kinetics of biogenic silica. Mar Chem 35:423–434

van Cappellen P, Qiu L (1997a) Biogenic silica dissolution in sediments of the Southern Ocean. I. Solubility. Deep Sea Res 44:1109–1128

van Cappellen P, Qiu L (1997b) Biogenic silica dissolution in sediments of the Southern Ocean. II. Kinetics. Deep Sea Res 44:1129–1149

van de Poll WH, Vrieling EG, Gieskes WWC (1999) Location and expression of frustulins in the pennate diatoms *Cylindrotheca fusiformis*, *Navicula pelliculosa*, and *Navicula salinarum* (Bacillariophyceae). J Phycol 35:1044–1053

Vasilescu M, Caragheorgheopol A, Caldararu H (2001) Aggregation numbers and microstructure characterization of self-assembled aggregates of poly(ethylene oxide) surfactants and related bloc copolymers, studied by spectroscopic methods. Adv Coll Interface Sci 89–90: 169–194

Vrieling EG, Beelen TPM, Gieskes WWC (1999a) Valve formation of diatoms: control by the intracompartmental pH inside the silicon deposition vesicle. J Phycol 35:548–559

Vrieling EG, Beelen TPM, van Santen RA, Gieskes WWC (1999b) Diatom silicon biomineralization as an inspirational source of new approaches to silica production. J Biotechnol 70:39–51

Vrieling EG, Poort L, Gieskes WWC (1999c) Growth and silica content of the diatoms *Thalassiosira weissflogii* and *Navicula salinarum* at different salinities and enrichments with aluminium. Eur J Phycol 34:307–316

Vrieling EG, Beelen TPM, van Santen RA, Gieskes WWC (2000) Nanoscale uniformity of pores in diatomaceous silica: a combined small and wide angle X-ray scattering study. J Phycol 35: 1044–1053

Vrieling EG, Beelen TPM, Sun Q, Hazelaar S, Gieskes WWC, van Santen RA (2003) Ultrasmall, Small, and Wide Angle X-ray Scattering analysis of diatomaceous silica: common features, interspecific differences. J Phys Chem B (submitted)

Vrieling EG, Beelen TPM, van Santen RA, Gieskes WWC (2002) Mesophases in (bio)polymer-silica particles inspire a model for silica biomineralization in diatoms. Angew Chem Int Ed 41:1543–1546

Wang L, Shi J, Gao J, Tomura S, Yan D (2000) Rapid synthesis and morphological control of self-assembly mesoporous silica materials. J Non-Cryst Sol 278:178–186

Wijnen PWJG, Beelen TPM, Rummens KPJ, Saeijs HCPL, de Haan JW, van de Ven LJM, van Santen RA (1991) The molecular basis of ageing of aqueous silica gel. J Coll Interf Sci 145:17–32

Wijnen PWJG, Beelen TPM, Rummens KPJ, van Santen RA (1993) The role of cations in the formation of aqueous silica gels. J Non-Cryst Sol 152:127–136

Yang H, Coombs N, Ozin GA (1996) Morphogenesis of shapes and surface patterns in mesoporous silica. Nature 386:692–695

Zaremba CM, Stucky GD (1996) Biosilicates and biomimetic silicate synthesis. Curr Opin Solid State Mater Sci 1:425–429

Zones SI, Nakagawa Y, Lee GS, Chen CY, Yuen LT (1998) Searching for new high silica zeolites through a synergy of organic templates and novel inorganic conditions. Micropor Mesopor Mater 21:199–211

# Subject Index

## T



## U

MIX
Papier aus verantwortungsvollen Quellen
Paper from responsible sources
FSC® C105338

If you have any concerns about our products,
you can contact us on
**ProductSafety@springernature.com**

In case Publisher is established outside the EU,
the EU authorized representative is:
**Springer Nature Customer Service Center GmbH**
**Europaplatz 3, 69115 Heidelberg, Germany**

Printed by Libri Plureos GmbH
in Hamburg, Germany